I0055859

The
Brain-Behavior
Continuum
The Subtle Transition
between Sanity and Insanity

The
Brain-Behavior Continuum

The
Subtle Transition
between
Sanity and Insanity

Jose Luis Perez Velazquez
University of Toronto and The Hospital for Sick Children, Canada

Marina Frantseva
University of Toronto, Canada

World Scientific

NEW JERSEY · LONDON · SINGAPORE · BEIJING · SHANGHAI · HONG KONG · TAIPEI · CHENNAI

Published by

World Scientific Publishing Co. Pte. Ltd.

5 Toh Tuck Link, Singapore 596224

USA office: 27 Warren Street, Suite 401-402, Hackensack, NJ 07601

UK office: 57 Shelton Street, Covent Garden, London WC2H 9HE

British Library Cataloguing-in-Publication Data
A catalogue record for this book is available from the British Library.

THE BRAIN-BEHAVIOR CONTINUUM
The Subtle Transition Between Sanity and Insanity

Copyright © 2011 by World Scientific Publishing Co. Pte. Ltd.

All rights reserved. This book, or parts thereof, may not be reproduced in any form or by any means, electronic or mechanical, including photocopying, recording or any information storage and retrieval system now known or to be invented, without written permission from the Publisher.

For photocopying of material in this volume, please pay a copying fee through the Copyright Clearance Center, Inc., 222 Rosewood Drive, Danvers, MA 01923, USA. In this case permission to photocopy is not required from the publisher.

ISBN-13 978-981-4340-60-1
ISBN-10 981-4340-60-X

Typeset by Stallion Press
Email: enquiries@stallionpress.com

Printed in Singapore by Mainland Press Pte Ltd.

CONTENTS

PREFACE

Since time immemorial, the myriad of efforts to understand the nature of mental life epitomize what is one of the most fundamental quests for knowledge. The research field of mental illness abounds with models reflecting the reality of troubled minds with various degrees of accuracy, ranging from unrestrained presumptions to the hard facts of neuroscience. Perhaps one of the most posed questions in philosophy and the natural sciences is the construction of a model about the relationship between mind and body. Are they separate or parallel? Or is there a sequence? And if so, which one is in charge? Does our brain orchestrate our behavior with the inevitability of an epileptic seizure or a migraine? Is there a controlling mind that does the conducting? This uncertainty contributes to the state of confusion that the study of the brain seems to cause, because compared with other organs in the human body, the brain seems to be regarded in a very different manner. Few people question the need for insulin once their pancreas is diseased and diabetes diagnosed. Yet, it normally takes a few relapses for the mental illness sufferer to finally surrender to the idea of taking regular medications. Why do we have this peculiar attitude toward the brain, apparently disregarding the fact that its pathologies are just as valid and real as those of the pancreas? This position, perhaps, is not entirely unfounded, for there is a fundamental difference between brains and other human organs: brains determine behaviors and possess the power of logical thinking and self-referentiality that no other organ does. Brains can monitor themselves, can correct any peculiar style of behavior, if only endowed with enough logical competence and will ... or so we think.

Ever since the times of Hippocrates, who was one of the first to recognize epilepsy as a "disease caused by the brain's disharmony," we have come to accept that we cannot control epileptic seizures by the power of our minds. Seizures are behaviors (indeed, the definition of epilepsy is based on the behavioral manifestation of seizures) produced by the brain activity resulting from some pathological process. On the other hand, we are fast to judge criminal psychopaths for their reckless behaviors, ample evidence linking antisocial behaviors with brain abnormalities notwithstanding. We assume that social predators must, and, in fact, can "easily" control their vicious habits by the use of proper thinking, acceptable logic, and the power of self-control. Yet, how much of a choice do they have in reality? Maybe, little more than when trying to stop an epileptic seizure? After all, seizures and antisocial behaviors are both fabrications of "diseased" brains, by-products of deviations from the considered normal neurophysiology.

The purpose of this monograph is to explore the perspective of a smooth continuum between distinct brain activities resulting in different behaviors, from seizures to mental illnesses and personality disorders. Many books have appeared on the topics of brain, mind, and behavior. So, why devote more words that may be lost in the already vast sea of words dedicated to these matters? This book is not proposing new theories of brain and behavior, its main purpose is to integrate different levels of description using frameworks that have been advanced and are well established in the natural sciences. In this sense, this volume describes, and integrates, what is already known, and endeavors to guide researchers in several neuroscientific areas to achieve a more fundamental comprehension and appreciation, absolutely needed in these days of massive data gathering, of an integrative approach to understanding brain function and behavior. The realization of a global understanding of brain function and behavior requires a flexible switching between levels of description, today more than ever. The current trend in the neurosciences and brain research is to develop sophisticated studies that benefit from an impressive arsenal of cellular and molecular methodologies. Thus, we have at our disposal an overabundance of empirical data generated under experimental conditions that, most commonly, try to isolate the phenomenon under scrutiny in supposedly controlled laboratory settings: reductionism at its best. In general in the biological sciences, the comprehensive understanding of high-level laws lags considerably behind the understanding of elementary processes, and the choice of the level of description determines to a large degree the nature of the understanding that is finally achieved. This book introduces a high-level perspective, searching for simplification among the structural and functional complexity of nervous systems by consideration of the distributed interactions that underlie the collective behavior of the system. Thus, the focus is on the nature of the interactions among the components of nervous systems: the dynamics of coordinated cellular activities that, of course, result from underlying neurophysiological events. We will see that, by focusing on the perspective of the dynamics of organized activities, with the added advantage that it can be made independent from the material particles that constitute a system, apparently diverse phenomena become conceptually closer.

The term "continuum" is used both with the connotation utilized in continuum theory (a branch of topology), and with that used in physics. The mathematical aspect denotes anything that goes through a gradual transition from one condition to another, without abrupt changes. In physics, theory of relativity for instance, the space–time continuum model explains space and time as part of the same reality rather than as separate entities. Our understanding of "brain–behavior continuum" denotes, first, the gradual transitions that most of the times occur in brain activities and behaviors, the progression from heath to disease. We illustrate with some examples how subtle alterations at the level of the analysis of electrophysiological recordings may gradually transition into robust changes in brain activities that sometimes are translated into a distinct pathological state. Second, it denotes, in the physics sense, that brain activity and behavioral actions are not separate and independent entities. Rather, it is a continuous transition between the microscopic (molecular), mesoscopic (cellular/network), and macroscopic (collective behavior, coordinated activity patterns) levels which constitute the brain and its activity that ultimately manifests itself into perceptions, emotions, and actions; which can, in their turn, alter the brain in a sort of feedback loop, affecting not only patterns of synchrony, but also the very molecular structures that have produced them (phenomena generally known as "neuroplasticity"). We argue that, due to the plastic nature of the brain and the fact that behaviors can alter brain structure and function, the boundaries between brain and behavior are rendered artificial, at least as it is seen at a certain level of description.

Our monograph is then a comprehensive study (overview) of the main current concepts in brain activity at the global, collective (or network) level, and an attempt to explain how normal neurophysiology can develop into brain pathological states, always concentrating at the network or collective level but paying attention too to the underlying molecular and cellular aspects that result in the specific pathologies here described. The emphasis is on a few crucial aspects that, we think, need to be understood and considered by a wide variety of medical practitioners and neuroscientists, for a global understanding of brain and behavior in these days of massive data gathering. Of particular importance, which is still underestimated by a vast majority of neuroscientists, is the notion of "small changes, large effect" and the determinism of brain activity resulting in specific behaviors. In this regard, the concept of free will has to be addressed, or at least how much freedom there is in that "will." This is discussed in the context of the possibility of control of some brain networks using other networks in various physiological and pathological states.

This book, if needs to be classified into some field, we would tend to place it into the neurophysics category: it is a neurophysics book about some clinical aspects, rather than a clinical treatise with neurophysical aspects. It is intended for an audience composed of neuroscientists (experimentalists as well as theoreticians and computational neuroscientists), psychiatrists, neurologists, neurosurgeons, biophysicists, and those interested in the workings of the brain and its relation to behavior. The main concepts can be understood by educated readers in general. Some very technical particulars are included for those interested in specific details because of the need for precision in these days when concepts from other disciplines are permeating neuroscience, particularly analysis methods and frameworks from dynamic system theory. The intuitive appeal of many of these theoretical constructs has attracted much attention from neuroscientists, psychologists, and clinicians, which has resulted in a metaphorical language that, while very useful, sometimes lack accuracy thus making these metaphors become extended a bit too far. Hence, for the sake of accuracy, sections in some chapters present detailed analyses that try to go beyond the metaphors, to demonstrate that a relatively rigorous application of dynamical frameworks can indeed be accomplished in brain science. Because of the interrelatedness of many of the aspects discussed in various chapters, no chapter stands alone, and, at least, it is recommended to go over the main concepts presented in Chapters 1 and 2 before reading any other section.

A general underlying uncertainty in most of the neuroscientific studies here considered is whether mental states, that determine behavioral outputs, can be reduced to neurophysiological phenomena. Many words have been devoted to this query, but this is not the place to address it. Perhaps better to be content with the formulation of some correlations between neurophysiological phenomena at several levels of description and behavioral manifestations. It is not without interest noting the fact that it is the brain that is trying to comprehend itself, as well as the nature of reality as perceived by the constraints of neural information processing. Francis Bacon (1561–1626) decreed that "Human understanding is like an irregular mirror, which distorts and discolors the nature of things by mingling its own nature with it" (Novum Organum, 1620). Physicists know quite well that the laws of nature deal more with our perception and knowledge of phenomena rather than with phenomena themselves. The contents of the brain, whether molecules or ideas, cannot be separated from the rest. The continuum proceeds at many levels. It is hoped that the careful consideration of the brain–behavior continuum will reveal that the pathological brain exists only in terms of degree, exposing the artificiality and fuzziness of the created boundaries between normal neurophysiology and neuropathologies. We apologize beforehand if some of the many works, empirical and theoretical, that have been commented throughout this monograph have been misrepresented, and if such a thing has occurred, a feedback form the original authors would be appreciated so that future editions can be amended.

QUOTES

"Science will teach man that he never has really had any caprice or will of his own, and that he himself is something in the nature of a piano key ... and that there are, besides, things called the laws of nature; so that everything he does is not done by his willing it, but is done of itself, by the laws of nature. Consequently, we have only to discover these laws of nature, and man will no longer have to answer for his actions, and life will become exceedingly easy for him."

F. Dostoyevski, *Notes from the Underground* (1864)

"If we ask whether we are free, the kind of answer we want may not be possible. A better question to ask is: do we make choices? The answer is certainly yes... Are our choices constrained? Yes."

Schall, J.D. (2001) *Nature Reviews Neuroscience* 2, 33–42

"Further conceive, I beg, that a stone, while continuing in motion, should be capable of thinking and knowing, that it is endeavouring, as far as it can, to continue to move. Such a stone, being conscious merely of its own endeavour and not at all indifferent, would believe itself to be completely free, and would think that it continued in motion solely because of its own wish. This is that human freedom, which all boast that they possess, and which consists solely in the fact, that men are conscious of their own desire, but are ignorant of the causes whereby that desire has been determined. Thus an infant believes that it desires milk freely; an angry child thinks he wishes freely for vengeance, a timid child thinks he wishes freely to run away. Again, a drunken man thinks, that from the free decision of his mind he speaks words, which afterwards, when sober, he would like to have left unsaid. So the delirious, the garrulous, and others of the same sort think that they act from the free decision of their mind, not that they are carried away by impulse.

As this misconception is innate in all men, it is not easily conquered. For, although experience abundantly shows, that men can do anything rather than check their desires, and that very often, when a prey to conflicting emotions, they see the better course and follow the worse, they yet believe themselves to be free."

Spinoza, *Letter to G.H. Schaller* (October 1674)

"El triunfo supremo de la razón… es poner en duda su propia validez."
(The supreme triumph of reason… is to cast doubt on its own validity.)

Miguel de Unamuno, *Del Sentimiento Trágico de la Vida* (1954)

ACKNOWLEDGMENTS

We are grateful to Wojciech Kostelecki, Vera Nenadovic, Roberto Fernández Galán, Siv Sivaloganathan, and Richard Wennberg, for reading and commenting on some parts of this volume, and to Delizia Ferri for her diligent help with administrative matters. Our thanks also to the original reviewers of our publication proposal for their suggestions and to Dr. Elena Nash of Imperial College Press for her assistance in all these matters. Our research on some topics here discussed is supported by the Natural Sciences and Engineering Research Council of Canada (NSERC), the Hospital for Sick Children Foundation, The Savoy Foundation, and the Fundação Bial.

CHAPTER 1

BASIC PRINCIPLES OF NERVOUS SYSTEM ACTIVITY AND FUNCTION

The purpose of this chapter is to introduce fundamental aspects that determine how nervous systems function and determine behaviors. It is not a review of basic neuroscience, information that is assumed the readers already possess. The emphasis is on the integration of the phenomena occurring at different levels of description into a more comprehensive, albeit less detailed, understanding of neural actions. While some specific results do appear and are commented in some detail in this and other chapters, an effort is made to put those data into context and to extract the fundamental relations among the observations. From the apparent immense complexity of the nervous systems as it necessarily appears at the molecular and cellular levels of description, simplicity is sought here by concentrating on the collective level of description, the coordinated patterns of activity of the cell ensembles that constitute the brain, synchronization as a particular aspect of this coordination. We follow the advice of J. Williard Gibbs (in a letter to the American Academy of Arts and Sciences in 1881): "One of the principal objects of theoretical research in any department of knowledge is to find the point of view from which the subject appears in its greatest simplicity." This "point of view," it is proposed, is the consideration of the relation among the activity in cells and in cell networks, descriptions of the collective activities. In more general terms, it is the relations between observations that interest us here, whether the observations are of cell spiking activity or about human behaviors is all the same. In the words written long ago by Henri Poincaré (1902): "The aim of science is not things themselves, as the dogmatists in their simplicity imagine, but the relations among things; outside these relations, there is no reality knowable." A system is not complex because it contains a vast number of constituents, but because of the manners the constituents interact. In nervous systems, the activity of one unit (a cell, a cell ensemble, a brain module, etc.) has meaning only as related to the activity of other units. Concentrating on the study of the distributed interactions among the constituents offers a high-level view on the system's collective behavior that underlies its activity and function. An advantage of taking this perspective is that diverse phenomena appear conceptually closer.

Many of these relations remain qualitative, which may deter investigators engrossed in quantification. However, qualitative fundamental relations offer, many times, deeper insight than precise quantifications; the attitude of Arthur Winfree (2001), one of the founders of the field of non-linear oscillations, is strongly advised: "My deeper motivation is a feeling that numerical exactitude is alien to the diversity of organic evolution, and pretence of exactitude often obscures the qualitative essentials that I find more meaningful." Hermann Weyl (1932) pointed the way to extract sense from the overwhelming mass of empirical observations that exists currently in neuroscience, when he said: "The assertion that nature is governed by strict laws is devoid of all content if we do not add the statement that it is governed by mathematically simple laws. Thus, simplicity becomes a working principle in the natural sciences. The astonishing thing is not that there exist natural laws, but that the further the analysis proceeds, the finer the details, the finer the elements to which the phenomena are reduced, the simpler—and not the more complicated, as one would originally expect—the fundamental relations become and the more exactly do they describe the actual occurrences." Finally, Albert Einstein (1936) was another sponsor of the wonders of simplification: "The aim of science is [...] a comprehension as complete as possible, of the connection between the sense experiences in their totality and, on the other hand, the accomplishment of this aim *by the use of a minimum of primary concepts and relations*". But these scientists were referring to physics, mostly, and perhaps the quest for simplification is not plausible in biological sciences. Can simplicity and generality be found in the basic laws of nervous system function and behavior? It is certainly possible because nervous systems and brains in particular are characterized by the abundance of cooperative phenomena in cellular activity, a source of a host of "emergent" states that are difficult to be understood by the individual operations of the units in isolation, not even mentioning the fact that it is not feasible to record the activity of all cellular elements simultaneously. Hence, neuroscience faces similar dilemmas as physics did, in that we cannot determine the behavior of all the constituents of the system but we can measure macroscopic variables that capture the collective, coordinated patterns of activity. But physics also includes thermodynamics, and the success of thermodynamics was characterized by overcoming these limitations, such that at the expense of losing a detailed description of the system's elements, a global description was attained. However, expert readers will object regarding the fact that most of the thermodynamics and statistical mechanics applies to closed systems at equilibrium, while nervous systems are not at equilibrium. Nevertheless, pattern formation in far-from-equilibrium systems has been and is being thoroughly studied and such framework may represent interesting resources to brain research for the understanding of the macroscopic spatiotemporal patterns of neural activity and their relation to behavior. The field of synergetics, that developed in the late 1970s and is being applied to neuroscience, follows the thermodynamical tradition of inspecting qualitative, macroscopic changes and searching for order parameters that can capture these transitions (Haken, 2002, 2006).

Charles Sherrington's classical studies on spinal reflexes led him and others to envisage the nervous system as an input–output reflexive device: the nervous system remains inactive until stimulated by the sensorium and then produces an output behavior after the processing of the sensory input. These ideas were challenged by early work on central pattern generators (CPGs), where investigators demonstrated that these relatively small neuronal networks, found in invertebrates and vertebrates, were able to produce rhythmic patterns of activity in the absence of sensory inputs. These notions are in contrast to the current dynamic conceptualization of the activity of brain networks. The rich internal dynamics of brain circuits is revealed by cognitive studies illustrating that, in addition to the processing of incoming sensory information, brains seem to generate information: e.g. in the responses to sensory inputs that depend on the initial, cognitive state of the subject, on past experiences (memories), and on expectations. "Information" is understood here as novel responses that cannot be accounted for by the incoming sensory stimuli. In this sense, dreaming during sleep can be considered generation of information, or, at least, a rearrangement of the stored information, in the absence of sensory stimuli. All these cognitive aspects are hard to explain based on linear stimulus–response paradigms. Here, cognition is equated to information processing by nervous systems and it encapsulates a continuum, rather than discrete processes, that starts with the decoding of sensory inputs, elaborations of these and their storage in memory, and their final transformation into motor outputs. Rather than disintegrating cognition into a variety of aspects including attention, perception, retrieval, and many others, we prefer to integrate those processes, following Ulrich Nieser's (1967) conception of cognition as "all processes by which the sensory input is transformed, reduced, elaborated, stored, recovered, and used". This is one aspect of the continuum, the gradual transition from one condition to another, without apparent abrupt changes, the gradual transitions that most of the times occur in brain activities and behaviors. But what may seem a gradual and subtle transition at one level may become an abrupt change at another. It is our main purpose to illustrate this point. Subsequent chapters contain a variety of examples of how either subtle or strong alterations at the molecular level or at the level of the analysis of electrophysiological recordings may gradually transition into robust changes in brain activities that sometimes are translated into pathological states. Nonetheless, there is a continuum in awareness (consciousness) as all of us experience it at every second of our lives, ignoring (or not being aware of) the predominantly discontinuous nature of perception, to start with. A myriad of cognitive experiments performed so far indicate that there are not sharp demarcations in this continuum. The continuum proceeds at many levels. It is hoped that the careful consideration of the brain–behavior continuum will reveal that the pathological brain exists only in terms of degree, exposing the artificiality and fuzziness of the created boundaries between normal and abnormal neurophysiology.

1.1. Main Principles of Nervous System Organization and Information Processing

For the purposes of achieving a better understanding of nervous system activity and function, as well as for the classification of a diversity of processes taking place, it is a common practice that three basic levels are usually considered:

– Microscopic level: this includes the molecular composition of intracellular and extracellular compartments, ion currents, membrane channels, and other microcomponents. To study and analyze details of the interactions among all these constituents of this level, biophysicists normally use systems of differential equations.
– Mesoscopic level: the description of the activity at this level is done by means of ensembles, in extensions that are large compared with intermolecular distances but small compared to macroscopic spatial patterns. These phenomena are probabilistic in nature, and thus stochastic differential equations are used to characterize processes at this level.
– Macroscopic level: characterizes the emergence and evolution of spatiotemporal patterns of activity in extensive cell networks. Mathematically, these phenomena are normally described in terms of time-dependent partial differential equations.

The study of these three levels is equally fundamental, because even though the microscopic level, obviously, underlies the other two, there are dynamic patterns, particularly the so-called self-organizing spatiotemporal patterns, that cannot be understood by any careful characterization of the microcomponents. To comprehend the emergence of the collective spatiotemporal patterns, one necessitates concepts that transcend the microlevel. As an advice to those interested in a comprehensive understanding of brain and behavior, we encourage them to gather enough knowledge to be able to switch comfortably between levels of description: become aware of the nature of the constituents, of the interactions among them, and of the distinct phenomena that can occur in the three levels, because what looks like one phenomenon at one level of description may become a different one at another level. This motivates the discussions that permeate this monograph on abrupt versus gradual change of activity, or whether a smooth continuum can be discerned in specific cases of deviations from the normal. Hence, the following chapters present discussions on a variety of neurophysiological and mental phenomena that at certain level may appear to result from gradual changes but at another level may manifest as abrupt transformations of activity. Just a sample: in epileptiform phenomena, the synchrony measured in cell networks as a seizure approaches appears slow and gradual in some cases, but the seizure manifests itself in the patient as a sudden change in behavior. As noted above, our main focus is at the macrolevel, yet emphasizing the integration among the three levels. Natural phenomena should be investigated at all possible levels of description, but it is fair to note that in biology, and neuroscience in particular, the lower levels (molecular/cellular) are thoroughly and intensively investigated, whereas the more comprehensive and theoretical efforts to create frameworks in which to interpret the low-level phenomena are not so well esteemed (McCollum, 2000); this general stance on theoretical neuroscience underlies present inconveniences in the peer review of theoretical/computational publications as clearly described by E. de Schutter (2008), current editor of *Neuroinformatics*, in one of his editorials. Reductionism, while it is fine to describe low-level phenomena, poses severe limitations if the purpose is to understand the collective behavior of the whole, and this may lie beneath the relative failure in treating some diseases by seeking very specific mechanisms of action of the drugs. Section 2.8 of Chapter 2 (Figure 2.16) is devoted to comments on the notion of specificity at different levels of description, where it is proposed that the realm of specificity in natural phenomena is restricted to a short time scale and to a microdomain in space. Please note that the claim is not that it is unnecessary to focus at low levels of description, but that a more balanced approach should be encouraged and valued in contemporary neuroscience.

Three basic principles of neuronal function are the following, derived from observations that date back to the early 20th century, basically from the work of Adrian and Zotterman (1926) in experiments performed on sensory nerves:

1. Action potentials (spikes) are all or none;
2. Spike firing frequency increases as the stimulus strength becomes larger; and
3. Adaptation of spike firing occurs to static stimuli.

Those were three seminal empirical facts, observations that, among others, give rise to the following assumptions (these, in fact, could be considered axioms in neuroscience) in the study of information processing by nervous systems:

1. Information processing takes place by the transfer of electrical gradients (potential differences between the intracellular and extracellular compartments) from cell to cell; this transfer is performed mainly by chemical interactions between cells. There is also direct electrical coupling between cells but this type of communication

is not considered so fundamental as the chemical. The differences in electrical potential are due to the movement of ions between the intra- and extracellular compartments, that is, there must be ionic gradients between interior and exterior for neural information processing to occur.

2. Neurons are considered to be the cells that process information, communicating via action potentials (spikes), according to point 1 above.

3. Reliable information processing requires stability in the transmission of spikes from network to network, rather than from neuron to neuron. Nervous system's main function consists in the propagation of activity from net to net.

These three "axioms" are substantiated by some empirical observations, such as the anesthetic blockade of synaptic transmission resulting in loss consciousness. Note, however, that these drugs will bind (and thus inhibit ion currents) not only to the receptors in neurons, but also those present in glial cells. Glial cells, more numerous in the brain that neurons, are mostly characterized by forming an extensively connected network because they are profusely connected by gap junctions. Gap junctions establish a direct contact between the cytoplasms of two cells, because the pore of the junctional proteins is large enough for ions and small molecules to pass. The term "syncytium" is used when referring to this wide glial net. Note that the gap junctional coupling affords a direct electrical communication with very little delay since ions move freely between the connected cells, as opposed to that of chemical synaptic transmission between neurons where there is a delay inherent in the release of neurotransmitter and binding to the post-synaptic receptors and opening of the channel for ions to move down their electrochemical gradients (the chemical synaptic delay is of the order of $\sim$1–2 milliseconds, and for this reason as well as due to the conduction velocity of spikes in axons, it can be said that we certainly live in the past, for the processing of any sensory input takes several milliseconds so what we perceive is the past). Yet, not much is thought of glial cells as information processors. There is definitely an impetus among researchers in efforts to frame glial activity in terms of their relative importance as modulators of synaptic function, among other aspects. It is suspected, for instance, that myelinating glial cells can influence synchronizing activity. Schwann cells in the peripheral nervous system and oligodendrocytes in the central form myelin sheaths that surround neuronal axons. Oligodendrocytes depolarize in response to neurotransmitters, even in response to the normally inhibitory transmitter gamma-aminobutyric acid (due to possibly high internal chloride). The depolarization of these glial cells can help synchronize action potentials traveling down neuronal axons because each oligodendrocyte myelinates many axons of different neurons, and axons will depolarize passively due to the depolarization of the surrounding glia membrane, thus enhancing the tendency to spike generation and propagation. Another type of glia, astrocytes, can influence neuronal synchronization and coordinated activity in general (Allegrini *et al.*, 2009). Those interested in neuron–glia interactions are invited to consult the special issue of the *Journal of Biological Physics on Neuron–Glia Interactions*, Vol. 35, No. 4, 2009. This renewed contemporary interest in glial cell function may still leave some neuroscientists unsatisfied as the absolute main emphasis continues to be placed on neurons, but at least there has been a considerable improvement compared with the importance that glia was given in the past when they were considered passive trophic elements of the nervous system. Nevertheless, despite the surge in interest to investigate glia and to bestow upon them a more prominent role in mental phenomena, the available empirical evidence indicates that glial elements act as modulators of neuronal communication (Tsacopoulos and Magistretti, 1996; Perea *et al.*, 2009). However, this notion of "gliotransmission" has been debated very recently. Gliotransmission has received support from all studies where pharmacological manipulations have been made, particularly manipulations that alter internal calcium levels in glial elements and a consequent effect on neuronal synaptic transmission is observed. However, in a recent study using transgenic rodent models (and thus no pharmacological interventions to alter calcium signaling were needed), no evidence whatsoever for the glial modulation of neural synaptic transmission was found (Agulhon *et al.*, 2010). In any case, the last word on this topic still remains to be heard, but for our present purposes it makes little sense to ask whether glia or neurons are more fundamental for mental activity, for what is clear is that there would not be any mind if glia or neurons were missing.

1.1.1. Functional coordination: Integration and segregation of cell ensembles

To those three first principles aforementioned, others could be added but these derive from those three and for this reason the word "axiom" was used, and also because in reality there is no critical proof of any of those statements and yet everybody accepts them, perhaps because it is not possible to resolve them with the current techniques. Let us not forget that there are other alternative proposals for brain information processing and consciousness in particular, based on quantal phenomena, these being still an object of debate (consult, e.g. J. Tuszynski's edited book *The Emerging Physics of Consciousness*, Spinger-Verlag, 2006). For example, another very basic principle of nervous system function is that the patterns of activity determine the different mental states and behaviors, with the added notion that the combination of the anatomical arrangement of cells (intercellular structural connectivity) and the particular intrinsic biophysical characteristics, establishes the

Anatomical arrangement/connectivity pattern $\Big\}$ | Anatomical, functional, and effective connectivity |

Intrinsic biophysical characteristics $\Big\}$

Environmental (sensory) perturbations $\Big\}$
Other modulators $\Big\}$ — Functional coordination

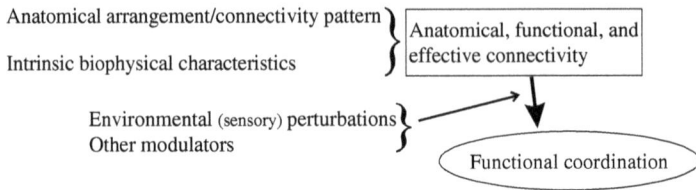

Figure 1.1. Structure and function in the nervous system are intertwined to produce coordinated activity patterns.

functional interactions that generate the diversity of activity patterns. This is represented schematically in Figure 1.1. That some type of correlated activity must exist at some level in the nervous system derives from axiom #3, the transmission of activity from net to net. Thus, two aspects of connectivity must be considered: the *anatomical* and the *functional connectivity*. One aspect determines the other only to some extent, for even in cases when cells are not structurally connected, still they can functionally interact. Consider, e.g. the electric coupling via field effects, the so-called ephaptic transmission, in which a neuron can "feel" the activity in nearby cells without structural contact. There could be, of course, anatomic contacts via gap junctions, in which case the electrical transmission was classically labeled electrotonic, and may be more efficient than just field coupling alone. It is the combination of the anatomical arrangement of cells with particular intrinsic biophysical characteristics and their functional interactions that will generate one or another activity pattern. But this is not the whole story, as two more things need to be added to the recipe for the emergence of neural activity patterns: the *perturbations* from the environment, that is, the sensory stimuli, and the *modulation* by hormones and other modulators within the nervous system. These two aspects can perhaps be considered control mechanisms, and their effectors (sensory inputs and hormones,) as control parameters (Bressler and Kelso, 2001). Subsequent chapters have detailed comments on control parameters that may drive the activity patterns in pathological directions. Peter Getting's (1989) research in the 1980s already pondered in depth the functional consequences derived from the combination of the anatomical arrangements and the individual cellular properties.

The term "functional coordination" appears in Figure 1.1, and is used in other chapters too. The more usual expression in the literature is "functional connectivity," but because there seems to exist some semantic controversies regarding the use of this term with another similar, "effective connectivity," and to avoid being trapped in semantic discussions on terminology that, in the final analysis, mean basically the same thing, in this book "functional coordination" is used. It is our opinion that this notion is more general and includes the ideas in the other two (this expression was proposed by Scott Kelso during one of the talks in a recent conference on brain coordination dynamics). Functional connectivity captures common statistical dependencies between units (cells or networks), whereas effective connectivity is invoked to capture causal interactions, that is, the directionality of coupling between units. Both of these ideas are encapsulated in functional coordination, which manifests itself in a variety of patterns of collective activity involving many cell networks. For similar reasons, we prefer to use "coordination of activity" rather than "correlation of activity" when talking about brain patterns of activity, because the former is a broader concept and includes the correlations. Coordination takes correlations one step further and entails a certain degree of order among those correlations in order to give rise to ordered, organized activity patterns. Although, as described in Section 1.3, it is the correlated activity that is normally analyzed from neural signals. From these analyses of correlations in activity, coordinated patterns emerge. But enough of semantic arguments. Let us rather concentrate on the importance of coordination in neural activity for information processing.

One can speak of several types of coordinated activity: between individual cells if action potentials can be measured in several neurons simultaneously, or between cell groups if local field potentials or other type of collective recordings like electroencephalography (EEG) or magnetoencephalography (MEG) are employed. Neuroimaging methods, like functional magnetic resonance imaging (fMRI), have been used too to derive coordinated patterns based on the signals measured using these techniques that rely on metabolic activations. A note of caution is here needed, as the physiological interpretation of the coordination between brain areas derived from these distinct recordings methods will necessarily show differences. Nevertheless, regardless of the methodology used, our concern here is to emphasize that these collective activity patterns are necessary for the integration and segregation of the information, the two most fundamental aspects for optimal (that is, adaptive) information processing (Tononi *et al.*, 1998). Integration because the different aspects of a percept (e.g. texture, color, smell) have to be integrated in order to perceive the unity in the percept; segregation because these different aspects are processed by distinct networks so each has to be relatively independent to be able to carry out their task, otherwise deviations may arise such as synesthesia, when colors, e.g. elicit sounds in the brain of the synesthetes. The perspective on cognition aforementioned in the first few paragraphs condenses these two aspects, both in terms of processing the sensory stimuli and in terms of elaborating adequate motor responses, the overt behavior that is the variable used in determining whether behaviors are normal ("healthy") or deviant. Coordination of activity, that is, *the organized exchange of information among cell ensembles*, can accomplish at the same time the integration of the activities between some ensembles and the segregation into several functionally connected groups of ensembles. An example of how this can be achieved can be found in the consideration of a particular aspect of coordination, the synchronization of neural activity and more specifically the synchronization between the phases of the neural oscillations. This aspect is treated in Section 1.4.3. The neuroanatomical connectivity present in brains seems also suited to integrate and segregate neuronal ensembles; for instance, in detailed studies performed in rhesus monkeys, Cavada and Goldman-Rakic (1989) reported that subdivisions of parietal cortex were connected to unique sets of frontal networks, providing evidence for structurally segregated cortico–cortical networks. On the other hand, the anatomy of integration is found in the multimodal cortical association areas that receive convergent inputs from many sensory regions. Note that glial cell networks cannot be easily segregated because of their extensive gap junctional coupling. The mammalian cerebellar cortex also displays signs of segregation of networks as has been inferred based on intracerebellar multiunit recordings.

Those observations seem to support a modular view of the brain. This perspective had predecessor in the 19th century most notably advocated by the neuroanatomist Franz J. Gall with his disputed and many times unaccepted phrenological theories. More modern histological staining works support the notion that nervous systems in general seem to be formed by modules, networks of interconnected neurons and glial cells, as already suggested by early 20th century investigators like Lorente de Nó (1938) using the remarkable Golgi histological staining procedure. He described these elementary units as cylinders or columns, and the functional activity of these units was confirmed later on by investigators like Vernon Mountcastle (1957) in his classical neurophysiological studies in cat somatosensory cortex, reaching a pinnacle in the precision of characterization of the cytoarchitecture and correlated physiological function in the Hubel and Wiesel's studies on the ocular dominance columns of the visual system (Hubel and Wiesel, 1963). Thus, in mammalian cerebral cortex, the basic module is considered to be represented by these (hyper)columns. In other brain areas, like basal ganglia for instance, the elementary modules identified with different staining methods (normally targeting particular enzymes like acetylcholinesterase) have been found to be elliptical or round, rather than column-like. In invertebrates, the basic module is the ganglion, and the emergence of brains in some phyla of invertebrates is thought to result from the fusion of ganglia. Thus, whereas invertebrate modules tend to be discrete hemispherical lobes that are basically visible without any anatomical staining, the mammalian modules are columnar or round compartments and normally require careful histological staining to be identified. Borders between parts are becoming diffuse in advanced brains, a continuum in neuroanatomy is emerging. For the present purpose, it does not really matter too much how the basic module is built or whether larger nervous systems have been enlarged by the addition of more modules, the important aspect is that there seem to be elementary networks of mutually interconnected cells that process information. While discrete modules are already identifiable in annelids (worms), larger modules perhaps appeared first in cephalic ganglia to process information of distinct sensory modalities (Leise, 1990), as these elementary interconnected networks of cells offer the possibilities of maintaining the activity among the cell networks due to their mutual interactions (that is, they help the integration of information), and also of separating the activity of one network from another (or the segregation of networks processing different inputs). As aforementioned, integration and segregation are two fundamental aspects for optimal (that is, adaptive) information processing. In this regard, R. Malach (1994) proposed that the cortical columnar arrangement serves a purpose, in that it maximizes the connectional diversity among the neuronal populations. The maximization of contacts aids the diversity in the sampling of inputs, which, in general, will favor adaptability of the organism. These related concepts of diversity, variability, and fluctuations in neural activity, and the consequent adaptive (or maladaptive) values, are of crucial importance in the understanding of why and how nervous systems developed and reached the present stage. Beyond anatomical considerations, in terms of function, the analysis of the spatial patterns of background brain activity, based largely on functional magnetic resonance imaging (fMRI), does not support a clear modularity in the human brain. What this probably means is that networks, or whatever areas are consider to form a network, are not rigid, and these are better thought of as transient clustering due to similar activities in those regions.

1.2. Developmental Considerations and the Fundamental Activity Patterns of the Nervous System

The very basics of the phenomena we know as life on this planet originated from a separation: the creation of a boundary between an internal and an external compartment. Compartmentalization of a connected metabolism was the fundamental aspect that gave rise to living phenomena. The separation of the two compartments by a lipid boundary necessarily created differences in the molecular composition between the inside and outside, molecular gradients appeared and because some molecules carry charges, electrical gradients emerged and thus excitability arose: ions could move between the intra- and extracellular milieu crossing the lipid membrane through particular proteins. Excitability meant that there is an opportunity to react to inputs. As organisms developed, and from the very basic goal of generating coordinated responses to stimuli, there appeared mechanisms that relied upon electrical activity: differences in transmembrane voltage that can be caused by a variety of means, normally involving ion movement through pores in the cell membrane (voltage-gated ion channels exist even in archebacteria). Electrical excitability thus appeared long before neurons or nervous systems emerged, and it serves its purpose from sponges to mammals. Some sponges (these do not have nervous systems), upon finding sediments for nourishment, generate "spikes" that last 5 s, extremely long according to neuronal standards (neuronal action potentials last 2–4 ms). Obviously, the need for fast reflexes required a more advanced electrical machinery that could respond in the millisecond time scale, and, not surprisingly, specialized cells, neurons, arose to be built into what we term nervous systems: ctenophores (comb jellies) and cnidarians (corals, jellyfish, and anemones) already have neurons able to fire fast action potentials and form net-like structures. Organisms thus became increasingly adaptable as their responses to external stimuli were faster and more accurate. In other words, as the sensorimotor transformations performed by nervous systems allowed a better control over the environment, more efficient behavior arose, as eloquently expressed by Roger Sperry: "...instead of regarding motor activity as being subsidiary, that is, something to carry out, serve, and satisfy the demands of the higher centres, we reverse this tendency and look upon the mental activity as only a means to an end, where the end is better regulation of overt response. Cerebration, essentially, serves to bring into motor behavior additional refinement, increased direction toward distant future goals, and greater adaptiveness and survival value. The evolutionary increase in man's capacity for perception, feeling, ideation, and the like may be regarded, not so much as an end in itself, as something that has enabled us to behave, to act, more efficiently." (Sperry, 1952). Certainly, brains did not evolve to philosophize.

To achieve this feat of improved control over the surroundings, what is needed is a network of connected cells that produces patterns of activity to guide the behavior. A consideration of earlier nervous systems and their development is always illuminating to understand the basic functions and the mechanisms of action of modern brains, and reveals that the basic principles of an organization and activity have remained constant throughout evolution. Let us consider the possible common connectivity and activity patterns that mammalian brains have with primitive nervous systems, by taking a closer look at the invertebrate central pattern generators (CPGs). The CPG is a good start to reflect on these matters because it arose early in evolution. A CPG is an interconnected net of neurons that produces a rhythmic pattern of activity (even in the absence of sensory inputs), which in invertebrates result in rhythmic movements in, for instance, swimming or crawling (this field is thoroughly reviewed by Marder and Calabrese (1996)). The connectivity patterns in these relatively small cell networks have certain common features. The common connectivity patterns, that serve computational roles to produce the periodic activity, can be described in the presence of reciprocal inhibition and excitation, direct electrical coupling between cells, and delayed feed-forward inhibition with feedback excitation. Computational models have shown that a combination of electrical coupling, inhibitory transmission, and intrinsic cell properties can generate flexible phase relations between spike firing of CPG neurons (Weaver *et al.*, 2010). This general structure of CPGs has been proposed to underlie, as well, the structure of the mammalian neocortex (Yuste *et al.*, 2005). In this proposal, the cerebral cortex is viewed as a "plastic," more adaptable CPG than classical CPGs. Indeed, the anatomical arrangement of neocortical circuitries shares many similarities with those aforementioned of the CPGs: reciprocal inhibition, mutual and recurrent excitation, and direct electrical coupling (even though this type is more restricted than chemical coupling in adult tissue, but very abundant in early development). At least one aspect of the structural arrangement of the CPGs is shared not only by cortical circuits, but also by many brain areas: reciprocal connectivity. In fact, we would say that this is commonplace throughout the whole brain. Reciprocal connections have been clearly anatomically identified in numerous areas including in mammalian frontoparietal cortices (Cavada and Goldman-Rakic, 1989) and between visual cortical areas (Anderson and Martin, 2009). But anatomy is not all; Yuste and colleagues also observed biophysical similarities between CPG neurons and cortical ones, including the feature known as bistability. Bistability in the level of "spontaneous" activity (this term needs to be interpreted with caution, for, are not all cellular activities *in vivo* spontaneous to some extent?) has been commented upon in several works, mostly referring to the so-called "up" (depolarized) and "down" (hyperpolarized) states in the membrane potential, these two being the "stable" states (but see Chapter 2 for stability considerations from the dynamical perspective). This phenomenon has been reported in cortical, thalamic, and spinal neurons, and seems to be another feature of importance in the resulting activity patterns. For instance, this switching between the two states, up and down, results in two different frequency of discharges in (rat) spinal neurons: a low level about 0.3 spikes/s corresponding to the "down" state and a higher frequency of $\sim$10 spikes/s associated with the "up" state, and, of interest when considering dynamical aspects of neural activity (Chapter 2) and the role of sensory perturbations in modifying the intrinsic dynamics, Monteiro *et al.* (2006) found that the transitions between these two states were, in some instances, induced by somatosensory stimulation of the animal. Bistability has also been inferred from the collective, macroscopic scale derived from EEG recordings (Freyer *et al.*, 2009).

Of course, there are main differences between neocortical activity and function and those of the CPGs, since the cortex, after all, evolved not to implement strict locomotor patterns like CPGs do, but to better exert fine control over these motor actions. Thus, the relatively stable limit cycle dynamical behavior (limit cycle and related dynamical concepts are explained in Chapter 2) of the CPGs has to be altered in the case of brains to endow a more flexible information processing, thus a great variety of dynamical behaviors can appear. One intriguing possibility could be here pointed out, derived from the fact that brains never rest. There is always substantial brain activity as has been shown in numerous studies where subjects were not performing any specific task. Even during sleep there is continuous neural activity, and some of these sleep patterns recorded in animals and humans are thought to represent consolidation of memories but these are still speculative ideas. Hence, would it be conceivable that, much like in the everlasting and regular CPG patterns of activity, the cortex is constantly exhibiting these not-so-regular patterns that look irregular and aperiodic because there is a myriad of interconnected cell networks? The "cortical CPGs" could be the small cell ensembles that (transiently) coordinate their activity perhaps displaying a regular rhythm, but because there are myriad, the summation of all these oscillations that is recorded as EEG appears irregular: the typical low-amplitude gamma band activity in awake states of the EEG. Perhaps, this is something to consider, as Yuste *et al.* (2005) put it: "we argue that Hebbian plasticity does not preclude the presence of predefined neuron classes and connections with predefined dynamics." These "predefined connections and dynamics" could be the tiny cortical CPGs, even though at a greater scale the cortex also has a CPG structure. Nevertheless and to conclude this topic of the comparison between the old and the new, from a global perspective as it is taken in this volume, it can be stated that the basic anatomical and functional connectivity patterns of the primitive nervous systems still remain in the highly developed nervous systems.

Vertebrates have CPGs too, those found in brainstem and spinal cord that generate regular oscillatory activity patterns involved in, for example, breathing and locomotion. The CPG's main function, the production of relatively stable rhythms, relies many times on the activity of pacemaker cells. However, some do not depend on pacemaker cells, or at least these have not been identified in some specific cases in invertebrates, such as the gastric mill rhythm of crustaceans that is generated by the stomatogastric ganglion. In the absence of pacemaking activities, thus, this gastric rhythm (that controls the three teeth that these animals possess in order to chew their morsels) is considered to represent an emergent network phenomenon that depends on the chemical and electrical contacts among neurons, and on modulators. Interestingly, this same ganglion produces other rhythms, such as the pyloric oscillation, that do depend on pacemaker neurons firing at a relatively stable bursting frequency. It can already be appreciated that much of nervous system activity at the cellular and network levels tends toward the production of periodic oscillations.

1.2.1. Rhythms: The primordial nervous system activity

Proceeding now with developmental considerations, the emergence of periodic rhythms in nervous systems and the extraordinary tendency of the constituent cells to synchronize their activity deserve a special paragraph. It has been long known that periodic rhythmic activities are prominent in the developing and in primitive nervous systems. Before proceeding, let us specify what is understood here as rhythmic or periodic oscillations because these phenomena are sometimes differentially conceptualized by clinicians and scientists in diverse fields. Periodic activity is taken as a regular activity, with a defined period that remains relatively constant, and is treated as synonymous of rhythmic. For example, the recordings in the lower panel of Figure 1.2 are periodic. It is a different matter to ask whether these rhythms are synchronized, and different types of synchronization that can be considered are presented in Sections 1.3 and 1.4.

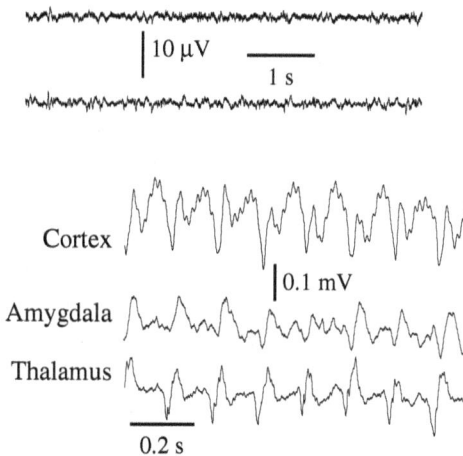

Figure 1.2. Upper and lower panels represent simultaneous intracerebral recordings from rats. The two signals in the upper panel were taken in the cortex during an awake state of the animal, while the three lower recordings, in the areas specified, were taken during an absence seizure. As noted in the text, these (bipolar) recordings are local field potentials that represent the collective activity in the area surrounding the electrode, and thus show synchronous activity, more evidenced in the lower recordings as high-amplitude signals (note the great difference in the amplitude scale between the first and second panel) which indicate that a pronounced synchronous activity has taken place. Or, more precisely, that post-synaptic potentials are arriving synchronously at the area surrounding the electrode, even though in this particular case of the spike-and-wave seizure, it is known that the "spike" (downward deflection) represents synchronous action potential firings while the "wave" represents post-synaptic activity. Nevertheless, the two upper traces, while low in amplitude as compared with the others, are also manifestations of synchronous activities but at higher frequencies (gamma range). Lower panel reproduced with permission from Proulx *et al.* (2006).

A major physiological reason for the commonality of periodic oscillations in neural tissue seems to be the presence of recurrent circuits. The emergence of spontaneous rhythmic activity is favored by reciprocally connected neurons since this arrangement endows the circuit with high levels of excitability and, as mentioned in the previous section, reciprocal connectivity is widespread. The notion of *recurrence* will appear over and over again throughout this book, sometimes in the guise of self-referentiality, other times as denoting anatomical or functional facets. These recurrent arrangements favor reverberation of activity, which was already postulated long ago by the likes of Lorente de Nó (in the 1930s) and Donald Hebb (in the late 1940s). What is then special in these anatomical arrangements to generate rhythmic patterns? Theoretical frameworks have been developed to understand how reciprocally connected cells can coordinate their activities (with synchrony as a specific aspect of coordination), and especially the case of inhibitory neurons reciprocally connected has been considered from a theoretical/computational stance for quite some time (Wang and Rinzel, 1992; van Vreeswijk *et al.*, 1994; Skinner *et al.*, 1994, just to cite a few among an extensive literature). These computer simulations using specific neural models first indicated, and we think this was before any clear experimental evidence was available, a tendency of inhibitory neurons to synchronize their activity even in the absence of excitatory inputs, that found experimental support in later observations demonstrating the importance of inhibitory potentials in promoting synchronization and shaping brain rhythms (Whittington *et al.*, 1995). The precise computational models vary in the form of the formulae used in the simulations of one or another type of activity, and this is not commented any further here. Suffice to mention that the resulting coordinated activity is due to specific membrane conductances (the hyperpolarization activated cation current I_h seems to be of particular importance as well as the cellular features of rebound responses) and their synaptic interactions.

This extraordinary tendency of neural activity to synchronize, regardless of its precise nature (inhibition, excitation, field coupling, etc.) is manifested in a multiplicity of phenomena observed *in vitro* and of course, *in vivo*, and it is this tendency that results in pathological situations when more than "normal" synchrony is present. The "normal" synchrony cannot be easily defined, let alone quantified, and this matter is treated specifically in each of the following chapters that deal with distinct diseases and behavioral deviations. Sections 1.4.1 and 1.4.2 do present preliminary comments. In addition to chemically mediated synaptic interactions promoting synchronous behavior, direct electrical coupling via gap junctions or field effects are also important to improve synchrony. In the case of gap junctional coupling, synchronization between neurons is facilitated by the similarity of the cells, and perhaps this is a reason why this type of coupling occurs mainly between same cell types (Perez Velazquez, 2003a). However, even completely different neurons, when artificially coupled via a computer, will tend to synchronize, as shown in the artificial "gap junctional" coupling between a pyramidal and a thalamic cell, two cells that have almost no chance of being coupled in this manner in the brain (at least it has never been reported). In these experiments, the computer simulated a gap junction (basically a resistor, using the dynamic clamp technique) while these two neurons were simultaneously patched in the *in vitro* brain slice, and, while same cell types (pyramidal to pyramidal) exhibited 1:1 synchronization in action potential firing with a certain strength of artificial coupling, the thalamic–pyramidal cell pair never reached 1:1 but was very close to it (Perez Velazquez *et al.*, 2001). But, in theory, the reasons underlying the tendency to synchronize activities in neural tissue stem from the more general phenomenon of synchrony in coupled oscillators: as it is known in physics, coupled oscillators that have similar intrinsic frequencies of oscillation will sooner or later synchronize. Further, essentially for the reasoning presented in this volume pertaining to the role of synchronous activities in the transition to pathology, it should be noted that there is no need for a large population of cells to become excited and finally synchronize their spike firing. Detailed *in vitro* studies have revealed that it is enough that a small and discrete neuronal population becomes active to originate synchronized neuronal bursting in whole cortical networks (Connors, 1984). This finding is of relevance to seizures, in particular, but in general to any other behavioral deviation that relies on extensive synchrony, for an undetectable hyperexcitability of a discrete neural net could underlie more synchronous activity resulting in hallucinations in schizophrenia, or in obsessive-compulsive behaviors, or autism, as discussed in the following chapters. In the final analysis, deviations from proper neurophysiological function could be the result of the neural tissue's tendency to synchronize.

The previous section introduced the pacemaker cell in the generation of rhythms. It is then conceivable to ask whether the various brain oscillations rely on pacemaker activities. Pacemaker cells should sustain stable periodic firing in the absence of inputs. However, individual neurons in the mammalian central nervous system cannot really be considered self-sustained oscillators. Here, the term "individual" is meant isolated, for the situation changes when neurons are allowed to interact in a network, in which case spontaneous rhythms do appear. But, going back to the individual cells, it seems that the great majority of central neurons do not manifest stable self-sustained periodic firing patters. Reasons for this are numerous, starting with constant perturbations like the bombardment by synaptic inputs, and continuing with some intrinsic features of the cells such as the phenomenon called spike-frequency adaptation (Fuhrmann *et al.*, 2002). This phenomenon is typical of, for example, pyramidal neurons, and occurs when the cell is subjected to a constant depolarizing input to make it fire such that it will tend to respond with irregular and aperiodic firing patterns, but if the input is maintained for a relatively long time (which in the neuron's time scale this means a few hundreds of milliseconds) the spike firing will diminish and eventually cease. No self-sustainability of firing is found here. These are all manifestations of the transience, variability, and nonstationarity of brain activity, three main phenomena with which any student of the brain should become acquainted, and that will reappear over and over again in subsequent chapters.

There are, although, few exceptions when cells look like intrinsic oscillators, classical pacemaker cells. These are found in the inferior olive nucleus, a major input to the cerebellar Purkinje cells, and in the septum, an area that projects to, and to some extent drives into rhythmic activity, the hippocampus (Stewart and Fox, 1990). Recently, instances of pacemaker-like neurons in cerebral cortex have been reported in juvenile tissue of rodents (Le Bon-Jego and Yuste, 2007), which is of interest because the immature mammalian brain tends to exhibit more stable rhythmic patterns than the mature one. But in general, for the majority of adult brain neurons, at least those of the cortex, their firing patterns are state dependent and tend not to exhibit stable rhythmic activity, with notable exceptions such as during sleep or the alpha band frequency that appears upon closing the eyes. The state dependency dictates that there are moments when cells indeed behave as intrinsic self-sustained oscillators, such as the neurons in the nucleus reticularis of the thalamus (NRT) as it was shown that the isolated (deafferented) NRT, with synaptic inputs from other brain areas severed, is still able to generate rhythmic activity in the 8–14 Hz frequency range (Steriade *et al.*, 1987). These oscillations are known as spindles and are characteristic of drowsiness and sleep states. However, when bombarded with inputs from connected brain areas, NRT cells may not sustain that particular activity and fire in a variety of patterns. Not only NRT cells but also all thalamic cells tend to fire in different patterns during sleep or wakefulness: in the former, nearly periodic bursting firing patterns are typical, while, in the latter, thalamic cells fire mostly in a non-bursting mode. Thus, considering that neurons display distinct firing patterns depending on the cognitive/behavioral state, perhaps it is conceivable to ascribe a "functional intrinsic frequency" of firing associated with specific brain/mental states, rather than a pure intrinsic frequency since this one depends on the functional state, on the context. In any case, brain oscillations are not completely intrinsic, but depend on the interaction between sensory, or internally generated, inputs and intrinsic biophysical membrane properties. These comments referred to the individual cells, because at the collective level, on the contrary, the landmark is precisely the emergence of stable periodic oscillations. Seemingly stochastic cellular activities can nevertheless result in coherent behavior at the macroscopic scale (El Boustani and Destexhe, 2010). Studies performed on sympathetic cells of the rat revealed relatively stable rhythms at the cell population level and that these collective periodic activities can exist even in the absence of mutual entrainment of its constituents (Chang *et al.*, 2000). These studies also reported an important feature in nervous (and many other) systems, that of transient phase locking between these sympathetic post-ganglionic neurons. This transience in coordinated activity is, as seen in subsequent sections and in the rest of the chapters, a fundamental aspect of brain function. Many other examples of collective patterns exist and have been observed both *in vivo* and *in vitro*, like the theta oscillation (4–8 Hz) that emerges even in the isolated *in vitro* rat hippocampus (Goutagni *et al.*, 2009). The remarks in Section 1.6 on the activities of the thalamo-cortical circuitry depict a classical example of this notion of how the combination of intrinsic membrane properties of individual cells and the anatomical and synaptic connectivity result in the emergence of a variety of periodic rhythms.

The last matter to note in this section is that the previous arguments revolve around individual cell activity and network phenomena that are observed with distinct recording techniques, and because these methodological considerations are important, the next section is devoted to recording methods.

1.3. Measurements of Nervous System Activity. Brief Notes on Recording Methodologies

A deep understanding of the nature of the signals recorded from nervous systems is crucial for a reasonable interpretation of what the activity observed means. There are different levels at which brain activity brain can be recorded, recall the three levels presented at the start of Section 1.1. One of the most basic methods is to measure electrical activity, for, after all, it was stated above that the most crucial cellular interactions occur via changes in electrical potentials. In this case, the time series that constitutes the recording consists of voltage values representing differences between intracellular and extracellular potentials. To measure at the single-cell level, intracellular recordings that allow access to the cytoplasm are optimal to reveal the excursions in the cell's membrane potential, these mostly representing synaptic potentials that are of low amplitude, normally smaller than 5 mV, action potentials (of high amplitude $\sim$100 mV), or subthreshold oscillations of the resting potential (subthreshold because these do not cross the threshold for action potential generation). These individual recordings are achieved using either intracellular electrodes that impale the cell, or the patch-clamp method, in which case the recording electrode, made of glass, fuses with the cell membrane and therefore allows access to the internal cytoplasm. It is normally hard to perform multiple intracellular recordings simultaneously with these methodologies, therefore to focus on the collective level of description other methods are required that allow for the measurement of global network activities. Collective activity can be recorded as field potentials by placing recording electrodes in the extracellular space in the tissue, and a grounding electrode somewhere depending on the specific preparation (*in vivo* recordings methodologies are detailed and reviewed by del Campo *et al.* (2009)). The recorded signal in this case represents a summation of several electrical events in the neighborhood of the electrode: synaptic activity, neuronal spike firing, and glial cell activity, all these of course nothing more than manifestations of the diffusion of ions between interior and exterior of cells. Normally, it is the synaptic potentials that contribute most to the signals, because these last longer than the spikes and thus the currents spread to reach the electrodes. Spikes will contribute to the field potential recording if they are synchronized, as it happens during seizures for instance (indeed it is known that action potentials underlie the "spike" in the spike and wave discharge shown in Figure 1.2, lower panel). Basically, the same can be said about scalp EEG or MEG recordings, the former records the spread on the scalp of the currents due to, mostly, synaptic activity, and the latter measures the magnetic fields associated with the tissue's electrical activity, for, as Oersted and Faraday taught us, electric currents have always an associated magnetic field.

Consideration of these techniques is important because when we deal with MEG or EEG signals we are really looking at activity from many brain areas, not only those right under the recording electrode but also from others, distant, that are sending synaptic input to the region under the sensor. The recordings in Figure 1.2 depict a classical example of waveforms that, in one case, are considered "synchronized" (lower panel in the figure) and in another case "not synchronized" (upper panel). Yet, the signals in the upper panel represent synchronous activity at γ frequencies (see Table 1.1 for the frequency ranges) arriving at the area where the recording electrode is positioned, of course synchronized to less extent than those of the lower panel as can be seen in the amplitudes of the waves. Classically, recordings in these relatively high-frequency bands were considered unsynchronized, and still this conception remains among clinicians, but, to emphasize once again, if there were not a trace of synchrony among the synaptic potentials arriving at the electrode, then only a flat line would be seen. Another matter is to examine whether there is synchrony between these two signals shown in the figure (or the three signals in the lower panel), for which synchrony analysis methods can be employed. The fact that there is synchronous arrival of synaptic potentials to one region does not guarantee, of course, that those are synchronized to the potentials arriving at another region.

Hence, the evaluation of what the signals represent is essential for studying and modeling neuronal synchronization patterns.

The recordings in Figure 1.2 are local field potentials representing collective events. The different recording techniques that allow one to visualize cell and network activity may result in ambiguities, if a good understanding of the methods and the physiology is lacking. One point to emphasize is that what may look like random activity at one level, it may appear as coherent at another level. This is why, as explained above, there is always a local field potential that can be recorded extracellulary as this is a collective phenomena due to the summation of many events (the synaptic potentials and spikes). This is almost a biological manifestation of the central limit theorem, which is really an expression of a most basic collective phenomenon and states that contributions of many random variables (post-synaptic potentials and spikes) lead to a collective behavior (the local field potential) which becomes simple, the individual details are washed out, or, in the words of Gnedenko and Kolmogorov: "In fact, all epistemologic value of the theory of probability is based on this: that large-scale random phenomena, in their collective action, create strict non-random regularity" (B.V. Gnedenko and A.N. Kolmogorov, *Limit Distributions for Sum of Independent Random Variables*, Addison Wesley, 1954). Figure 1.3 is a simple scheme to illustrate these points. A last word of caution relates

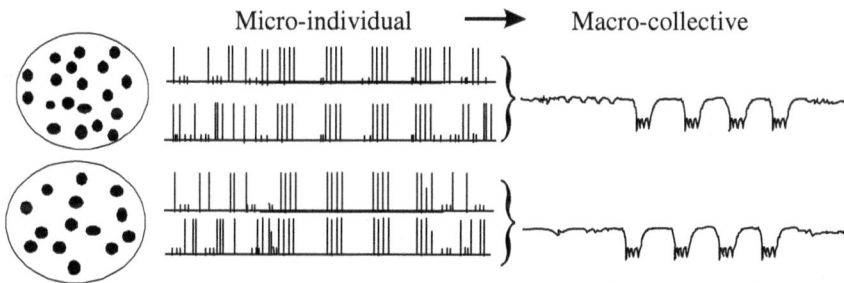

Figure 1.3. Neurophysiological signals at the micro- and macrolevels serve to understand the transition from the microscopic to the collective organization of activity. Two cell sets are represented, and two hypothetical intracellular recordings from two neurons in each set are shown, each cell receiving synaptic potentials (the short spikes) and firing action potentials (long spikes) in either a nonsynchronous or synchronous manner. The macroscopic recordings, local field potentials, represent the summation of the individual activity in the cells. So, the synchronous firing interval (located around the middle of the traces) is equally evident from both types of micro- and macro-recordings, but during desynchronized activity there is still a summation of events that results in coherent activity in the field potential signal, and while measurements of synchrony may not detect any in one type of the recordings, it may be detectable in the other depending on what is used to assess synchronization. For instance, there are synchronous synaptic events and spikes in the cell recordings that may not be captured in the synchrony analysis between the field potential traces.

Table 1.1. Main division in the frequency domain of the waves in EEG recordings, and behavioral states in humans in which each band is more prominent.

Frequency band name	Frequency band width	State associated with bandwidth
Slow oscillation	<0.5 Hz	Slow wave sleep
Delta	0.5–3.5 Hz	Slow wave sleep
Theta	4–7.5 Hz	Drowsiness
Alpha	8–12 Hz	Relaxed state, closed eyes
Beta	13–30 Hz	State of alertness, sensory-motor processing
Gamma	30–100 + Hz	Alertness, cross-modal sensory-motor processing

to the aforementioned fact that with macroscopic recordings we are never sure to what extent neurons are firing spikes, hence, since reliable spike transmission from net to net is the basis for the transfer of information in nervous systems, to record field potentials from two connected sites only informs us as to synaptic inputs that are arriving at those areas. It is not known how many cells need to fire spikes in each ensemble to activate another and thus transfer information.

1.4.   Collective Activity in Nervous Systems: Coordination Dynamics

The first brain recordings using scalp electroencephalography, attributed to Hans Berger and published in 1929, revealed a mixture of waves that were subsequently described in the frequency domain using the Greek alphabet, represented in Table 1.1 below. Translated to the time domain, frequencies become periods and then the different bands can be seen as doubling of the periods, for example the theta band at 4 Hz means a period of $1/4 = 0.25$ s, and the next one is called alpha with a frequency of 8 Hz and thus a period of $1/8 = 0.125$ s, one-half of the period of the theta wave. Period doubling is a dynamical regime that is discussed in Chapters 2 and 3 in the context of epilepsy. The classification presented in the table is, naturally, completely arbitrary as it was conceived by investigators in efforts to correlate some activity patterns with some behavioral situations. For instance, during sleep lower frequencies are more abundant and thus they had to be baptised with such names as delta. This terminology makes easier the communication between clinicians, but the general advice is to avoid getting conceptually trapped in this alphabet: there is not a one-to-one relation between a frequency band and a specific behavior, although it is correct to say that some frequencies seem to correlate better than others with specific situations. Tort *et al.* (2010) put this consideration in a clear fashion: "it is overly simplistic to create monolithic characterizations of the function of rhythms in particular frequency bands." Related to this matter, a very commonly asked question is what the functional role of each frequency band is on cognition, to which a few words are devoted in the final section of this chapter. Now, just a consideration is advanced, in that the fascination with gamma frequencies being present in recordings during wakefulness and associated with almost any possible task that can be imagined, has really nothing mysterious to it: it is just the result of myriad of sensory stimuli arriving to the cortex via subcortical structures that make cells fire action potentials in a wide variety of fashions, hence the multiplicity of gamma rhythms from 30 Hz up. Further to demonstrate that there may not be anything too specific about this range, consider that cells also fire at gamma frequencies during deep sleep, as shown in Figure 5.1(B) of Chapter 5: that cell fires high-frequency spikes on top of a slow bursting oscillation, whether or not these spikes can be detected at the level of scalp EEG or other techniques is irrelevant to the fact that cells do fire in almost any possible pattern during any possible behavioral state.

At the level of scalp EEG recordings, as mentioned in Section 1.3, what is measured is the collective activity representing a summation of synaptic potentials, and to a lesser degree neuronal spike firing and glial cell activity. Hence, the aforementioned waveforms are manifestations of collective, synchronized activity because if the potentials were not synchronous then the summation would be minimal and the recording would be a flat line. But this is obvious. What is equally obvious, derived from the observations of brainwaves, is that synchronous activity is omnipresent, for there is never a flat line in any healthy neurophysiological recording (but cessation of activity occurs in pathological conditions like some late stages of ischemia and, of course, brain death). We would like to note a common misconception that was alluded to in the previous section: viewing the EEGs during awake states, that consist of mostly gamma frequencies, and thinking that these are desynchronized, a view that comes from old since the early times when it was shown that stimulation of the reticular formation caused "EEG desynchronization," is not correct, for high frequencies can be as synchronized as low ones. See, for instance, the gamma synchrony observed in intracerebral recordings in cats during anesthesia or wakefulness, although it is true that this coherence of fast oscillations was more confined to short-range spatial scales than the large-scale synchrony found for low frequencies, at least during unconscious states (Steriade *et al.*, 1996). As emphasized in Section 1.2.1, of note is the extraordinary tendency of brain cell ensembles to synchronize. Indeed, this is perhaps the principal characteristic of nervous systems, whether brains, spinal cords, or invertebrate ganglia, the tendency is to produce rhythmic synchronous patterns of activity. It is this tendency, as described in the rest of the chapters for several pathological conditions, that underlies the deviations in behavior that are labeled as pathological. But let us not forget that, in the final analysis, it is our perspective that classifies and distinguishes behaviors (and thus brain activity) as healthy or unhealthy. From a more natural viewpoint, perhaps unhealthy, or pathological, behaviors are those that do not have an adaptive value, that is, those that are detrimental for the survival of the individual. It could be a matter of debate whether our subjective division between normal and deviant brain function coincides and maps the non-adaptive behaviors. Chapter 7 concludes with a discussion on this and similar themes.

The observation of the brainwaves by the early brain explorers of the beginning of the 20th century prompted a natural interest in synchronous neural dynamics, but it has been recently that a surge of interest has emerged along with the idea that coordinated activity in brain cellular networks is fundamental for information processing. This notion originated with proposals already advanced in the mid-1900s by the Russian psychologist Alexander Luria (1966), who envisaged that the dynamic interplay between brain areas is the essence of brain function, followed later by the ideas of Livanov (1977) who had already articulated many of these matters on synchrony with his theory of the spatiotemporal organization of brain function. Seminal contributions were also those of von der Malsburg (1981), and many others (Damasio, 1989; Singer and Gray, 1995). These proposals that attributed a fundamental importance to brain correlated activity were substantiated by early empirical evidence of correlations in the activity between cortical cells (Li, 1959). More recent empirical evidence indicated that synchronous brain oscillations depend on the sensory stimulus to be processed, as, for instance, was shown in cat cortex using cross-correlation analysis of unit and local field potential responses to visual stimuli (Engel *et al.*, 1990). This study also presented evidence for a constant phase relation of cell firing, that is, the activity was phase locked, and this was interpreted as evidence for the role of synchronous activity in the formation of cell assemblies, each ensemble processing one aspect of the stimulus. Synchrony is invoked to bear on aspects of neurophysiology ranging from perception and cognition to consciousness (Singer, 2006), and the investigation of the mechanisms and functions of synchronous activities is becoming a very popular trend in neuroscience, notwithstanding some relatively serious methodological considerations in the analysis of this type of coordinated activity (for current trends in brain synchrony analysis and concepts, see Perez Velazquez and Wennberg, 2009).

The general concept is that synchronization of neuronal activity supports the temporal correlations in cell action potential firing that seem to be necessary for neuronal ensembles to activate one another, to ensure the stability of the transmission of information in the form of action potentials (Abeles, 1991; Harris, 2005), see axiom #3 in Section 1.1. The very basic and most fundamental reason why synchronous discharges of spikes in a group of neurons is absolutely necessary for nervous system function is because, normally, more than one spike is needed to generate another spike in the receiving cell. Thus, a group of neurons firing in synchrony increases the probability that the train of spikes will continue to be propagated in a connected network, because the post-synaptic potentials generated by the synchronous spike volleys in the receiving cells will summate and depolarize the neurons enough to reach threshold to fire action potentials, and thus the chain proceeds to other connected networks. As a recent illustration of this phenomenon, the work of Bruno and Sakmann has revealed that single thalamic neurons cannot activate a layer 4 cortical cell because of the "weak" thalamocortical synapse that generates a small post-synaptic potential. However, although these individual connections are weak, the abundant convergence of thalamic axons onto cortical neurons and the synchronous thalamic discharges produces reliable activation of the cells in layer 4 of the cortex, the main target of thalamic axons (Bruno and Sakmann, 2006). This manner of activating a connected set can be considered, in dynamical terms, forcing, or, more precisely, periodic forcing if a periodic oscillation was enforced by a driver network. This will come up again when discussing, below, the mechanism of neural synchronization, because forcing, rather that the adjustment of frequencies of weakly coupled cells, seems to be more prevalent in the brain (these ideas are extensively discussed by Perez Velazquez (2005, 2006)). General evaluations of the stability of neural transmission can be found in Abeles (1991). The stability of transmission ensures that the same output is given to similar inputs even in the presence of noise. Notably, it has been shown that loops in the structural arrangement of the units, by feedback and feed-forward paths, facilitate the stability of transmission (Majumdar, 2007). These loops that facilitate what can be considered a synaptic reverberation are related to the concepts of reentrant excitation that are invoked to play a major role in, for example, epileptiform phenomena as discussed in Chapter 3, and in memory (Wang, 2001). Some have argued against the dynamic stability of reverberatory cell assemblies (Milner, 1996), and some theoretical studies have been devoted to address the conditions for the stability of the reverberatory activity with regards to the attractor theory of brain dynamics and the formation and stabilization of memory traces (Tegnér *et al.*, 2002). In more general terms, the anatomical loops can be thought of constituting a mode of self-referentiality that is a major feature of the brain, not only anatomically (most areas are reciprocally connected) but also functionally, and mentally (as it relates to self-awareness and consciousness). Perhaps, all this is a vestige of the origins of the nervous system, as was described when discussing the structural organization of the CPGs in Section 1.1. In this sense, it can be said that nervous systems exploited early the advantages of self-referentiality.

To reiterate, a nervous system's main function consists in the propagation of activity from net to net. Those who are versed in neurophysiology will have probably read about synfire chains (Abeles, 1982), sequences of synchronous activity propagating from network to network that are anatomically linked to each other in a stepwise manner. Experimental evidence for synfire chains has been obtained using MEG while subjects evaluated speech processing (Pulvermüller and Shtyrov,

2009), with multiunit recordings in the vocal control nuclei of birds that produce songs (Kimpo *et al.*, 2003) and in the rat somatosensory system during the typical whisker twitching in these animals (Nicolelis *et al.*, 1995), and several others too numerous to be cited here. Thus, these works represent substantial empirical evidence indicating the importance of the serial triggering of activity in connected cell assemblies. While this is mostly accepted by neuroscientists currently, it seems that the implications of this phenomenon have not permeated our scientific community, for what it entails is that there is no start or end to the chain, and from this, a host of consequences derive. The last section of this chapter includes some reflections on the implications of the serial triggering of activity for brain function and behavior.

These cellular and network processes underlying collective actions can be linear or nonlinear, and for our present purposes this distinction is not fundamental. A multitude of studies have treated the problem of linearity versus nonlinearity in nervous system activity, trying to resolve how it can be best conceptualized. As usual in science, avoiding dichotomies and answering "in both ways" will probably be most accurate. The ever-increasing popularity of non-linear system analysis may be confounded with the previously posed query, which is of different nature, in that one thing is the consideration of whether to use a linear- or non-linear methodology to analyze some neurobiological signal, and another thing is to conceptualize the system as linear or not. These are two different aspects. A linear method can be used to analyze a non-linear system. Nonetheless, pondering on this dichotomy is not needed for the purposes of the notions that follow in the next chapters, for what is linear (nonlinear) at one level may be nonlinear (linear) at another. Proportionality (or additivity) and independence are the two main conditions for linearity. If the total output response of the system to an action from several factors is equal to the sum of the results of the values of each separate factor, that indicates independence: the total force on a car resulting from two people pushing the car is the sum of each person's effort. If the response to the action of each separate factor is proportional to the factor's value, this indicates proportionality: you push the car with double the force and the car is dragged double the distance. But one can already notice in this trivial example that the fact that the individuals are nonlinear (as most physiological processes are, thus bodies can be considered non-linear systems) does not guarantee that the whole system (car plus people pushing) behaves nonlinearly. Some neurobiological aspects are clearly nonlinear, like synaptic transmission because synaptic potentials do not add linearly; as well, neurons possessing a threshold for spike generation (all-or-none response) presupposes another nonlinearity. Many intrinsic biophysical aspects are thus nonlinear, but, as aforementioned, this does not mean at all that the global activity is nonlinear! Strict views on linearity and nonlinearity will not match the variety of neurobiological phenomena. Hence, the interest in the coordinated activity in nervous systems prompts the study of its dynamics, and this is the province of coordination dynamics.

1.4.1. Fluctuations in coordinated activity

So synchrony is needed, but, to what extent? It can be envisaged that too much synchronous activity involving extended brain areas for a long period of time may not be adequate to process information. Recall the previous comments (Section 1.1.1) on global integration and local segregation of networks that has been invoked in adaptive information processing of stimuli. Temporal correlations facilitate the segregation and integration of cell ensembles so that each set of connected networks processes one aspect of the stimulus and the different aspects are integrated in, perhaps, cortical association areas, but this correlated activity should be transient so that flexible formation and dissolution of cell ensembles takes place. Current theories of brain function propose that the coordinated integration of *transient* activity patterns in distinct brain regions is the essence of brain information processing, giving rise to consciousness. Hence, fluctuations in the synchronous activity should be present when brains are processing information, perhaps not so much when they are sleeping. But, if the flexible formation of neural activity patterns requires transient coordination which is manifested in a certain variability in the measurement of synchronization, is there any upper (or lower) bound to it? When is too much or too little synchronization unhealthy? This characteristic of fluctuations in neural coordinated patterns should encourage the investigation of not only the magnitude of synchrony, which is the main focus in most studies, but also the variability in the spatiotemporal patterns of organized activity. Some works are emerging analyzing variability and fluctuations in neural activity, addressing the general hypothesis that lack of variability in synchrony is associated with disease. Thus, measures such as complexity of brain signals and a wide assortment of equivalent notions have appeared in the literature, and the general trend seems to support the hypothesis. Conditions in which higher than normal synchronous brain activity occurs are normally associated with loss of consciousness, such as during seizures (Blumenfeld and Taylor, 2003). However, this is state dependent, because during sleep there is too an increase in coherence compared to wakefulness (Achermann and Borbély, 1998) but in this condition the body is not processing information, at least not sensory information, and thus it is reasonable for the brain networks to be highly synchronized. It all depends on the context. Subsequent chapters do discuss and present illustrations of the lack of variability in coordinated activity associated with diverse pathologies like epilepsy (Chapter 3), Parkinson's disease (Chapter 4), traumatic brain injury (Chapter 5), and autism (Chapter 6), and specific results are mentioned for each particular pathology that will be addressed.

Related to this topic, a frequently asked query is what the roles of the different frequency bands on brain function are, and why gamma band activity is associated with wakefulness (and REM sleep). The answer can be found in the need for variability, in the requirement for the existence of many different temporal correlations to integrate and to segregate activities. During wakefulness, there is a myriad of networks working in parallel and thus a wide variety of frequencies are necessary to unite specific networks and separate their activity from others that are processing distinct aspects of the stimulus. On the other hand, during non-REM sleep there are no incoming inputs from the periphery so there is no need to integrate and segregate and thus no need of multiple frequencies. Schroeder and Lakatos (2008) similarly propose that during continuous vigilance in wakefulness low-frequency oscillations are suppressed because they entail long periods of low excitability hence detrimental to sensory processing. But the matter may not be the oscillations per se, but the synchronization of those oscillations in widespread brain regions, because a long period of synchrony in extended neocortical areas will impair efficient information processing.

The discussion in this section prompts the consideration of the concept of metastability in brain dynamics, which is, in fact, the most fundamental notion that is used throughout this book to explain mental activity in the healthy and diseased condition. The next section is, then, perhaps the most important to understand in order to make sense of subsequent chapters.

1.4.2.   Stable, unstable, and metastable activity of nervous systems

Since the nervous system can be considered a dynamical system, the activity of cell ensembles should exhibit stable or unstable steady states. Chapter 2 presents dynamical aspects in detail, but for the present purposes it suffices to consider the importance of certain activity levels in brain cell ensembles, particularly those that represent synchronous, or, using the more general term, coordinated actions. Section 1.2 introduced the CPGs as creators of very rhythmic and synchronous activity (limit cycles in dynamical terms), but while this may be pertinent for those animals that rely on a very structured pattern of motion, it may not be optimal for vertebrate brains to process information in a fast and flexible manner. Can therefore unstable synchronous states be more prevalent than stable ones in advanced brains? Perhaps, the transient stabilization of the unstable states manifests as synchronous oscillations giving rise to the wide spectrum of brain rhythms or to pathological states (Figure 3.9(C) of Chapter 3 and Section 2.5.2 of Chapter 2 present evidence suggesting this notion). If we think of these coordinated patterns as representing steady states or equilibria in nervous system activity (at least for the time being before reflections on these topics are presented in succeeding chapters) then the ephemeral stabilization of equilibria may be so short that in fact it may never really occur, and what we observe are in actuality metastable states. The notion of metastability applied to neuroscience is not as strict as the original conception in physics (most of the ideas of physics translated to the biological domain lose some stringency, but no need to panic). The concept of metastability in coordinated patterns of brain activity is related to an idea already advanced in the mid-1900s by the behavioral physiologist Erich von Holst (1939) who postulated that the basic nervous system activity pattern was the result of the interaction between a tendency of the oscillators (cells) to maintain a steady rhythm and the tendency to draw and couple their oscillation to that of other connected oscillators. These two tendencies create an infinite number of variable couplings and form a state of relative coordination. On the other hand, the CPG activity would be close to absolute coordination. It was emphasized in previous sections that extensive and long-lasting synchronous activity in our brains should be avoided, lest we fall into unconscious states such as during seizures or (slow wave) sleep. Conceivably, this is what the thalamocortical circuitries achieve due to the intricate anatomical connectivity and the diversity of constituent cells. These factors contribute to the expression of the two tendencies that von Holst noted, the tendency of cells to express their intrinsic independent activity and the tendency to couple together, and the normal state of affairs for an awake neocortex is to avoid falling into the absolute coordination pattern, rather to drift around the synchronous states so that what we normally measure when analyzing brain synchronization is relative coordination. However, the tendency to couple together and to synchronize, as expounded thoroughly in this volume, results in states of almost absolute coordination that are considered pathological. Metastability in nervous systems and behavior is discussed more and more often in current times, after its early exponents proposed the notion of metastable brain states a couple of decades ago (Kryukov *et al.*, 1990; Kelso, 1995).

It is not trivial to experimentally capture metastable coordination, these ephemeral states (in fact, these are not states at all). In terms of transient synchrony, this has been detected in numerous studies using macroscopic recordings (EEG, MEG) of individuals performing a variety of behavioral tasks, and specific experiments are treated in different sections in each chapter. Are these transient brain patterns of synchronization manifestations of metastable dynamics? Are there methods to quantify metastability? There are precise quantifications of stability and instability derived from dynamic system theory to which Chapter 2 is devoted. Chapter 3 presents a possible indication of metastable states in epileptiform activity. However, because in reality metastability implies that there are no states, or that the manifestation of stability is extremely transient, the techniques of dynamic theory, that are fit to analyze stable or unstable states, may not be pertinent in this case. Recently, a proposal to sort of characterize metastability has been advanced. Two characteristic times, a dwell time for how long a determined collective tendency persists and an escape time when ensembles are expressing their individual activity, have been proposed (Kelso, 2008). These time factors depend on the specific situation and have to be operationally defined, the author advises. The ratio of these two times is a dimensionless number that provides a measure of the "quality of metastability" in the words of the author, and can be conceived as being conceptually similar to other dimensionless number in physics such as the Reynolds number, the ratio of inertial forces to viscous forces, because this represents a competition between diffusion (the inertial force, the tendency to express individual characteristics) and convection (viscosity, the tendency to couple together in an organized pattern). The analogy is of interest and can be extended to a similar number also used in the study of turbulence, the Rayleigh number, representing the ratio of convective forces and diffusive (thermal and momentum) processes. From a very global perspective, the "competition" (if it can be called like this) between stochastic processes such as diffusion and cooperative collective phenomena permeates natural phenomena, not only related to the living but also to the nonliving, and constitutes a central theme in pattern formation (Perez Velazquez, 2009). This is, again, von Holst's principle put in different words. Another approach to determine metastability could be to somehow measure the spatiotemporal dynamics of the transient formation and dissolution of cell assemblies, or, using the words of Greenfield and Collins (2005), the "turnover of neuronal assemblies." However, our current methodologies may be far from achieving this experimental feat, in spite of the state of the art imaging methods. Theoretical efforts have been devoted to this problem as well. A phenomenological model for the characterization of collective phenomena in the mammalian brain has been proposed, based on the quantification of neural assemblies via synchronization. Essentially, it consists in detecting the sequence of synchronized states and evaluating the rate functions that describe the transition of the network to burst again, which depends on the elapsed time after the last burst, and can be extracted from experimental data (Deppisch *et al.*, 1994). In general, any convenient metric of cellular interactions can be useful in this regard, particularly those metrics that can be framed in terms of dynamic changes in correlations. The problem of quantifying metastability is somewhat similar to the problem that will appear in other chapters about the computation of variability in brain activity: to say that certain degree of variability is "healthy" is synonymous to investigate metastability.

The notions of metastability and multistability are much related and it may not be easy to define and differentiate in experiments. Indeed, these two notions are so intertwined that any separation may not be feasible, at least experimentally, even though theoretically and using the expressive powers of words, some classification can be endeavored. Conceptual frameworks that encapsulate these ideas are presented in Section 2.2.3 of Chapter 2 (Figure 2.7 in particular). Suffice to say for now that multistability, and more specifically bistability, has been theoretically and experimentally addressed in numerous publications in context as varied as epileptic dynamics (Takeshita *et al.*, 2007; Perez Velazquez *et al.*, 2007b) or in multiarticular movements (Chow *et al.*, 2009). Multistability and metastability are concepts of fundamental importance to comprehend brain and behavior and appear in all the following chapters regardless of the disease or deviation considered. Their usefulness in characterizing neural activity and behavioral patterns is not restricted to any particular level of description. Thus, at microscopic levels, one can talk about how spike firing in two neurons can phase-lock with different frequency ratios resulting in a complex multistable structure of the synchronization regime, which has been determined in computer simulations (Rulkov, 2002) and in brain recordings (Perez Velazquez *et al.*, 2007d); or, at macroscopic levels, these concepts are useful in discussions about how brains operate as "mental chains" of switching thoughts and behaviors of individuals (Kelso, 2008).

This section is closed with an afterthought that was already alluded to in Section 1.2. The developmental discussion in that section introduced primitive nervous systems as generators of very regular activity, as fits "primitive" animals. However, it is clear that the emergence of advanced nervous systems required flexible processing of information so that the anatomy and physiology of these newer brains allow for novel dynamic patterns not governed by too regular activity, but by stochastic, metastable dynamics. At the same time, we not only possess CPGs that produce strict rhythmicity, but also the "resting" state of the brain is characterized by a more periodic activity ("resting" here should not be taken literally as doing nothing, as the brain never really rests), such as the slow wave sleep oscillations or the alpha oscillation upon closing the eyes. In addition, the brain recordings from neonates or preterm babies show higher-amplitude waveforms as compared to the recordings of adults, indicating more synchronous activities in parts of the brain (recall that the fact that high-amplitude potentials are recorded in one site is an indication of a barrage of synchronous synaptic inputs arriving at that area, hence some networks have to be synchronized), and seizures tend to happen more frequently in the newly born and young children than at any other stage in life, occurring in preterm neonates more frequently than in full-term babies. Along the same lines, it has been known for many decades that the developing nervous system shows more spontaneous rhythmic activity than the adult tissue (O'Donovan 1999), this activity appearing most frequently as highly correlated bursts and in some cases it has been attributed to intrinsic membrane conductances that "pace" some specific cells favored by gap junctional coupling (Strata *et al.*, 1997). All these observations on immature, developing, and primitive nervous systems support the notion that periodic neural activities constitute the main characteristic pattern of any nervous system, the brain included. Perhaps, those aforesaid periodic brain dynamic patterns during behavioral situations that do not require important cognitive concentration and information processing (today these are investigated within the field of the so-called brain default network) are vestiges from past evolution, a neural relic that manifests itself as soon as given the chance due to the tendency to synchronization. In reality, this seems to be a general physiological tendency, take for instance the more periodic spontaneous fluctuations in the size of pupil diameter that appear as soon as alertness diminishes (this is regulated via the brain stem reticular system), so when given the chance (less central control in his case), more rhythimic oscillations emerge. Schroeder and Lakatos (2008) wrote: "We propose that the brain is biased toward either a 'rhythmic' or a 'continuous' mode of operation, depending on the dynamics of task demands." Conceivably, taking the dynamical perspective that is presented in the next chapter, it can be pictured that the stable states afforded by the dynamics of primitive neural systems are transformed into a variety of unstable states, or metastable, by a different dynamics of the modern brains conferred by the thalamocortical circuitries (among other factors). This new dynamics could be characterized by sensitivity to small perturbations, either external or internal inputs that destabilize long-range synchronization states with the consequent arising of metastable dynamics, so that efficient information processing can take place, thus offering possibilities that primitive systems do not have. This is how the "predefined dynamics" of Yuste *et al.* (see Section 1.2) can coexist with other dynamics in a state space of considerable complexity, as the examples depicted in Figures 2.7 and 7.3 of Chapters 2 and 7, respectively.

1.4.3. Coherent oscillations: Phase relations

Oscillatory activity entails periodic rhythms and specific phase relations between interacting rhythms. The synaptic inputs from different connected cells that arrive at one particular neuron will summate depending on their timing of arrival, that is, their phase difference in the case of oscillatory inputs, and also depending on the integration properties of the dendrites that are dictated by the intrinsic biophysical features particularly of the membrane, which is normally called in neurophysiological terminology intrinsic membrane characteristics and include a variety of features; among the most crucial for determining the cell's excitability (the response to synaptic inputs) are the membrane time constant, input resistance, and resting potential. But, here, the focus is on the network, collective activity, thus relations among variables are most suited for our present purpose. If phase relations are important, conceivably, the difference in the oscillating phase could be useful as a parameter to capture some brain dynamic aspects. The next chapter introduces the concept of order parameter and emphasis is placed on the relative phase: the phase difference between two oscillations. Results involving the relative phase are presented in most of the following chapters that deal with distinct pathologies and behavioral deviations. To understand the reasons for this emphasis, let us look briefly at the crucial importance of phase relations in neuronal information processing.

Figure 1.4 depicts a very simplified representation to illustrate the importance of phase relations in neuronal activity. In-phase synchronization of the membrane potential oscillation and the arrival of the post-synaptic potentials, as represented in the figure, will favor the contribution of all these factors to the propagation of the neural signals as spike trains. This is trivial: spikes that arrive at the peaks of excitability of the post-synaptic cell will have more probability to make it fire an action potential.

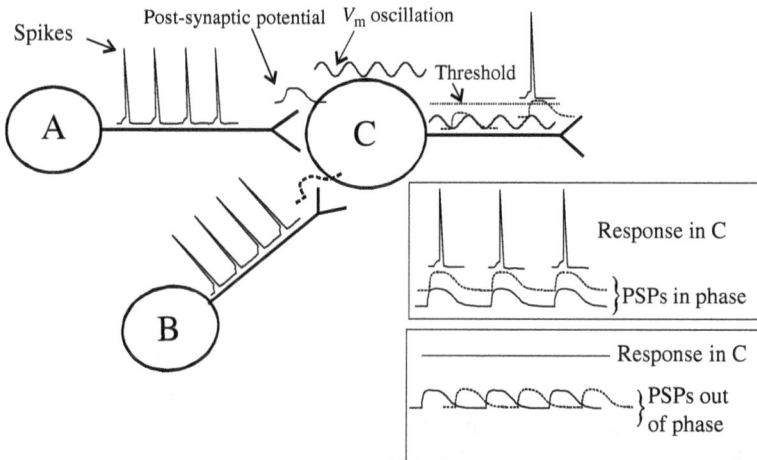

Figure 1.4. Phase relations between post-synaptic potentials (PSPs) determine neuronal output. Neurons A and B send a train of action potentials (spikes) along their axons to neuron C, generating PSPs that represent a slight depolarization of the membrane potential (V_m) that has some intrinsic oscillations (V_m oscillation). The threshold for spike generation in C is represented by the dotted horizontal line. If the depolarization caused by the PSP coincides with a trough of the V_m oscillation in C, no spike will be fired, but if it coincides with the crest, the PSP will add to that depolarization and a spike will be produced along the axon of C, as shown on the right-hand side. In an analogous manner, the phase relation between the PSPs from cells A and B will determine whether neuron C will fire spikes or not: if the PSPs are in-phase (represented in the upper part), the summated depolarization may be enough to cross the threshold for spike firing (normally the addition of PSPs is nonlinear), but if they are out-of-phase there will be little summation and thus the cell will remain silent.

As can be appreciated from this concept represented in Figure 1.4, flexible patterns of phase relations can dynamically modulate the functional connectivity between cells in spite of their constant anatomical connection. Differential effects of stimulation at distinct phases of an oscillation (the alpha cycle in this case) on reaction times of individuals performing tasks and on the brain-evoked potentials were noted as early as the 1960s (Dustman and Beck, 1965). The importance of phase relations in synaptic plasticity and in the determination of spike firing patterns has since been substantiated in several experimental studies (Volgushev *et al.*, 1998; Fries, 2005; Womelsdorf *et al.*, 2007). The term "phase-of-firing coding" has been introduced to refer to the encoding of information in spike trains relative to the phase of a background, ongoing oscillation of the local field potentials, and it has been shown that this phenomenon contributes to synaptic plasticity mechanisms needed for learning and memory (Masquelier *et al.*, 2009). Early evidence was presented by Pavlides *et al.* (1988) in a study exploring how phase relations within the hippocampal θ rhythm determine long-term potentiation of synaptic pathways. Perhaps, one of the first examples demonstrating not only the decisive role but also the flexibility of phase relations in dictating synaptic plasticity mechanisms was the study performed initially *in vitro* (similar experiments were subsequently done *in vivo*) using the hippocampal slice preparation where synaptic potentiation or depression could be elicited depending on the phase relations between two synaptic pathways: as soon as the interaction in-phase that potentiated synaptic responses was changed to out-of-phase, synaptic depression was elicited (Stanton and Sejnowski, 1989). More recently, other studies presented further support for the impact of the relative phase of incoming synaptic activity to a group of cells that are synchronized in the β-band oscillation, in that the phase relation between the synaptic input and the ongoing oscillation modulates the gain of the synaptic input that the group receives (van Elswijk *et al.*, 2010). Not only in the brain but also in the insect nervous system, phase relations are crucial. For instance, phase changes in cellular activity within their CPGs regulate some reflexes that these animals experience (Proctor and Holmes, 2010).

The evaluation of coordinated activity can be done at different levels, from temporal relationships between neuronal spikes (Gerstein and Perkel, 1972) or between EEG signals, to correlations in metabolism measured in neuroimaging. The study of phase relations, specifically, uses the so-called phase synchrony analysis. While most studies consider only the 1:1 phase synchrony, other frequency ratios do occur and, in this fashion, cells are still able to coordinate their activity but do not remain completely 1:1 entrained, and thus allows for a more efficient transfer of information. This is, once again, another sign of the transience in coordinated activity. Admittedly, the use of the term "synchronization" in neuronal studies may not be too adequate, as it does not represent precisely what is meant by synchronization in the classical sense of physics: the adjustment of rhythms between weakly coupled oscillating systems (Pikovsky *et al.*, 2001). Since in a synchronized regime the phase differences of the oscillators remain bounded, synchronization can also be understood as the constancy of the instantaneous phase difference between two oscillators, but the strict constancy of a phase difference that can be observed in mathematical simulations is rarely found in measurements of neurophysiological variables due to noise and non-stationarity among other factors. Consequently, the precise physico-mathematical description of a synchronized state is substituted by a statistical one in which phase differences do not need to remain constant but may fluctuate at a significantly low variance, and this is what is known as phase locking and normally what is measured in the studies of phase synchrony, encapsulated as a synchrony index (Mormann *et al.*, 2000). Even a more general description of coordinated activity can be used: any relation between neurophysiological observations—measured by any reliable method that represents cellular activities—in separate brain areas, which may or may not be directly interacting. Thus, the term synchronization can be used to describe statistical constancy of phase differences between two signals or, in a wider sense, to describe all sorts of correspondences among signal parameters. These are very broad concepts of brain coordinated activity but they encapsulate the classical description of synchronization in the case of directly interacting areas, as well as instances where one brain region sends synchronous inputs to separate areas, in which case a correlated activity in these receiving areas may be detected but is not dependent on a direct connectivity between them (they are connected via a third region). For all these reasons, we recommend use of the term "coordination" rather than "synchronization," in neuroscience. Synchronization would be an aspect of coordination. There are many technicalities and limitations in these analytical methods that have been discussed in other works (Perez Velazquez *et al.*, 2009).

1.4.4.   On the importance of coordinated cellular activity for brain information processing

Previous sections have considered the importance of some aspects of coordinated activities, like phase relations, in cell function. Now, we consider the significance on behaviors. Consider the somewhat trivial task of reaching for and grasping an object. Sensory information about the spatial properties (location, shape) of the object has to be integrated with the spatial properties of the effector systems (arm, hand) using some sort of self-centered coordinate reference frame. Neuroimaging studies indicate that parietal and frontal areas of the brain neocortex compute such sensorimotor transformations and are particularly involved in this grasping task. Without some sort of coordinated cellular activity among those brain regions, the accomplishment of this task would not be feasible. Cognitive processes involve not only serial stages of sensory signal processing but also massive parallel information processing circuitries, and therefore it is the coordinated activity of brain cell ensembles that is the manifestation of cognitive processes.

Different coordinated activity patterns in nervous systems have been associated with distinct behaviors, as has been observed in mammals and, perhaps more clearly, in invertebrates because some invertebrate ganglia have few cells such that they can be easily identified and recorded, and the connectivity between and within ganglia is well known and amenable to manipulations and recordings, so that the possible synfire chains, mentioned in Section 1.4, can be scrutinized in more detail than in complex vertebrate nervous systems. Consider the mollusk *Aplysia*, one of the preferred invertebrates for studies of the nervous system. It was reported that several behaviors, including respiration, spontaneous gill contractions, and the reflex withdrawal, were associated with activity in the same neurons; however, the activity patterns were different, suggesting that it is the distributed organization of the activity in several cell ensembles that best correlates with behavior (Wu *et al.*, 1994) and, naturally, a causal link is prompted: these organized patterns cause the behaviors. But, perhaps, one of the clearest evidence for the role of neural coordinated dynamics as synchronous discharges in sensory processing has been obtained in studies in locusts, where MacLeod *et al.* investigated whether the responses of a neuron set downstream from the odor-activated sensory network depended on the synchronization of the projection neurons, which is one of the first sets of neurons in the chain of odor discrimination in insects. These investigators took advantage of the fact that the synchronization in this set depends on inhibitory (GABAergic) transmission, so they suppressed it with a $GABA_A$ receptor blocker, picrotoxin, that induced desynchronization in the projection set. As a result of this manipulation, the downstream neuronal set showed degraded specificity of responses to specific odorants (scents like cherry and hexanol were used), such that it could not discriminate the spike trains evoked by different odors (MacLeod *et al.*, 1998). Likewise in honeybees, desynchronizing the projection neurons results in impaired odor discrimination of similar odorants (Stopfer *et al.*, 1997). Obtaining evidence for the association of behavioral changes with transitions in the coordinated activity of cell ensembles in the vertebrate brain is a more challenging task and requires novel methods. This is currently being investigated with state-of-the-art optical technologies. Early attempts used simultaneous recordings from various neurons, experiments that are difficult to perform and to interpret because the number of cells that can be sampled is not too large compared with the groups that can be assessed by the modern optical measurements. Obviously, to study coordinated activity, the larger the number of cells whose activity can be monitored, the better. In any case, the electrophysiological evidence based on multiple cell recordings generated a debate whose end is not near, in that while some studies found supporting evidence for cell synchronization that was correlated with behavioral states, others reported accidental synchronization, for example in monkey cortex during forelimb movements, accidental in the sense that the synchrony measured was not related to movement execution (Murthy and Fetz, 1996). To make matters even more interesting, other studies also showed not an increase, but a reduction of synchronization during behavioral performances, for example in monkey visual cortex processing visual stimuli (Cardoso de Oliveira *et al.*, 1997). However, reductions in synchrony constitute a pattern too, nobody ever said that only the increases in synchronization are the important patterns, even though it is fair to note that most research focuses on the increases of the magnitude of synchrony. Other studies evaluating multiunit activity in the antennal-lobe neuron nets of moths have revealed that, among four possible coding mechanisms including instantaneous cell firing rate or total network synchrony, it is the pattern of ensemble synchrony and the mean firing rates that best correlate with the behavior of the animal in response to odorants (Riffell *et al.*, 2009). These few examples serve to depict the ongoing debate over the functional role of neuronal synchronization. Nevertheless, two recent studies have used optical measurements based on calcium imaging in mice and fish to probe the activity of cell ensembles, thus using intracellular calcium levels as a surrogate for cell activity, a reasonable approach. These methods allow one to assess activity in large cell ensembles, provided the tissue can be properly loaded with fluorescent calcium indicators, which is not a trivial task either. The study in rodents by Komiyama *et al.* (2010) reported temporal correlations in neuronal groups that increased during learning of a motor task appropriate for mice: a lick/no-lick task which is equivalent to the go/no-go tasks for humans. Using calcium imaging as well, Niessing and Friedrich (2010) found, in the fish brain, that while the pattern of the global population activity in the olfactory bulb of zebra fish was insensitive to changes in odors, these transitions in odor sensory input were associated with changes in the coordinated response among small neuronal ensembles, thus supporting the notion that the brain coordinated activity is the biophysical manifestation of behavioral outcomes.

It should not be forgotten that the distributed, organized coordinated activity patterns emerge from the structural (anatomical) connectivity and the biophysics of cells and their contacts, so different organisms will exhibit unique collective activity patterns in the nervous system that relate to particular behaviors relevant to those individuals. This consideration is important when trying to extrapolate observations in one animal and its nervous system to other animals and their distinct neurophysiology. Since the focus of this monograph is on human behavior, the next paragraphs provide a very specific example of how a distorted brain coordinated activity results in cognitive impairment in patients with a specific syndrome.

1.4.4.1. *A specific example on the importance of brain coordinated activity for information processing*

The impaired behavior during a continuous epileptiform activity, the Landau-Kleffner syndrome, serves as indication of the importance of coordinated cellular activity in brain information processing. In general, consciousness is, to some extent, lost during epileptic seizures (Gloor, 1986), but it can be argued that, of course, a paroxysm is a very considerable departure from the normal physiological cellular activity, thus unconsciousness is a most probable event in these conditions. However, there are more subtle changes, unlike seizures, in coordinated activity that result in cognitive impairments. There are studies that have dealt with alterations in brain synchronization associated with a variety of deviations characterized by different styles of information processing, like autism and schizophrenia, which are reviewed in Chapter 6. However, the following example, taken from the epilepsy literature, is chosen now because of the specificity of the cognitive dysfunction, the abnormal brain synchrony patterns, and the recovery of the function that occurs after the aberrant synchronized activity is restored to normal. The condition was originally described by Landau and Kleffner (1957), and thus took their names, as a "syndrome of acquired aphasia with convulsive disorder." It is associated with abnormal electroencephalic patterns that may result in seizures. Patients lose the ability to speak while their EEGs show brief bursts of activity and this occurs without loss of consciousness or awareness; only occasionally do patients develop intermittent partial and/or generalized seizures that involve a transient loss of consciousness. These abnormal bursts tend to manifest as temporal or temporo-occipital spike and wave discharges, the most typical being during sleep as continuous 1.5 to 5 Hz spike and wave discharges. The outcome of this syndrome varies among individuals. What is important for the present illustration is the existence of an altered brain coordinated activity (the abnormal bursting pattern) associated with a very specific behavioral alteration (aphasia), and the interesting phenomenon is that in some cases where the abnormal bursting is controlled with antiepileptic drugs, the patients recover the faculty to speak. Thus, in many cases, a correlation was found between increase in EEG discharges and aphasia, or, translated to our vocabulary, between abnormal coordinated activity and impairment of flexible behavior. Morrell *et al.* (1995) reported cases where multiple subpial transections of the cortex were performed to abolish epileptic discharges in 14 children with acquired epileptic aphasia who had been unable to use language to communicate for at least two years; sustained improvement was obtained in 11 of them after the surgery that reduced or abolished the epileptiform activity. The language disorder, however, may never resolve in almost half of the patients (Mantovani and Landau, 1980). However, the fact that, at least in some patients, aphasia recedes in correlation with a normalization of the EEG patterns is suggestive of the importance of the right type of brain coordinated activity needed to process information (in these cases, to be able to speak). Evidence from evoked potential recordings also exists indicating the ability of single spike-wave discharges to interfere with auditory sensory processing in this syndrome (Seri *et al.*, 1998). As always when discussing these topics about behaviors in the epilepsies, the question arises as to whether EEG epileptiform activity is a manifestation of underlying abnormalities of the brain (the speech areas in the case of Landau-Kleffner syndrome), rather than the cause of the abnormality (aphasia) (Holmes *et al.*, 1981). It is entirely possible that in critical times of development, abnormal interictal discharges interfere with the underlying maturational processes resulting in inappropriate connectivity (Gordon, 1997). But this is not the place to discuss the perennial question of what comes first, seizures or brain damage, or equivalently, whether seizures are causing brain damage or they are just manifestations of an already damaged brain. These queries have been addressed in other publications (Cortez *et al.*, 2006). For the present purposes, suffice to note that between 30 and 50% of the epilepsy cases have been reported as symptomatic, that is, caused by a known etiology that alters what was once a normal brain (Hauser and Hesdorffer, 1990). In the rest of the cases, there is no apparent brain damage to which seizures can be attributed. In any event, there is evidence that epileptiform activity causes cell death (Frantseva *et al.*, 2000) and that molecular injuries such as oxidative stress initiate epileptic events (Frantseva *et al.*, 1998). Considering the current views in neuroscience and complex system theory, the question posed above may not make much sense in the context of a complex, open dissipative system like the brain: cause–effect relations become blurry and hard to differentiate in far from equilibrium conditions. But this does not deter researchers, particularly those devoted to the reductionist spirit, to seek the understanding of epileptiform events as if these were clear-cut phenomena, almost negating that epileptic variety arises not only from internal elements but also from environmental factors.

Another observation that underscores the importance of maintaining the precise coordination of activities in brain networks, in this case in setting up memory traces, is the retrograde amnesia for a period of up to approximately 15 s preceding an absence seizure (Figure 3.1 of Chapter 3 shows typical waveforms of these ictal events). This phenomenon, which has been known to occur since the mid 20th century (Jus and Jus, 1962), is due to the disruption of the normal activity in the thalamocortical networks by the paroxysm, normal activity that may be needed to lay down memory traces. Persistent retrograde memory impairment is also a well-established side effect of electro-convulsive therapy where generalized tonic-clonic seizures are artificially induced in order to treat psychiatric illnesses. This also occurs after impact to the head in traumatic brain injury, but in this case the consequence of the injury is an almost immediate cessation of neural activity due to spreading depression and related phenomena, so it stands to reason that memories cannot be laid because there is no functional synaptic activity at all. As noted above, normally, loss of information processing (loss of consciousness) occurs during ictal events, and because these are characterized by a more extended and enhanced synchrony, which by the way also happens during slow wave sleep, then one can already envisage one pattern of activity that alters in a fundamental manner brain information processing: the aforementioned (Section 1.4.1) long-lasting and extensive synchronization of brain areas. To end this discussion, let us briefly mention that alterations of brain coordinated activity have been reported in cognitive dysfunctions such as Alzheimer's disease and Down syndrome (Babiloni *et al.*, 2010).

1.5.   Brain Activity as Coupled Oscillator Phenomena in Stochastic Settings

From the arguments in the previous sections it should be apparent that the tendency toward neural synchronization is quite robust in general. But the external world is very noisy, constantly bombarding the nervous system with stimuli, so the question becomes to what extent can one find stable synchronization. As observed, most of the synchrony in wakefulness is transient, which already can be explained by this relentless external perturbation. Nonetheless, synchronization can perfectly occur in the presence of noise, although in this setting it is limited to short time intervals as already described in the leading work of Stratonovich (1963, 1967). Synchronization in stochastic systems does take place and has been studied for a long time (Anishenko *et al.* (2002), provide a clear account of the ideas, methods and mechanisms underlying stochastic synchronization). The study of synchronization of chaotic oscillators (systems that oscillate but have the properties of chaotic dynamics as described in Chapter 2) requires more general definitions of the concept, and thus the notions of generalized, phase and full synchronization arose (Afraimovich *et al.*, 1986; Pecora *et al.*, 1997). Chaotic phase synchronization, for instance, can be characterized using diffusion coefficients of the phase difference. The fact that stochastic, or random, activity can synchronize may seem paradoxical, nevertheless it is known that two coupled systems governed by chaotic dynamics can synchronize their chaotic trajectories, even though the coupled systems still remain chaotic: it is only their trajectories that are locked to each other. Even more interestingly, these chaotic networks can split into subsets of synchronized units (Kestler *et al.*, 2008), which may be revealing when translated to brain activity patterns, as, for instance, the complementary neurophysiological evidence presented by Le van Quyen and colleagues studying the fine structure of synchronization during seizures. In this study, intracranial recordings taken in patients revealed a temporal pattern of synchronization that starts as massive synchrony at one specified frequency but that splits up into several subclusters, each one being synchronized at distinct frequencies, and ends up at seizure offset in the frequency dispersion seen before the ictus (Le van Quyen *et al.*, 2006). This "convergence and divergence" (in the words of the authors) of the dynamics of brain synchronization links with other theoretical studies and supports the notion of noise-induced synchronization in brain networks, limited to short time intervals (except in cases of pathologies) and splitting into sets of connected cells. Specific levels of noise can induce synchronization, as shown in modeling studies of neuronal networks (McMillen and Kopell, 2003; Zorzano and Vazquez, 2003), even though clear empirical observations of this phenomenon are still lacking, as, to our knowledge, only computer simulations have clearly shown that noise creates and stabilizes synchrony. The robustness of noise-induced synchronization has been addressed theoretically, because this is the only manner to obtain quantitative measures of the stability of synchrony in the presence or absence of noise, as experimental manipulations to show this phenomenon are much more involved and of difficult interpretation perhaps precisely because of the inherent noise in neurophysiological systems that tends to destroy synchronous phenomena. Having said this, some results can always be compared between experiments and theory, and Figure 1.5 shows the remarkable similarity between the statistics of the intermittent phase slips that occurs in brain signals and the theoretically determined statistics. The right-hand panel comes from an MEG recording performed in our institution on a subject and the phase difference (Ψ) between two MEG sensors was determined using the analytic signal concept. The modeled system, on the left-hand side, consisted of coupled oscillators driven by stochastic processes (Teramae and Tanaka, 2004). In both cases, the synchrony (defined as 0 phase difference, that is, the horizontal segments in the time course of Ψ) is transient and shows numerous intermittent phase slips. The distribution of the time intervals between the phase slips obeys a similar exponential decay in the experiment and in the model, and sometimes the exponent was found to be close to $-3/2$ in the experiment, as shown in the figure. In Chapter 3, Section 3.4.4.2, it is detailed that intermittent bursts are typical of stochastic processes driven by noise sources and that the power-law decay with a $-3/2$ exponent is associated with the dynamical regime known as intermittency (Čenys *et al.*, 1997).

While a classical view regards noise as a degrader of information processing, a change in perspective seems to be taking place currently: noise and fluctuations in brain signals are studied in great detail and it is already known that this stochasticity is not detrimental to brain function, but indeed advantageous in phenomena like stochastic resonance, and in general necessary to avoid becoming trapped into long-lasting and stable synchronous states. The constructive nature of noise continues to be explored in themes related to the default brain activity, psychophysical stochastic resonance, and others. In any event, the similarity in the results obtained from the studies shown in Figure 1.5 suggests that the synchronization of the oscillating phases in brain activity is a typical manifestation of coupled oscillator phenomena in a stochastic setting.

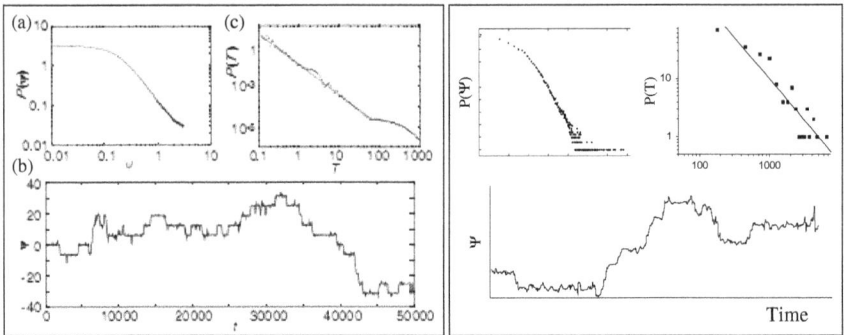

Figure 1.5. Noise-induced phase synchronization in models and in brain signals. The panel on the left comes from a computer simulation of the synchrony between two oscillators driven by noise, and the panel on the right from an MEG recording in an individual. Inside the panels, the following are shown: the time evolution of the phase difference between two modeled oscillators or two frontal MEG sensors (Ψ), the distribution of the phase difference ($P(\Psi)$), and the distribution of the interphase slip intervals ($P(T)$). A full phase slip occurs when one signal advances with respect to the other a complete circle (360° or 2π radians), and the time intervals between these slips are then calculated to construct the diagram on the right-hand side of each panel. In the case of the MEG signals, the phase difference was computed at 35 ± 2 Hz. Note the similarity in all the three plots between the model and the experiment, particularly the distribution of the interphase slip intervals (represented by dots) obeys a power-law decay with an exponent of $-3/2$ (-1.5), as determined from the slope of the straight line that fits the experimental points (it was -1.42 in the experiment, with $R = -0.93$). Left-hand panel reproduced with permission from Teramae and Tanaka (2004), *Phys. Rev. Lett.* 93, 204103, copyright by the American Physical Society.

1.6. Thalamocortical Activity as an Illustration of the Combined Individual and Network Levels Resulting in Neural Rhythms

As a very clear illustration of the importance of the combination of network and intrinsic cellular properties in the determination of spatiotemporal patterns of coordinated activity, a look at the thalamocortical system is of value. The networks of the thalamus and neocortex display a variety of rhythms, and the dynamics of the whole system is fundamental in determining the levels of consciousness. The following description is very schematic and does not consider many details, because the main purpose is to see how patterns of activity emerge from the intrinsic cell properties, the anatomical connectivity and the nature of the synaptic contacts, and this can be best explained without too much detail. The basic thalamocortical circuitry, schematically shown in Figure 1.6, consists in a loop where thalamic relay cells (these are the neurons in all the thalamic nuclei except for the reticular nucleus) send axons to the cortex (mainly to layer IV), and in turn cortical pyramidal neurons send their projections back to relay neurons and also to the thalamic reticular neurons. The neurons in the NRT are inhibitory and send their projections to the relay cells. This is the basic neuroanatomy of the circuitry. It is very simplistic because the three main networks, the cortex, relay

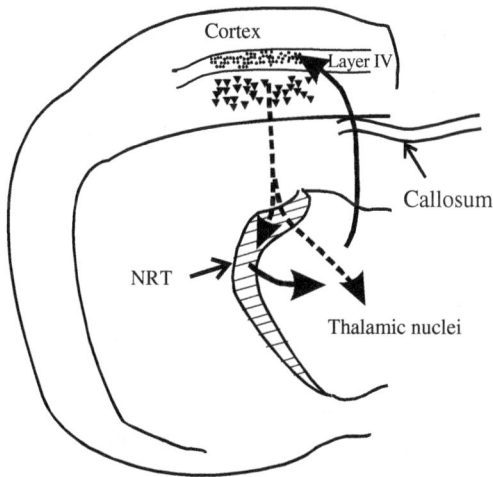

Figure 1.6. Schematic depiction of the three basic connections in the thalamocortical circuitry, shown in a brain section of one hemisphere. The reticular nucleus (NRT) output is inhibitory, whereas the others are excitatory. The cortex receives the major thalamic input in layer 4, composed of spiny stellate neurons, while deeper layers (composed of pyramidal cells) project back to the NRT and to the rest of thalamic nuclei.

thalamic nuclei, and the NRT, can be further subdivided into other "oscillator parts." As to the essential cellular intrinsic characteristics that determine thalamocortical activity patterns, these are: the depolarization conferred by low-threshold calcium spikes (T-type calcium currents) in the thalamic cells, the hyperpolarization-activated cation conductance I_h (which is a non-inactivating sodium/potassium current), and after-hyperpolarizing potentials carried out by a diversity of potassium channels. The basic synaptic contacts are all excitatory (glutamatergic) except for the inhibitory projection of NRT cells to relay neurons mentioned above. From these cellular and synaptic properties and the recurrent anatomical connectivity between thalamus and cortex, the main thalamocortical rhythms arise.

The low-threshold calcium conductance, commonly known as the T-type conductance, present in relay and reticular neurons is crucial because it remains inactive at resting membrane potential, deinactivates by hyperpolarization, and is activated (the calcium channel opens) at the onset of the rebound toward depolarization from the hyperpolarized level (Jahnsen and Llinás, 1984). The rebound response results in a calcium spike of considerable amplitude and duration that can make the neuron fire action potentials if the threshold for spike generation is crossed (Deschenes *et al.*, 1984). Therefore, the presence of the T-type calcium current means that, for thalamic neurons, more hyperpolarization does not necessarily mean less activity, since these channels will open from hyperpolarized potentials. So, if thalamic cells are depolarized as normally occurs during wakefulness, the T-type channels have a lower probability to open, but if the cells become hyperpolarized as occurs during sleep, these calcium conductances are prone to be deactivated and generate low-threshold calcium spikes that depolarizes the cells and can make them fire action potentials that ride on the calcium spike, thus the net effect is to enhance excitability.

As a consequence, the thalamic reticular neurons are particularly important as they are inhibitory and can cause deep hyperpolarization of relay cells because of the synchronous burst-firing in reticular neurons. This robust hyperpolarization of relay cells can de-inactivate the T-channels and can promote calcium spikes and bursting, with the possibility of starting synchronized activity in the whole thalamocortical loop because the bursts will be relayed to the cortex via the axons of the relay neurons, and the cortex in turn feeds back onto relay and reticular neurons so that the cycle can thus continue (some functional manifestations of this loop are described by Steriade *et al.* (1993), and an in-depth description of the interaction between intrinsic and extrinsic mechanisms in the generation of rhythmic activity can be found in the volume by Destexhe and Sejnowski (2001)). These factors, along with others, can create a variety of thalamocortical rhythms. One schematic situation is depicted in Figure 1.7, to describe how two different rhythms, at 6 and 10 Hz, can appear in thalamic neurons.

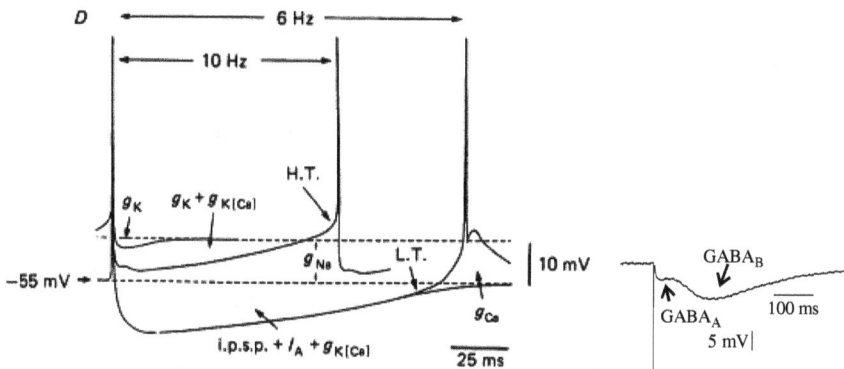

Figure 1.7. Schematic depiction of the origin of two oscillations in thalamic cells. The traces represent recordings from a thalamic neuron. The action potential is followed by an after-hyperpolarization mediated by voltage-dependent (g_k) and calcium-dependent potassium currents ($g_{k[Ca]}$). When the membrane potential is relatively depolarized (upper dotted line), no activation of the low-threshold calcium current (LT) occurs, and the potential is brought back to the resting level by a slow sodium conductance (g_{Na}) and, if the cell fires another spike (if the potential crosses the threshold while recovering), it may result in oscillatory firing at the specific frequency of the oscillation dictated by the time courses of those ionic mechanisms (10 Hz). On the other hand, if the cell starts at a relatively hyperpolarized potential (lower dotted line at -55 mV), excitatory input from the cortex can trigger a low-threshold calcium spike (g_{Ca}). If the cell receives IPSPs from reticular neurons, the hyperpolarization is more profound and longer lasting, so the frequency of the oscillation will depend on these other ionic contributions from the synaptic potentials and membrane properties, and thus slower rhythms (6 Hz shown here) will materialize. Reproduced with permission from Jahnsen and Llinás (1984). The inset to the right is a whole-cell recording from a neuron to demonstrate the different time course of the two inhibitory potentials, the short mediated by GABA$_A$ receptors and the long by GABA$_B$ receptors. The different lengths of the hyperpolarization caused by these IPSPs will cause the LT rebound burst to appear with different delays (more delayed in the case of GABA$_B$-mediated IPSPs) which will determine different frequencies of the oscillation.

The synchronous activities in these circuitries express a variety of frequencies, the illustration of some of the causes underlying some of these oscillating frequencies can be inspected in Figure 1.7 that is used here for the sake of simplicity because there are many other factors contributing to the assortment of these rhythms. For example, during wakefulness the thalamic neurons are kept depolarized by cholinergic projections from the brain stem, and therefore the probability to express a rebound response mediated by the T-type current is low. On the other hand, at the start of and during sleep, the cholinergic input diminishes and thalamic cells hyperpolarize. This hyperpolarization promotes rebound responses that will generate oscillations at different frequencies depending upon synaptic activity in the network, but normally will be lower frequencies (longer periods) than those found in awake states, like the one shown in the figure (with a long IPSP) that can generate rhythms of about 6 Hz. If the IPSP were longer, for instance if it were mediated by $GABA_B$ receptors that open potassium channels and can last 200 or 300 ms (see inset in Figure 1.7), then the low-threshold rebound with the burst of spikes on top can produce a ~3 Hz oscillation (one every 300 ms), that is the δ frequency range. The longer-lasting IPSPs thus imply a slowdown of frequencies, and that is why the typical sleep EEG patterns are slow waves (except during REM episodes). The rebound response of the thalamic cell is transmitted to the cortex and, because these rebound bursts contain many spikes and tend to be synchronous among thalamic cells, the input to cortical neurons is large enough so that the chain of activity will continue, and the loop can self-sustain until something occurs, for instance, a depolarization of the thalamic cells by cholinergic inputs that will decrease rebound firing and thus break the loop, and as a result the individual either wakes up or starts to dream in a REM episode. It can be appreciated how the combination of intrinsic conductances, connectivity and synaptic activity contribute to the emergence of distinct frequencies on synchronous oscillations, which is the main point to be made in this brief digression.

This thalamocortical circuitry also furnishes an example of the progression from "normal" thalamocortical rhythms to paroxysmal spike-and-wave discharges (SWDs). This has been extensively studied in the feline penicillin model (reviewed by Gloor and Fariello (1988)) and in anesthetized cats (reviewed by Steriade (2001), and McCormick and Contreras (2001)). These transitions stress the fact that paroxysms, in general, are but an extension of normal, physiological brain activity; indeed, cellular studies have determined that epilepsy-related thalamic cellular activities are very similar to those found in normal activities (Pinault, 2003). At the risk of belabouring, the propensity to synchronous activity in nervous systems can never be enough emphasized, and the thalamocortical loops are an extreme exponent of this tendency, one reason why we fall asleep every day as explained above. The maintenance of thalamocortical synchronized activity is not too difficult: after a set of cortical or thalamic neurons sends a relatively strong, synchronized input to the connected cells, the aforementioned intrinsic and network features will engage the loop into synchronized activity, it does not matter where the chain starts. This is demonstrated by the fact that one single intracerebral electrical stimulation either to the thalamus or to the cortex is able to generate a SWD in the whole system, as reported in experiments using pharmacological rodent models of absence seizures (Proulx et al., 2006; Perez Velazquez et al., 2007b). This aspect is of interest to clinicians, who have devoted much effort to answer the question of where absence seizures start, some in favor of the thalamus and others of the cortex as the site of initiation of an SWD (Seidenbecher et al., 1998; Meeren et al., 2002; Pinault, 2003). In the end, it may not really matter where the seizure is initiated, because it will be maintained by the thalamocortical loop. Extending laboratory scenarios to patients with absence epilepsy, we can expect that in any "hyperexcitable" brain, any extra activity in thalamus or cortex will be amplified resulting in a spike and wave discharge. Section 3.4.3 of Chapter 3 contains more on this theme.

But then, how difficult it is to end thalamocortical synchronized activity? This is again of interest in the context of absence seizures. A variety of factors are in play to halt "hypersynchronous" activity. Basically, anything that promotes depolarization of thalamic cells will contribute to the ending of the paroxysms because the T-type calcium spikes will be less likely to emerge with the consequent reduction in bursting firing which means a lesser input to the connected cells. In this regard, and for those interested in details, a specific action of the cation conductance, I_h, in thalamic neurons has been advanced (Bal and McCormick, 1996), as this mixed Na^+/K^+ current slowly but continuously depolarizes neurons, which will inactivate the low-threshold calcium spike. As a final consequence of this event, fewer and reduced calcium spikes will result in fewer bursts of action potentials in thalamic neurons, reduced input to neocortical areas, and the termination of the paroxysm.

Let us end this section with a phenomenon in absence epilepsy that seems paradoxical. As noted in Chapter 3 (Section 3.5), drugs that normally are beneficial in other seizure types are not effective in absence epilepsy. Thus, apparently paradoxical is the observation that drugs that promote inhibitory transmission enhance absence seizures, when they should decrease the propensity to seizures as they do in other seizure types. The explanation to this apparent paradox lies in the consideration of the intrinsic properties of thalamic cells that were mentioned in the previous paragraphs, specifically the rebound bursting response after hyperpolarized states. In general, whatever the mechanism that promotes hyperpolarization of thalamic cells (and inhibitory transmission does) will enhance the probability of developing a synchronized activity, call it a seizure or a sleep oscillation. Not surprising then is the fact that much research on the thalamocortical network and absence seizures has been done in cats, not because cats have any special propensity for SWDs, but because the animals develop spontaneous SWDs under ketamine/xylazine anesthesia, a natural consequence of the highly synchronized thalamocortical activity that this anesthetic mixture causes due to the neuronal hyperpolarization. Observations derived from this model are extensively reviewed by Steriade (2001), and we refer interested readers to this publication.

1.7. Local Versus Global Activity

The theme of local and global activity pervades most of the discussions in the following chapters. As in any other complex system, the problem of how to go from a local description to a global understanding of the system's activity, how the dynamics of the individual are related to that of the collective, is a major question that arises, prominently, in brain research. After all, the abnormal collective states in brain dynamics associated with diseases and behavioral deviations emerge from the individual constituents, the cells and their environment. The attempts at interpreting the collective (macro) dynamics from the local (micro) led to no few debates in the history of science, perhaps the most famous is the 19th century controversy on the irreversible dynamics of systems composed of reversible processes, that involved characters the likes of Boltzmann and Poincaré. There is no such a risk in neuroscience, provided nobody takes a strong stance on any level of description.

The comments in previous sections are clear indications of the general interest in understanding neural collective activity from the individual components, even though this is not a trivial task and many investigators may find it more comfortable to remain focused on just one level of description. The methods of dynamic system theory, as explained later, certainly allow for a global understanding, however most of the analysis that is normally performed reveals only local properties. For instance, the typical linear stability analysis (see Section 2.1, and more specifically on local versus global considerations, Section 2.3.1) represents a local approximation. Hence, this apparent dichotomy also exists in physics, in the analytical methods that are being applied to brain research.

In general, local neural activity is discontinuous, or discrete. From the quantal nature of neurotransmission to spike firing, discontinuity pervades the microscopic level. Yet, mental life has a strong feeling of continuity. Just as the brain interprets sensations as continuous, and in general this is how space–time is perceived, brains also perceive mental life as continuous which underlies the conception of the self (Chapter 7). Not only it is known that sensation is discontinuous (e.g., microsaccades in the eyes imply a discontinuity in retinal transfer of information to visual cortex), but also movements are made in step-wise fashion and the apparent continuity that is perceived when performing the actually discrete slow limb movements will be mentioned within the context of motor coordination and Parkinson's disease in Section 4.2.2 of Chapter 4. This continuous manner to construct a reality seems to be a general feature of the brain.

Each of the following chapters contain specific jumps between local and global levels, as they are relevant to the particular topics covered in those sections. The broad, common idea is centered around the emergence of abnormal macroscopic brain states in terms of the magnitude and extension of synchronization as a result of local alterations in the excitability of individual cells and networks. Physics handled this problem of local versus global using the allied concepts of competition and dissipation, as insightfully described by Alan Turing (1952) studying pattern formation in chemical systems. In his case, the competition was between diffusive gradients of reactant molecules, and dissipation took place in the energy involved in the localized chemical reaction. The situation may not be so simple in the brain, where we have to contend with the anatomy, the intercellular contacts, and the cell or network coherent activity, nevertheless simplification is always feasible and global views on competition, cooperation, and dissipation (but unlike in Turing's reaction, the sources of dissipation are numerous in the case of brain networks) in the nervous system represent an starting point to address these subject matters. To start with, it is known that, in general, competition is extensive in the nervous system, that between excitatory and inhibitory inputs, the former tends to be long range and the latter is more often short range, thus presenting, interestingly, similar features to those in the Turing's chemical reaction where the difference in diffusivity, long- or short-range, between reactants gives rise to pattern formation. This difference is the central determinant for pattern formation in reaction–diffusion phenomena. Possibly, nervous system dynamics can be captured in reaction–diffusion paradigms. Generally, the spontaneous organization of collective activities arises from the exchange of information between the system's components at the local scale and results in global patterns, and this applies to anything from physical, non-living systems, to animal and human crowds (Moussaid *et al.*, 2009). These are the roots of critical phenomena, where the competition between the ordering tendency due to the microscopic interactions and the disordering tendency due to stochastic internal and external fluctuations, determines critical points in a system's behavior. Different microscopic factors can thus give rise to similar macroscopic phenomena. With regards to brain activity, perhaps an aspect to consider in terms of interactions between constituents (the cells) is the aforementioned phase relation in synaptic inputs (Section 1.4.3), so one can envisage how there is an "ordering" tendency due to the intercellular interactions that tend to synchronize their activity, and a disordering perturbation from external sensory stimuli. The competition between these two tendencies may result in crises, or bifurcations in the terminology of dynamic systems. Chapter 2 discusses at length the topic of dynamical bifurcations in nervous systems, which, along with criticality, is becoming an important research area in neuroscience. The reason why it is thought that criticality is so prominent in neural activity is because critical phenomena result from the progressive build-up of interactions, from the cooperation and competition inherent in cellular activities. Perhaps with these views, we can start to obtain a crude glimpse at how organized, macroscopic brain states of activity can arise from the microscopic noisy patterns.

In closing, it should be considered that, while the global characteristics are determined by the local interactions, the reverse also occurs. In physics, especially in relativity, the global properties (of the space-time) determine local properties. This is the case too in brain dynamics (Haken, 2002, 2006), and in the brain and behavior

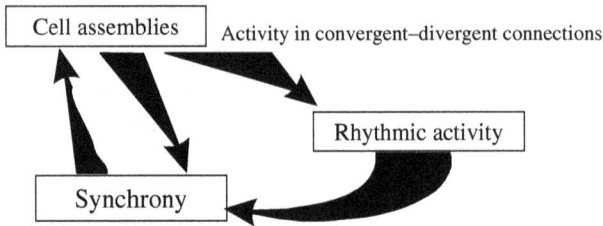

Figure 1.8. Simple scheme indicating that these three phenomena are aspects of a common theme. The activity (spike firing) in cell ensembles originates synchronous activity in receiving sets and possibly rhythms, depending on the specifics of the connectivity and biophysical features of the cells. The synchronization need not be rhythmic. In turn, synchrony, at different spike firing frequencies for instance, serves to create (and thus segregate) transitory cell assemblies.

relation due to the loop between the one and the other, as exposed throughout this book. Hence, perhaps the title of this section may not be too adequate, for it used the term "versus," as if local and global aspects were contraries. Rather, local and global levels are complementary and form what Kelso and Engstrøm would consider a complementary pair (Kelso and Engstrøm, 2006).

1.8.   Concluding Remarks

The foundation of neural information processing lies in the interactions between cells or between cell ensembles. This chapter has described basic activity patterns of cell ensembles of the nervous system and how these arise from the intrinsic cellular properties and the structural and functional characteristics of the intercellular contacts, noting the extraordinary tendency toward synchronization. It is proposed that this tendency underlies the deviations in behavior that are labeled as pathological. The interactions between oscillations, the interference pattern if you will, that occurs at the cell and ensemble levels along with the consequent synchronization, or phase locking, between the oscillations, establish temporal windows fundamental for the exchange of information. Synchrony, in this sense, is not a state but a process. Figure 1.8 describes the loop between the cellular activity that gives rise to synchrony and oscillatory activity, suggesting that these three phenomena are hard to disentangle. Synchronization phenomena pervade not only nervous systems but also almost all aspects of the living and nonliving, perhaps because it is an optimal manner to exchange information in any system (discussed by Perez Velazquez (2009)). It has also been emphasized the need for the consideration of the spatiotemporal distribution of the organization of the activity, the coordinated activity patterns that emerge from the microscopic, cellular interactions, and the fundamental importance of the relations among the activity in cells and in cell networks, descriptions of the collective activities, for one cell's (or network) activity has meaning only in relation to the activity in other cells (or networks). Without some level of coordination within brain networks, even the simplest of tasks would not be possible. Cognitive processes involve serial stages of sensory signal processing and parallel information processing networks, therefore the crucial relevance of coordinated activity of brain cell ensembles. Much has been talked during decades about the different frequency bands apparent in EEG and other neurophysiological recordings, particularly with regards to the role of those bands in cognition. While it is true that the information processing that a brain carries out at one time point is represented in the temporal distribution of those recorded activities, how to decipher this message needs the consideration of the spatiotemporal collective activity. There are advantages in using the paradigm of coordination dynamics in that it offers a tremendous reduction in the number of parameters due to the correlation in the individual activities. This is one reason why the reductionist paradigm, in our opinion, is bound to fail in the understanding of brain function and its relation to behavior, for there are thousands of parameters that need to be described at the very local and reduced level of description. The relative phase in neural oscillations has been stressed as a possible collective variable to use in the developing of a metric for neural interactions and information processing. Thus, phase synchronization is seen as an aspect of brain coordination. One goal of the following chapters is to illustrate the sometimes gradual, sometimes abrupt, transitions between brain dynamic patterns and between behaviors, attempting an integration of several levels of description. What may look like one phenomenon at one level, it could be a different phenomenon at another.

The flexible formation of neural activity patterns requires transient coordination, that is, ephemeral synchronization. Metastability, the relative coordination as the result of the tendency of cells to express their intrinsic independent activity and the tendency to couple together, has been noted as a fundamental framework to understand brain and behavior. Highly synchronous activity is entirely appropriate for worms to crawl and brainstems to govern breathing rhythms, but not for the cortex to process sensory information in fast and flexible manners enough to be adaptive. This reminds us of a Zen koan about a centipede that was asked how it could manage the coordination of 100 legs and how it knew what leg goes first and which one second, third…, upon which the animal reflected on this with the result that it could not walk anymore. The moral of this tale is that some things are best left without central control, to let the emergent pattern arise. One wonders whether there is any fundamental difference between the emergent activity of a centipede's neural system and that of people's brains, besides the fact that the former results in a highly structured motion and the latter in mental activity. It was described in Section 1.2 regarding how primitive nervous systems specialize in the production of relatively stable periodic activity to achieve stable stereotypical gaits, but more refined processing of sensory inputs and motor coordination required the presence of additional circuitries and metastable dynamics, afforded, we think, by the thalamocortical networks. Brain development brought about the emergence of distinct dynamic patterns such that the feeling, or perception, of agencies and selves arose. While fine control of motor output requires awareness, many aspects of our behavior are still as "automatic" and mindless as in the worm's case. In the final analysis, is it not all prescribed in the neural patterns of activity? This also reminds us of a common question in neuroscience related to why the brain uses this or that rhythm, frequency band, or synchrony pattern to achieve one or another behavioral goal. We would like to reframe these types of questions considering that the brain oscillations, the collective activity patterns, are not "used" by the brain (or by the self), but rather are the source and manifestation of cognitive processes themselves. In other words, the coordinated patterns of activity are the intellectual dispositions and features of the individual, the self, there cannot be a separation between these activity patterns and the mental state of the individual. Neural patterns are considered to cause behaviors, but possibly it would be equally accurate to say that the behaviors are the patterns themselves. To say that brains use glutamate as neurotransmitter may be accurate, but brains do not use oscillations, for these constitute the brain itself, the mental world of each individual. There is no need to dichotomize in excess.

A brief point that needs to be clarified before proceeding with the following chapters is what is understood by pathologies or diseases when discussing brain function and behavior, for all those chapters are dealing with deviations from normality that are labeled pathological. The World Health Organization (WHO) defines normality as a state of complete physical, mental, and social well-being. Absence of a mental disorder constitutes mental well-being, and, in DSM-IV terms, mental disorder is a behavioral or psychological pattern associated with distress. Yet, distress, particularly determined by a mental health professional, may be a somewhat subjective and ambiguous entity. Does portrayal of normality depend on our local, restricted perspective on brain and behavior? Everybody would agree that epileptic seizures are disturbing deviations from normality, but not everyone will consider subtler deviations as pathological. Consider, for instance, high-functioning autism (Chapter 6, Section 6.1). These individuals live more or less normal lives, tend to have decent jobs, albeit they present some deviations in behavior (ranging from perception to social interactions) that, nevertheless, do not prevent them from enjoying their lives. Other individuals in the autism spectrum disorders, those classified as low functioning, are more impaired. Yet, the presence of subtle behavioral deviations prompts the labeling of a condition, high-functioning autism in this example, as if anything that falls away from the center of the Gaussian distribution, the "normal" behaviors, has to be classified one way or another. The continuum between health and pathology makes the border between these notions quite nebulous, the specific separation points being arbitrarily defined. It is our local perspective that classifies and distinguishes behaviors (and thus brain activity) as healthy or unhealthy. Thomas Szasz, in his book *The Myth of Mental Illness*, suggests that normality can only be defined in terms of what the persons do and do not do, and that belief in mental illness is similar to believing in witchcraft or demonology. Does our subjective division between normal and deviant brain function coincide with non-adaptive behaviors? Sight should not be lost of the fact that the collective states, the configuration of activity, or whatever nervous system dynamic patterns are called, acquire sense only with respect to the environment, to adaptive behaviors. In some sense, then, pathological states or mental deviations depend on context: in a world without air, nobody would consider deafness a disease (but of course ears would not have developed in such a planet!).

A concept that appeared some decades ago and that may be useful is that of dynamical diseases, which is discussed in several chapters with regards to some deviations in brain function. This notion developed as an extension of that of "periodic diseases" (Reimann, 1963), where the symptoms recur at quasi-regular intervals, and the current notion is based on dynamic system theory frameworks (Chapter 2 is devoted to this aspect). The signature of a dynamical disease, according to Glass *et al.* (Glass and Mackey, 1979; Mackey and Milton, 1987) is a change in the qualitative dynamics of physiological processes as some physiological control parameters modify their normal values. Of course, any disease is caused by an alteration of a physiological parameter, however the distinction between a dynamic disease and a viral infection, for instance, is that the former is based on alterations of more or less periodic processes that define certain characteristic physiological rhythms while the latter does not represent a modification of a biological oscillation, it just occurs at a time point and then goes away when the infection disappears. In a dynamical disease, the temporal patterns of physiological variables change, and because there are plenty of biological rhythms (perhaps, to the extent that almost all physiological processes are rhythmic) thus deviations from the normal that perturb the rhythm can be considered dynamical diseases. Later chapters deal with specific cases and the parameters that change and are responsible for the alteration of particular rhythms. However rhythmic some activities look, there are always fluctuations. Sections 1.4.1, 1.4.2, and 1.5 have introduced evidence for the value of the intrinsic variability in neural actions. This may indicate that a purely deterministic description of nervous systems (or any complex system for that matter) is not possible due to the presence of noise (internal and external) and stochastic fluctuations, therefore stochastic approaches seem more suitable for the characterization of brain function. Sections of the subsequent chapters deal with the notion of randomness in brain function and in behavioral manifestations.

We have seen how brain rhythms are manifestations of synchronized cellular activity, ultimately depending on neuronal firing. The concept of synfire chains helps understand the nature of the, most of the times, forced oscillations propagating down neural nets. But, as alluded to in Section 1.4, there are implications for brain function and behavior of this serial triggering of neural activity. Gilles Laurent said: "Our thinking generally ignores the fact that, with the exception of motor neurons, a given neuron is never an end-point or its response an end-product [...] Thinking about sensory integration in these active terms (considering 'responses' not only as products but also as ongoing transformations towards some other goal) may be helpful [...] to understand some brain operations" (Laurent, 2002). There is no start or end to the chain. A sensory input could be considered a start, but what this input does is basically to add itself to the ongoing, background neural activity already present: sensory processing is basically a modulation of this ongoing activity (Buzsáki, 2006). Further, while it can be conceived, as Laurent does, that one end can be at the nerve motor endings contacting the muscles, is it not true that motor actions generate a sensory feedback? Then, is that an end or a beginning?

Readers will have noticed that the question of the neural code has not been mentioned in this chapter, perhaps an expected topic to treat, and yet the neural code is all that has been talked about here. Codes imply agents interacting and a decoding mechanism, maybe a program, a set of rules, it does not matter. The agents in the previous narrative have been cells and cell ensembles; the interactions, differences in electrical potentials carried out by ions; and the decoding program, the intrinsic biophysical characteristics.

A word that has appeared numerous times in this chapter, and is continuously used in the following chapters, is "state". Whereas considering cell membrane potential "states" may not be ambiguous, talking about brain or mental states is certainly unclear. What constitutes a brain state? The persistent activity present in a subset of cells and networks during cognitive performance? The electrical activity patterns of the cortex, whole brain (not possible to measure, anyway)? Cognitive scientists live on the notion that mental states can be associated with neuroimaging data. But, what is a mental state? Or, how about behavioral states? The notion of state in dynamics, which is clearer (Chapter 2), may be hard to map onto a neurophysiological configuration. Section 2.6 of Chapter 2 contains words on this subject, which is left for the time being, lest we get involved into a heavy discussion of the certainly problematic question of the identification of brain or mental/behavioral states.

One question is now asked to end this introductory chapter, and the rest of the book is devoted to finding an answer. Nobody seems to be proud of performing complex metabolic reactions such as glycolysis, but yet many are proud of, say, playing a musical instrument, or learning a trade, or obtaining a record in a marathon. What is the underlying reason behind this difference in attitude? It is the difference between mental life and other "lives," like the "metabolic life" of our cells, and this difference arises from our assumption that the mind (brain) can exert control over itself and thus regulates the behavior, achieves goals by making efforts, sometimes by suffering, all apparently accomplished by the power of the will. Hence, control seems to be the basic ingredient from which our special attitudes on brain and behavior emerge. The question posed is, how much control does a brain really have? Or, what parts of the brain, if any, have control over which others?

CHAPTER 2

THE DYNAMICAL PERSPECTIVE ON NERVOUS SYSTEM ACTIVITY

Chapter 1 has introduced several levels of description at which the activity of nervous systems can be scrutinized and conceptualized. Let us state the nature of the problem to be faced: as detailed in Chapter 1, brains are composed of a very large number of cells that interact in multiple manners and with synaptic contacts that are numbered in the billions. Behaviors, the manifestations of brain activity, result from the activity patterns in that immense cellular conglomerate. The question is how to make sense of this activity to capture the essence of the relation between brain function and behaviors. Can simple laws be found at some level of description, perhaps at the macroscopic level? Physics was once faced with an equivalent challenge during the quest for the understanding of systems composed of billions of interacting molecules. Thermodynamics offered aid: giving up a detailed, microscopic description, afforded a view on macroscopic behavior captured by physical laws that were independent of the specific nature of the constituents of the systems. This approach entails some abstraction. As commented upon in Chapter 1, neuroscientists are encouraged to practise a certain degree of abstraction in their reflections on neural phenomena.

While very successful in science, the reductionist approach, or the notion that phenomena are understood by breaking them down into its constituents and by characterizing these parts with great precision, fails in the comprehension of complex systems. The presence of dynamical bifurcations (also known as phase transitions or critical states) is but one reason why the reductionist approach fails, as bifurcations (explained below in Section 2.3) correspond to a large-scale behavior of the system determined by cooperativity among the units and the scaling up of interactions. The complexity of these interactions near bifurcations renders the knowledge of individual cell properties basically futile in determining the collective behavior. Knowing in extreme detail the nature of neurons and glial cells and their synaptic properties will not help in the understanding of the difference between, for example, a nonpathological neurophysiological phase transition and that of seizures. Of course, detailed descriptions of cells and networks are essential and should be pursued, but to resolve questions at the collective level of description requires a more holistic approach.

This chapter presents a global perspective on the characterization of the nervous system activity patterns (especially brains); a description based on dynamical system theory. Why this framework in particular? Some nervous system activity seems to be *robust*, occurring for a long period of time like that of the central pattern generators (CPGs) for instance; clinicians address the (pathological) *persistence of states* like depression; brain scientists talk about *transitions* from one cellular activity pattern to another, and they wonder how to *control* specific pathological activities; cognitive scientists speak about how individuals *switch* behaviors. These are common words that are used in brain research. In addition, this is in fact the language of dynamics, words and concepts that are the foundations of dynamic system theory. It offers a common language to capture the behavior of complex systems, regardless of the nature of the particular constituents. This common language is much easier to grasp using nontechnical words, and it is intuitive to a large extent. But it can also be made mathematically precise to characterize, for example, transitions between brain activity patterns, between behaviors, etc. It is this power of being relatively intuitive that has attracted enormous attention from biologists in general, and neuroscientists and clinicians in particular. However, this power has also resulted in these intuitive concepts being pushed too far in the biological and clinical fields, using metaphors that have been stretched to limits resulting in flawed concepts or methods. How far these metaphors can be stretched is a matter of discussion in the present and next chapters. Cautious application of dynamical methods can be extremely useful in the understanding of physiological processes. Time series analysis, for instance, can reveal patterns that are not evident by just visual inspection of raw physiological recordings. However, a noncareful application leads to misunderstanding.

In addition to offering a common language for the description of complex phenomena, dynamical system theory also offers conceptual frameworks to comprehend brain activity and its relation to behavior and the environment from a relatively abstract viewpoint. It facilitates the integration of the several levels of description used to study complex systems, and thus we think it is important for a global comprehension of brain and behavior. Equally interesting, conceptual approaches and analytical methodologies to process physiological signals are derived from this theory, techniques that differ from the more classical approaches to data analysis (such as power spectra) in what has been termed, in general, nonlinear time series analysis (Rapp, 1994; Kantz and Schreiber, 1997; Lehnertz, 2008).

Thus, the aim of this chapter is not to present a comprehensive overview of dynamical system theory, but rather to emphasize the particular notions and methods that will be needed in the following chapters to conceptualize the brain–behavior dynamics. This presentation is done from the intuitive as well as from the more formal perspective, which in some parts is quite technical. These technical sections may be skipped, however, readers are encouraged to go over them because, as aforementioned, there is a risk in not comprehending the essence of these physical/mathematical concepts and it is felt that a precise understanding of the concepts talked about here is really necessary currently, as the metaphors and analytical methods used sometimes create severe misunderstandings in the neurobiological literature. As a representative example of misinterpretation, some studies create a state space of a particular neural system with graphs presenting a cloud of data points which is called an "attractor." While this could be an attractor, surely a cloud of points does not constitute one. It has to be demonstrated using typical attractor characteristics. Thus, a formal and also intuitive description of what could in principle be considered attractors in state space is presented in sections below. Dynamic system theory is rooted, of course, in mathematics, commonly employing systems of coupled differential equations, an approach that has been the mainstream in science basically since J.C. Maxwells' system of equations describing electromagnetism demonstrated that reality (or some aspects of it) can be understood in terms of differential equations. However, these facts should not lead to an exaggeration of numerical exactitude. Indeed, it is many times the qualitative characteristics that are more fundamental for a full comprehension of dynamic phenomena, as eloquently expressed by A. Winfree in his words reprinted in page 2 (Chapter 1). Winfree's "qualitative essentials", more than exact quantitative results, will help outline unifying frameworks that are vitally needed in present day neuroscience, as W. Freeman put it: "The construction of unifying hypotheses requires a more poetic, holistic turn of mind and more tolerance for ambiguity and uncertainty that reductionists care for" (Freeman, 1995).

2.1. Basic Preliminaries. Brief Introduction to Important Concepts of Dynamical System Theory

Let us start with some, perhaps trivial, facts about the dynamics of nervous systems. These are open systems, exchanging matter/energy with the external environment. Brains, in particular, consume a large amount of energy. This fact signifies that the system is far from equilibrium, as nonequilibrium systems are maintained by a flow of energy (flow which may or may not be at steady state), so thermodynamics teaches us. Open systems plus nonequilibrium conditions and a source of noise (fluctuations) constitutes a recipe for interesting phenomena to occur. Namely, the formation of patterns, and in the special case of nervous systems we can talk about either activity patterns or structural (anatomical) patterns.

An overview of the concepts that is used in the following chapters is thus presented here, and is not intended to be a full, comprehensive review of dynamical system theory. For those interested in a deeper treatment, a recommended introduction to differential equations and dynamical systems is Perko (1991), and introductions to dynamical system theory can be found in the classic monograph by Bergé *et al.* (1984), which is centered on turbulence phenomena as many notions of dynamic theory developed fundamentally in this field. Guckenheimer and Holmes (1983) and Anishchenko *et al.* (2002) are recommended too, the latter introducing stochastic approaches. Specifically related to neuroscience, the monograph by Hoppensteadt and Izhikevich (1997) focuses on the application to neural models.

Let us start, for precision's sake, with the definition of a dynamical system. Mathematically, it is an evolution operator for the system's state variables, and the number of these variables constitutes the system's dimension. In simpler words, it is an entity whose state evolves according to a (deterministic) dynamical law, given by the evolution operator. The operator normally consists of a set of differential equations defining the temporal evolution of the state variable(s). From the experimentalist's perspective, it means that the concept of a state of the system can be defined at each time point and that an operator can be found that approximately describes the spatiotemporal evolution of the system (to be precise, to model the spatial and temporal aspects requires partial differential equations). If the operator is characterized by linear properties (recall Section 1.4 of Chapter 1 about the linear and nonlinear characteristics in nervous system function) then the dynamical system is linear, otherwise it may be nonlinear. The temporal evolution of dynamical systems is represented by trajectories in a state space, and this space is normally constructed from the variables of interest. The state space is the set of all possible states of a dynamical system, each state corresponding to a unique point in the space. States, sometimes called phases, of a system can be described as the spatiotemporal manifestations of the system and the variables that capture the basic properties of the states are called order parameters. The state space is just the space spanned by the values of those variables according to the evolution operator.

In this book, the system will normally be the brain. Hence, our problem becomes the reconstruction of the dynamical system that represents the brain, including its physiological and mental characteristics. Having considered this definition of a dynamical system, the two main aspects that need attention are:

1. the choice of the variables that will be used to characterize the system and thus define the appropriate state space in which to follow the dynamics, and,
2. the construction of the evolution operator (normally differential equations) that define the temporal evolution (the trajectories in the state space).

These two aspects deserve some consideration that is presented in detail in Section 2.5.

A linear system has the general form

$$dx/dt = \mathscr{F}(x) = Ax, \tag{1}$$

where x is a system's variable that can be measured, or it can be a collection of variables that constitute a matrix (in this simple case it is a vector, a matrix with only one column):

$$\begin{bmatrix} x_1 \\ x_2 \\ . \\ . \\ x_n \end{bmatrix}$$

dx/dt can be the matrix (vector) of the time evolution of the variable(s) and A, the matrix of coefficients describing the temporal evolution of the system given its current state. In general, $\mathscr{F}(x)$ is a vector function and defines a velocity vector field in the state space, and thus $\mathscr{F}$ is normally referred to as a vector field. So, a full system of coupled equations with several variables will look like something like this, in matrix form:

$$\begin{bmatrix} dx_1/dt \\ . \\ . \\ dx_n/dt \end{bmatrix} = \begin{bmatrix} a_{11} & \cdot & a_{1n} \\ a_{21} & \cdot & a_{2n} \\ . & . & . \\ . & . & . \\ a_{n1} & \cdot & a_{nn} \end{bmatrix} \times \begin{bmatrix} x_1 \\ . \\ . \\ x_n \end{bmatrix}.$$

Once the equations that describe a system are found (or created), the next step is to solve them. This means to find out the time evolution of each of the "x" variables, the values they take with time: $x(t)$ for all times "t" considered. This step involves solving the system of differential equations. The fundamental theorem for linear systems dictates that the unique solution of the above system of equations with certain initial value $x(0) = x_0$ at $t = 0$ is given by

$$x(t) = e^{At}x_0. \tag{2}$$

Hence, it can already be appreciated that the matrix A will be fundamental to characterize the states of the system. For instance, a large positive value of A will translate into exponential growth, while a negative value will dictate an exponential decay. Hence, the matrix e^{At} is an operator that encapsulates the global information of the set of solutions to Eq. (1) above, as it determines where the point $x(t)$ is positioned at time t. This operator describes, thus, the flow generated by the vector field Ax. This flow is just the set of all possible solutions to (1). An interesting aspect is that some solutions play a fundamental role. For instance, solutions that are invariant generate invariant subspaces. Invariant sets are of primary importance in dynamical systems. The term "invariant subspace" refers to any solution that starts in a subspace and does not leave it: it remains trapped in it. These subspaces are called the stable, unstable, and center subspace and are determined by the eigenvectors (and associated eigenvalues) of the matrix e^{At} (those familiar with matrix algebra will understand this notion, but this is not needed for the present purposes). This means that solutions that lie on the stable subspace are exponentially decaying, those lying on the unstable subspace are exponentially growing (as mentioned in the beginning of this paragraph), and those in the center subspace are neither growing nor decaying: these either oscillate or remain constant (a technical point: this is true provided the absence of multiplicity of the eigenvalues of the matrix). Fundamental invariant sets of importance are fixed points and periodic orbits (limit cycles), commented upon below. In technical terms, the values that dictate the evolution are the eigenvalues of the matrix e^{At}. Just one brief comment to aid those unfamiliar with matrix algebra: the eigenvalues represent the factors by which a matrix transforms an eigenvector, because matrices act on vectors changing the vector's magnitude and direction. Those vectors that are not altered in direction (only direction reversals are possible) are the eigenvectors, and thus these are important vectors to consider. These properties of matrices are fundamental for the dynamic behavior of the system, and are important in physics and applied mathematics. Readers interested in these mathematical aspects are encouraged to consult introductory textbooks on linear and matrix algebra.

Now, if the function $\mathscr{F}(x)$ in Eq. (1) is nonlinear, complications arise as most times an analytical solution, like that one shown in (2), cannot be found. For this reason, nonlinear systems are studied in the linear approximation, that is, using the mathematical trick of linearizing the system about the invariant sets. Pioneers in the study of nonlinear dynamical systems include Henri Poincarè in the 19th century (Poincarè, 1899) and A.N. Kolmogorov in the early 20th century (Kolmogorov, 1957). In current times, the lack of analytical solutions may not be an insurmountable obstacle because with the aid of computer simulations which can numerically solve differential equations, trajectories of nonlinear systems can be visualized in state space regardless of the nonlinearities. This was originally exploited by E. Lorenz with his now landmark study on weather prediction, a study that necessitated computer simulations to solve his system of three coupled nonlinear differential equations, the celebrated Lorenz system (Lorenz, 1963). This original study, while it took a good number of years to be appreciated by the scientific community, is considered to be the rebirth of nonlinear dynamics, as it exposed chaotic behavior (and in this manner "chaos theory" was born) and introduced the computer for obtaining numerical solutions of the system's equations, which has become the standard today in the study of similar nonlinear systems.

One of the first features to investigate in nonlinear systems is the presence of equilibrium points, also termed fixed points, those in which the variables do not change with time, that is: $dx/dt = 0$. Thus, solving the equation $\mathscr{F}(x) = 0$ will reveal the fixed points of the system, x_{fp}. Once the fixed points are found, the next step is to perform linear stability analysis: linearize the system at the fixed point and thus study the *local* behavior of the dynamics about those points. The aspects of local and global dynamics will be a recurring theme throughout this monograph, both in theoretical considerations and in practical neurophysiological matters, and in fact has already appeared in Chapter 1. The importance of linearizing the system is to obtain a linear system of equations that is easier to solve than the original nonlinear system. These concepts are related to the study of manifolds (Wiggins, 1994) as mentioned below, where geometric objects such as tangent spaces to manifolds are defined so that these locally reflect the structure of the perhaps very complex manifolds. The method to perform linear stability analysis is standard and explained in many works (Guckenheimer and Holmes, 1983), and consists of calculating the Jacobian matrix of first partial derivatives of the function $\mathscr{F}$. This Jacobian amounts to a matrix of derivatives of the function $\mathscr{F}$ when one variable is allowed to vary and the others are held constant, hence the name partial derivatives (recall that $\mathscr{F}$ depends on several variables and thus the derivatives are evaluated for each variable). Intuitively, the Jacobian describes a linear transformation between, for instance, two different coordinate frames and can also be thought of as describing the amount of "stretching" applied by a particular transformation between frames of reference. Readers may ask now, what can really be known about the solutions to the original system (1) by just studying the solutions to the linear system about the fixed points? Two fundamental theorems in dynamic system theory provide an answer: the Hartman-Grobman theorem and the stable manifold theorem. Without dipping into technicalities, the main message of these mathematical results is that, topologically, the local behavior of the full nonlinear system near fixed points is determined by the linear system's behavior, when the derivatives present in the

Jacobian matrix are estimated at the fixed point: $D\mathscr{F}(x_{\mathrm{fp}}) = \left(\frac{\partial F_i}{\partial x_j}\right)_{x_{fp}}$ for each of the i components of $\mathscr{F}$ ("i" is the number of equations that make up the system of differential equations) and each of the j variables.

As can be seen, it is fundamental to evaluate the Jacobian matrix in these investigations. These methodological aspects provide the readers with an indication to the topics needed to be mastered by those interested in getting their hands dirty in the application of dynamical system frameworks to neuroscience: a decent knowledge of differential calculus and matrix algebra is essential. The two aforementioned theorems, thus, allow us to forget the complex original system and instead to study the linearized system about the fixed points to determine properties like their stability. Analysis of epileptiform activity based on these ideas is presented below and in Chapter 3. A pictorial representation of what the linearization means and particularly the essence of the Hartman-Grobman theorem is shown in Figure 2.1. In the figure, a saddle is represented, which is a type of equilibrium point (see Figure 2.2 for the different types of equilibria in the plane). The original nonlinear system has an attracting (stable) manifold, W_{s}, and an unstable manifold, W_{u}. These are fundamental terms that are encountered in the next chapters. In general, trajectories that tend toward the fixed point as time advances constitute the stable manifold, while those trajectories that move away form the equilibrium point constitute the unstable manifold. These manifolds are also known as separatrices, because they separate areas of the state space having distinct qualitative features: look in Figure 2.1 at the different direction of the vector flow in each region separated by W_{s} and W_{u}. The linearization transforms the system around the equilibrium into another that is simpler, that is, linear. Note that now the stable and unstable manifold (E_{s} and E_{u}, respectively) are linear, so they become the aforementioned stable and unstable subspaces that were introduced when discussing Eqs. (1) and (2). There is no center subspace in this case as the equilibrium is hyperbolic. The important

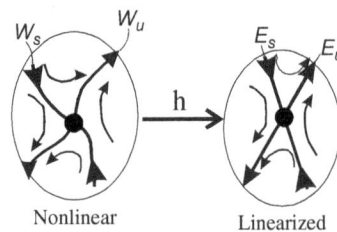

Figure 2.1. Linearization of the system around a saddle: schematic depiction of the Hartman-Grobman theorem. The function h defined near the equilibrium point (black dot in the center) maps orbits of the nonlinear flow characterized by stable (W_{s}) and unstable (W_{u}) manifolds to the linear one.

Stable node   Unstable node

Stable focus   Unstable focus

Center   Saddle

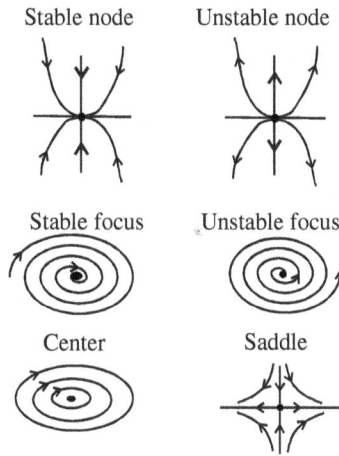

Figure 2.2. Types of equilibria in two dimensions. From top to bottom and left to right, attracting (stable) node with all trajectories converging toward the fixed point; repelling (unstable) node, with trajectories moving away; attracting (stable) focus, with trajectories spiraling toward the fixed point; repelling (unstable) focus; center, trajectories are neither attracted nor repelled; and saddle, with the stable (trajectories moving toward the node) and unstable manifolds. All but the center are hyperbolic equilibria, as discussed in the text.

role played by the stable manifolds of fixed points has been used in some studies to investigate transitions between multiple equilibria, as these manifolds act as separatrices between dynamical regimes. These studies necessitate accurate tracking of the manifolds, which is the difficult part and depends on the individual study. An application of these methods based on tracking separatrices to weather prediction can be found in Deremble *et al.* (2009), where transitions between weather regimes were derived from an atmospheric model.

Some properties of the Jacobian matrix will determine the nature of the fixed point, particularly its eigenvalues determine the local behavior at the equilibrium points. This is not the moment to go into many formal details, but it is important to remark that the two main classes of fixed points are hyperbolic and nonhyperbolic. Hyperbolic systems have an important characteristic: they are structurally stable. In dynamic system theory, structural stability means that small or moderate perturbations do not cause qualitative changes in the system's behavior. Or, equivalently in topological terms, that small perturbations yield topologically equivalent systems. Structural stability of dynamical systems was advanced by Andronov and Pontrjagin (1937) and it is a fundamental property of a dynamical system dictating that the qualitative behavior of the trajectories in state space is unaffected by "small" perturbations. How small is "small" is an aspect to be determined in each situation.

When the term "perturbation" appears in dynamic theory it usually denotes a small one, because a big perturbation can destroy the system, plus other more theoretical reasons. It should be noted that this notion of structural stability considers perturbations of the system itself rather than perturbations of the initial conditions on trajectories as for example Lyapunov stability refers to. The various concepts of stability are discussed in Section 2.2.2. This idea of structural stability can be of use to the analysis of physical systems. Specifically, are nervous systems structurally stable? Each chapter addresses this query considering the specific aspects that are being treated, but in general a partial answer can be advanced in the present chapter, although this depends on the level of description. At the behavioral level, more precisely at the psychological level, people feel (perhaps "perceive" is a better term) a unity in cognition and behavior and thus create the so-called "self," and this feeling seems unchanged during an entire lifetime (however, some special dissolution of the self do exist in pathologies and is discussed in subsequent chapters). Perhaps this feeling of self is structurally stable, if such a concept can be described in this manner. However, many personality traits are variable and unstable depending upon the external inputs the person is receiving. Even when the same external influences are almost identical, the subjective feelings are completely different if the person is in one or another mental disposition. The internal influences of the brain itself are promoting this variability in the response to external input: the vision of food can elicit different responses depending on whether the individual is satiated or hungry. The consideration of structural stability at the cellular/network dynamic level is a matter discussed in each chapter because it all depends on how this activity is measured and how the state space representation is constructed.

Nevertheless, these concepts may have some consequences when discussing natural systems, and specifically nervous systems to which dynamic theory is applied, for, if brains were hyperbolic (hence structurally stable), could they change their dynamics in a flexible manner to become adaptive? Perhaps, structural stability is not a desired property for brains, or for some brain functions, as these have to function in a way that allows alterations in their activities with a variety of small and large inputs, so that the activity results in enough flexibility in the behavioral manifestations and thus of adaptive value. We discuss this theme when introducing attractors in the next section, and will be a point that pervades discussions in the following chapters. The Hartman-Grobman theorem introduced above applies only to hyperbolic equilibria: the dynamics near a hyperbolic equilibrium is essentially the same as that of the linearized system about that fixed point. If the equilibrium happens to be nonhyperbolic, the linearization says nothing about its stability or instability. The nature of the fixed point is known after calculating the aforementioned eigenvalues of the Jacobian matrix, when these are purely imaginary the system is termed nonhyperbolic. Hyperbolic equilibria may still have imaginary eigenvalues but with non-zero real part that are used to determine the stability, because it is this real part which determines the rate of contraction (if negative real part) or expansion (if positive), recall that the Jacobian can be thought of as describing the amount of "stretching" applied by a particular transformation determined by its eigenvalues. Nonhyperbolicity implies that the system has a center subspace, one of those invariant subspaces mentioned above and depicted in Figure 2.2, and, unlike the stable and unstable subspaces where stability is properly defined (with positive or negative real parts of the eigenvalue), nothing can be said on the center subspace since it has purely imaginary eigenvalues (zero real part). To catch a glimpse at what the equilibria represent, Figure 2.2 depicts the main fixed points in two dimensions, details of each can be found in the legend of the figure.

2.2. Geometric Approach

The geometric representation is a convenient manner of visualization of the system dynamics. Since this is a crucial component of some material in the following chapters, an adequate understanding of the fundamentals of this approach and the information derived from it is necessary. The evolution of the dynamical system can be followed in a state space as mentioned in the previous section (also called phase space), where its phase portrait is graphed. It was previously described that Eq. (1) defines the motion of the variable $x(t)$ in state space along the solution curves determined by the differential equation, recall that $\mathscr{F}(x)$ is a vector field. More formally, a solution of the system's equation is a trajectory whose values define the states that the variables of the system go through.

Early thinkers in the study of trajectories in state space include H. Poincaré (Poincaré, 1885), while Andronov *et al.* (1937) are considered to be the first to introduce methods for the analysis of dynamical oscillatory behavior based on their graphical state space representations. However, it is fair to note that very early efforts at attempting a geometric representation of dynamics were done in the 14th century by the French cleric Nicolas de Oresme and later on by the likes of Galileo. The geometric perspective allows for the visualization of the structure of the dynamics and continually reinforces the fact that seemingly simple systems (defined by relatively simple differential equations) can possess solutions of amazing complexity. Figure 2.3 shows the very famous butterfly-like phase portrait of the Lorenz system of equations, a relative simple system of three-coupled differential equations but displaying a large variety of dynamical behaviors, chaos in

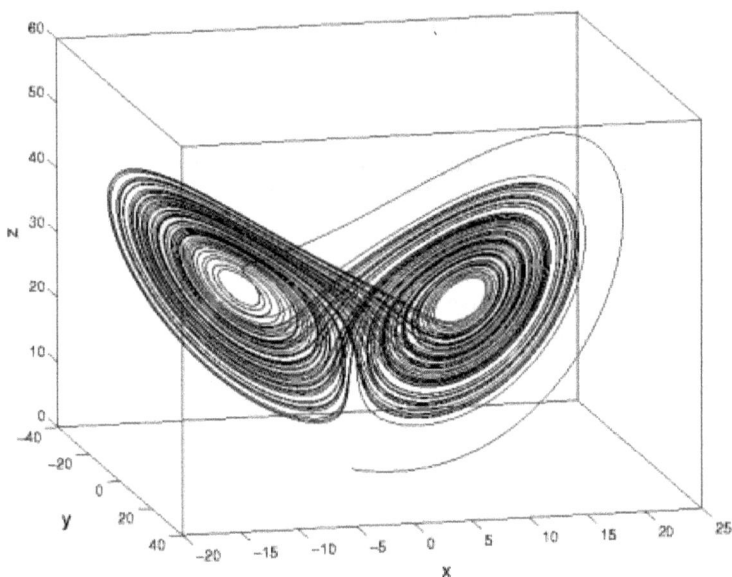

Figure 2.3.   The Lorenz attractor.

particular. Other renowned phase portraits include those of the Rössler and Duffing equations.

A look at some of these phase portraits reveals enough complexity to be amenable to detailed study, and at this point, a geometric trick developed by H. Poincaré comes to help. The method is based on the creation of a section transversal to the flow, known as the Poincaré section, to reduce the dimension of the system (by one), and thus the continuous flow shown in Figure 2.3 becomes discretized. This is schematically portrayed in Figure 2.4. It is important to understand now what this method entails and the knowledge that is gained by its application because the following chapters make abundant use of this technique to characterize brain dynamics. The method relies on finding a surface of a section (Σ in the figure) that is crossed transversally by the trajectories of the original system. Every point of crossing defines a point in the now reduced state space, the two-dimensional (2D) section Σ, since the original dynamics lived in a three-dimensional (3D) space. Instead of the butterfly of Figure 2.3, one would see a set of points representing the crossings of each trajectory with the section. By focusing on the dynamics on Σ there is no need to know the detailed structure of the orbits. It is important to note that our new representation is not continuous, like the original trajectories, but discrete. A discrete dynamical system is defined, basically, as iterates of

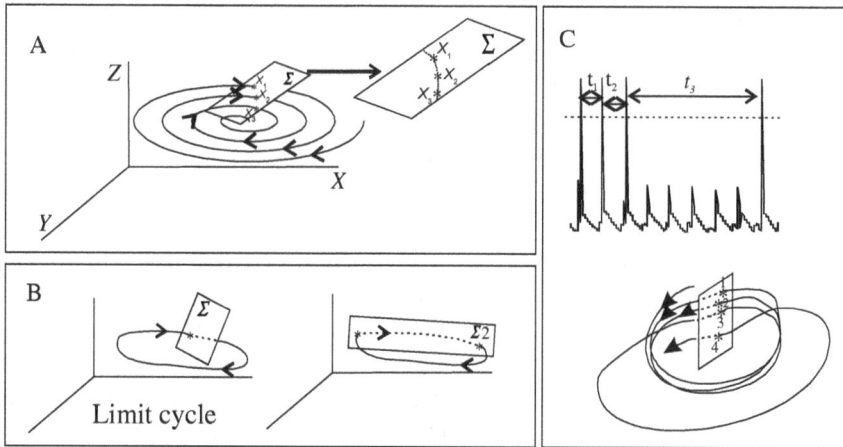

Figure 2.4. Illustration of the method of reducing the dimensionality of the system using a Poincaré section (Σ) and its relation to the analysis of experimental time series.

any diffeomorphic mapping function of the flow that represents the continuous dynamical system (Perko, 1991), and thus it can be represented as iterations of a function like:

$$x(k+1) = F(x(k)). \tag{3}$$

The equation is solved for $x(1)$ starting with an initial value $x(0)$, and then $x(1)$ becomes the initial value for the second iterate $x(2)$, and the same for $x(3)$, $x(4)$, etc. In this way, a new value $x(k)$ is successively obtained from $x(k-1)$ for k iterations. This, in fact, defines a mapping function, $F(x(k))$, which is often called a *first return map* because it specifies each data point given the previous. Chapter 3 takes advantage of first-return maps to study epileptiform dynamics. To stress it once again, one advantage of working with these maps is that the analysis is simplified. For instance, to determine the stability of the aforementioned fixed points one can perform the typical linear stability analysis on the discrete map (of one fewer dimension) rather than on the continuous flow. Simplifying is rarely a sin.

For simplicity, only three crossings of the section were shown in Figure 2.4(A), which define three points in that section. A crucial concept to be understood, as can be readily seen in the figure, is that the relation among those three points is determined by the original trajectories and how these cross the Poincaré section. The Poincaré map is then the mapping function that defines the relation between those points on the section, and is represented by a general equation like (3) above. In other words, the map is produced by considering the successive intersections of the trajectory with the Poincaré section. There are theorems that establish results that refer to the existence and continuity of the Poincaré map, but, while these are important mathematical results, there is no need to understand them in depth for the purposes of the subsequent discussions in the following chapters. The great advantage of this method cannot be underestimated: it shows that there is no need to consider, or to know for that matter, the original system's trajectories in a high-dimensional state space, rather it is enough to concentrate on the relations of the points on the section as these represent the original system's dynamics. This is a crucial concept that has importance in experimental studies of neurophysiological signals and related computer simulations, as the basic idea underlying this approach is that time-delayed phase portraits generated from experimental time series of one variable maintain the essential characteristics of the original, multidimensional phase space. Some of these first-return maps can be very simple. Imagine, for example, that the original system has strictly periodic activity and thus it is represented by a limit cycle as shown in Figure 2.4(B). Then, there is only one point on the surface of the section Σ, or two points if section Σ2 is chosen. It is clear that depending on the section, the number of crossings will be different, however it is important to realize that, in the final analysis, these are all equivalent. Here, then, we are confronted with the question of how to choose an adequate Poincaré section, and in particular the question of whether it always exists can be asked. Even having at hand the system's equations, it is frequently not trivial to analytically construct the mapping from the original differential equations, since it requires knowledge of the flow: if the equations are too complex to yield general solutions, the flow is unknown and the computation of the Poincaré map is not feasible. There are perturbation and averaging methods to approximate these maps, and readers willing to go into heavy mathematical details are encouraged to consult Chapter 4 in Guckenheimer and Holmes (1983). Numerical methods to compute accurate Poincaré maps have been proposed (Tucker, 2002). Nevertheless, sometimes the map does not exist, but there are systems in which it is well defined, specifically those systems that are periodic or have a more or less periodic rhythm. The reason for this, in simple words, is that in strictly periodic dynamics, at each period the point returns to itself where it was in the previous period, as shown in Figure 2.4(B), and a section can always be constructed transversal to the flow so that at each period the trajectory intersects the section once. It is conceivable, therefore, that these concepts can be applied to physiological systems since almost all of them are periodic or quasi-periodic.

This method of studying the dynamics on Poincaré sections is of fundamental importance for experimental purposes. Let us then consider some illustrations of experimental time series. Imagine that the orbits in Figure 2.4 come from an experimental system. For instance, it may represent a neuron's membrane potential, and every time the orbit crosses the section there is a spike generated (as if it were a threshold crossing for the firing of an action potential). Further, imagine that the recording methodologies do not allow one to measure either the x or y variables. However, spikes can be easily detected in neurophysiological traces. Then, a first-return plot of interspike intervals, plotting one interval versus the next, can be graphed and this will constitute an analog of the Poincaré map. Consideration of Figure 2.4(C) reveals this intimate analogy between the theory and the experiment: the trace shows four action potentials and five synaptic potentials recorded in a neuron and the dotted line represents a threshold that can be used to detect spikes and therefore the interspike intervals t_1, t_2, and t_3 can be calculated. The bottom graph could represent the dynamics of this neuron's membrane potential and the section used to detect the crossings of the trajectories, where each of the four crossings defines a spike. It can be seen that the time intervals between the spikes above are directly related to the time intervals between the section crossings below (it takes longer for the last orbit to cross the section at point number 4 since its excursion is the longest). Hence, the threshold used for spike detection can be freely interpreted as a "Poincaré section." This is just a qualitative, not too accurate, illustration, but it serves to appreciate that we can know a good deal of the neuron's behavior by just studying the spikes time series, which is encapsulated on the surface of the section. Related to this, the reconstruction of the dynamics from interspike intervals was proposed by T. Sauer (1994, 1995) in the mid-1990s, and this will be introduced in Section 2.5.1 where there will be discussion on the fundamental concept of embedding. These types of graphs are sometimes called recurrence plots as the notion is conceptually close to the concept of recurrence in dynamical systems, that, while it may seem obvious today, required the intellect of Poincaré himself to be revealed.

The relationship between the discrete mapping functions and the original continuous flows determined by the differential equations can easily be revealed by considering again the linear Eq. (2): $x(t) = e^{At}x_0$ and recalling that the matrix e^{At} defines the flow. Then, for a fixed time $t = \vartheta$, and letting $e^{A\vartheta} = P$, then, P is a constant coefficient matrix and sets up the difference equation $x(t) = Px_0$ or equivalently: $x_{n+1} = Px_n$, which is nothing more than Eq. (3) above. If the flow is nonlinear, the same ideas apply and will give rise to a nonlinear map. Main results from the theory for flows also apply to mapping functions, particularly the aforementioned Hartman-Grobman theorem, which means that one can work with maps and perform a linear stability analysis similar to the one described for flows. For those attracted to abstraction, there is a particular beauty in the comprehension of the essence of the relations between geometry–topology and dynamics. An appreciation of the profound connection between dynamical systems and differential geometry and topology can be gained by consulting works like Burns and Gidea (2005).

The topological approach is heavily involved with manifolds. The term manifold, that appeared previously in Section 2.1, will show up again in subsequent chapters and it is a crucial concept in dynamical system theory. The concept of manifold plays an important position to study the structure of the dynamics and the stability of its states. Wiggins (1994) provides a decent reading to understand the intricacies of normally hyperbolic invariant manifolds applied to dynamical systems. To recapitulate from previous sections, an invariant manifold is, in general terms, a surface in the state space with the property that trajectories starting on it will remain on the surface. Hence, it can be conceptualized as a collection of orbits that form a surface. Manifolds have been used in theoretical biology and particularly in neuroscience, as will be seen in next section and chapters, and have had fruitful applications in a variety of fields. Those readers interested in chaos should pay particular attention to the study of manifolds, as one mechanism that originates chaotic dynamics is precisely the intersection between stable and unstable manifolds. A description of the current interest in the determination of these manifolds in the study of epileptic dynamics will be presented in Chapter 3.

2.2.1. Asymptotic behavior of dynamical systems and the concepts of attractor

Asymptotic behavior refers to the long-term evolution of the solutions of the dynamical system, as time goes to infinity. In simple words, run the computer simulation for a long, long time and see where the trajectories end up. A principal characteristic of dynamical systems is the presence of regions in state space that attract trajectories originating in other regions. These limit sets are termed attractors. We have already encountered stable manifolds or subspaces, which attract nearby orbits. Asymptotic behavior thus refers mostly to the motion on an attractor. The concept of attractor is fundamental for the discussion in the next chapters, and because its use may sometimes be appropriate in neuroscience even though there is a tendency of taking it too lightly to the point of misrepresenting the concept, it is worth investigating in detail so that its meaning is not pushed too far in the metaphor landscape. The previously discussed invariant sets in Section 2.1 are crucial in the study of asymptotic behavior of dynamical systems. There are several manners to define an attractor, all of which are conceptually similar. In plain words, it is a set (a region) of state space to which trajectories converge. In other more empirical words, the dynamics of the system approaches a certain "activity" and then it is trapped into this particular activity pattern, only to escape if perturbed enough. Attractors thus determine the system's long-term behavior. Many physiological processes seem to be cyclic, which leads one to think that these can be represented by limit cycle attractors. Whether or not this notion is applicable to neural tissue, will be a matter discussed in Sections 2.2.3 and 2.6. The emphasis is now on the concept of attractor itself. And, for the sake of precision, here is a sample of a technical definition of attractor: an attracting set with a dense orbit (Guckenheimer and Holmes, 1983; Perko, 1991). Attracting sets are characterized by having a domain, or basin, of attraction, which is the region around it where all orbits are trapped and fall toward the attractors. Conversely, there are also regions of state space that repel the trajectories, and are termed repellors; basically, these sets are the opposite of attractors.

The two simplest attractors are the fixed point (a static equilibrium point) and the limit cycle, which represents periodic motion like that shown in Figure 2.4(B) (periodic equilibrium of the oscillation, to be more precise). Continuing with the final discussion on manifolds of the previous section, it can be said that these are the basic types of invariant manifolds (the torus is the third basic manifold, see Figure 2.5). In subsequent sections and chapters, examples of possible attractors in nervous system activity are introduced. At this point, it can already be conceived that a neuron that does not fire spikes, and thus has an almost constant membrane potential, is in an attractor state (the inactive state), while those neurons that fire periodically, like in the case of the so-called pacemaker neurons in central pattern generators, may reside in a limit cycle attractor. If the synchrony between two cells is analyzed, the representation can use a torus (Figure 2.5), rather than a limit cycle.

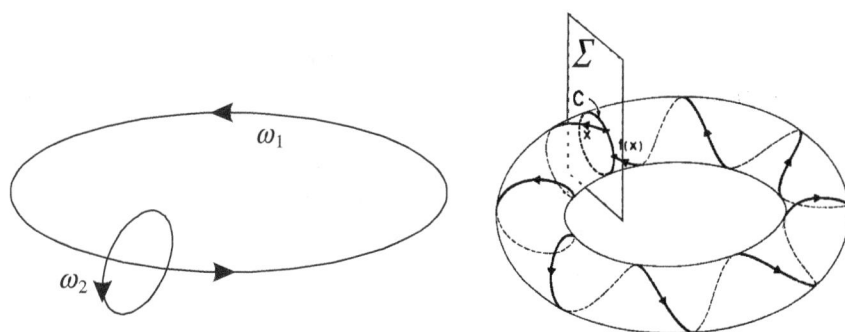

Figure 2.5. The 2D torus represents the mathematical (or geometric) representation of synchronization phenomena. It consists of the orbits representing the periodic motion of two oscillators with frequencies ω_1 and ω_2. The Poincaré section (Σ) cutting the torus allows us to visualize how a trajectory starting at x intersects after one revolution at $f(x)$, thus creating a Poincaré mapping. Modified with permission form Bergé *et al.* (1984) (copyright Wiley-VHC Verlag GmbH & Co.).

A great deal of theoretical research into attractors has focused on the "robust" hyperbolic attractors, that is, those that are structurally stable, while nonhyperbolic ones are structurally unstable and thus dynamical bifurcations, or phase transitions, can occur. It was mentioned above that the local behavior of hyperbolic systems is almost linear, and thus the linearization approach about fixed points is used to study characteristics of local equilibria, according to the aforementioned Hartman-Grobman theorem (Figure 2.1). The structural stability of hyperbolic equilibria implicates that these cannot display bifurcations. However, the neuroscientific literature is full of studies implicating dynamical bifurcations, as explained in the next section where it will be seen how bifurcations result in pattern formation and in most of the interesting aspects of nonlinear dynamics. Without bifurcations, life is rather boring as the activity remains stuck to an attractor and there is not much possibility for pattern formation. It was also briefly commented above that brain dynamics need to be flexible enough to be adaptive, that is, to respond to stimuli big or small. Thus, perhaps it can be concluded that hyperbolic equilibria are less interesting from the perspective of adaptive, open dynamical systems; in the words of Hoppensteadt and Izhikevich: "locally hyperbolic weakly connected neural networks are not interesting" (this is their Theorem 5.1 in Hoppensteadt and Izhikevich, 1997; this monograph presents in detail the mathematical aspects of dynamic theory applied to neural networks, and may be of interest to those who wish to go deep into these matters). In general, most "attractors" found in physical systems are not structurally stable, as these natural systems have many types of activities and, normally, display an abundance of pattern formation. However, as aforementioned, the study of hyperbolic attractors has been fundamental from the theoretical perspective to guide the thinking about more complex dynamics. Is it thus recommended to those interested in this topic to study in detail the concepts and methods of hyperbolic attractors that can be found in the several monographs indicated above, before jumping into more complex attractors such as the chaotic ones.

In brief words, three main types of attractors were known before chaos was discovered: the equilibrium state (fixed point), the stable periodic oscillation (limit cycle), and the stable quasiperiodic oscillation (represented by a torus). Two of these were represented in Figure 2.2. The torus provides a representative example of a geometric manifold, as shown in Figure 2.5, which in two dimensions has a doughnut-like shape that, topologically, consists of the product of two circles representing the periodic motion of two oscillators on perpendicular planes. It is used as a convenient representation of synchronized states and is very useful in modern neuroscience for the reasons detailed in Chapter 1 regarding the current interest in the study of synchronous activities. The presence or absence of frequency locking between two oscillators can be geometrically represented on the surface of the torus, since the combination of any pair of oscillators draws a trajectory on the toroidal surface. If the trajectory (the solution to the equations of the system) is a closed orbit, there is frequency locking between both oscillators. This will happen if the frequencies of the oscillators are commensurable, that is, if the ratio of the frequencies is a rational number. On the contrary, if the ratio is an irrational number, the oscillators cannot phase lock and the trajectory of their combined motion will eventually cover the whole toroidal surface without returning to the same point: the combined motion is said to be quasiperiodic. With the aid of a Poincaré section, represented in Figure 2.5, the periodic or nonperiodic dynamics can be visualized better. In the case of 1:1 frequency locking, there will be only one point, as the orbit always comes back to the same point, while in the case of 2:1, for instance, there are two rotations along the main axis in the time it takes to complete one along a perpendicular circle (the transversal section of the torus), and in this case there will be two points on the Poincarè section. As can be seen, other ratios can be visualized on the torus. It was discussed in the previous chapter that most synchronization studies in neuroscience focus on the 1:1 frequency locking, however, more complex ratios have been noted (Perez Velazquez *et al.*, 2007d). Synchronization patterns have been found, in theoretical studies, to represent stable attractors of network dynamics (Kestler *et al.*, 2008), so the torus will re-appear in some more or less metaphorical descriptions in subsequent chapters to stress the stability of many synchronous regimes found associated with diseases. In general, these notions are of importance throughout this volume.

2.2.2. Strange attractors and deterministic chaos

The discovery of chaotic dynamics brought about the concept of strange attractors that have been widely popularized in modern times. These attractors correspond to nonperiodic solutions to the system's equations, and are characterized by a complex geometrical structure, thus the name of "strange." The structural complexity is the reason why the dimension of a strange attractor is not an integer number, and this is termed fractal dimension. Since the early work in this subject, particularly by Ruelle and Takens (1971) who introduced the strange attractor as the representation of deterministic chaos, strangeness and chaotic dynamics have always been closely associated with each other. Note that this is not always true: there are systems displaying nonchaotic activity and yet they possess strange attractors, the so-called strange nonchaotic attractors (Prasad *et al.*, 2003; these concepts are reviewed by Anishchenko *et al.*, (2002)). Of note, one study attributed neocortical activity to a strange nonchaotic attractor (Mandell and Selz, 1993). Not only this, but systems can also display chaotic dynamics and have (chaotic) non-strange attractors. To understand the reason for this, a reference to the main properties and the two main descriptions of attractors is necessary.

An attractor can be defined probabilistically or topologically. The geometric (topologic) interpretation is by far the most common among neuroscientists and biologists in general, and was given in Section 2.2.1. Nevertheless, the geometry alone only shows the regions of state space that are visited by the trajectories, but says nothing about what regions are more visited. Consider the Lorenz attractor shown in Figure 2.3. Just by inspecting the geometrical characteristics it cannot be known whether all parts of the attractor are visited with same frequency, or whether there is a region most frequented by the trajectory. Thus, attractors also carry probability measures that can be used to answer the aforementioned questions, and this is the subject of Ergodic theory (Mañé, 1983). Experimentalists have used both descriptions of attractors, the probabilistic and the geometric, and this will be a theme of discussion in the next section. The probabilistic definition of attractors will be important in the following chapters when the possible neurodynamics of brain circuits is considered, particularly in notions like chaotic itinerancy. Perhaps the most used definition from this perspective is the one attributed to J. Milnor, who defined an attractor in a certain manner (too technical to be discussed here, but see Milnor (1985)), with a main difference from the topological description: some orbits have a certain probability to escape from the attractor, something forbidden in the topological definition as was discussed above. Thus, a geometric attractor can be considered a Milnor attractor (one in which there is never an escape), but not vice versa.

Two of the most used statistical properties of attractors are the dimension and Lyapunov exponents. The former is a measure of the geometry, while the latter considers the dynamics. Confusion arises among experimentalists working with physiological recordings when it is found that the reconstructed attractor from the experimental time series has a non-integer dimension, which is known as fractal dimension (the complexities of attractor reconstruction from time series are treated in Section 2.5), and thus this is taken as a "proof" of chaotic dynamics. There are many introductions to the currently very popular concepts of fractals and chaos (Mandelbrot, 1977). For the present purposes, suffice it to say that when the dimension of some complex attractors is not an integer, these are termed strange attractors, as mentioned above. However, it is worth emphasizing again that while chaotic dynamics is normally associated with strange attractors (thus with fractal dimensions), there are systems displaying nonchaotic activity that display strange attractors too. Hence, a fractal dimension means little with regards to the underlying dynamics. It just means that the attractor has a complex geometrical shape, that is all. Still, there is usefulness in the estimation of the dimension as it indicates the complexity of the signal. For instance, some studies will be mentioned in Chapter 3 that have used the correlation dimension to reveal relative differences between neurophysiological time series in epileptiform activity. This relative comparison within the same time series is reasonable, but ascribing chaotic dynamics to a fractal dimension is not.

The three types of attractors mentioned above, even the strange attractors, are hyperbolic (but consider the intricacies of chaotic attractors that can be classified too as quasi-hyperbolic or quasi-attractors, Anishchenko *et al.*, 1998) and, recalling the comments on hyperbolicity a few paragraphs above, these are structurally stable: do not bifurcate. While it was mentioned before that the dynamics of systems exhibiting hyperbolic points or attractors is not very interesting, the activity is not always predictable, and this is a characteristic of the strange chaotic attractors. For the term chaos, rather than a geometric property (as the characteristics of an attractor), is a *dynamic property*: the exponential separation of the trajectories on the attractor is the main feature of chaos. Figure 2.6 depicts this situation. The trajectories are completely determined by the system of equations (this is why it is called deterministic chaos) but these present an extreme sensitivity to initial conditions, such that a minute change in the initial value leads to completely different trajectories on the attractor. Thus, the behavior is not predictable because it is not feasible to know the initial conditions with extreme precision, as Lorenz showed with his weather model. The exponential divergence of trajectories in the attractor is characterized by the Lyapunov numbers (or exponents), these being a very used measure to assess chaoticity. These numbers are closely associated with the eigenvalues of the linearization matrices talked about before, and for our present purposes no more details are needed, perhaps only to mention that chaos needs at least one positive Lyapunov number. Quite a few studies have determined Lyapunov numbers from

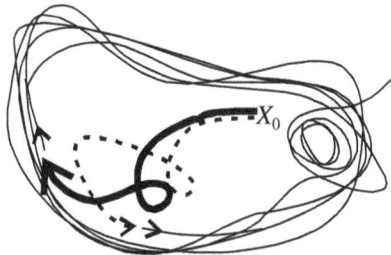

Figure 2.6. Separation of orbits within a chaotic attractor. The divergence of two solutions (trajectories) in state space is shown, for slightly different initial conditions near X_0. Note, however, that both trajectories remain bounded within the attractor, never leaving it.

experimental time series, most particularly in epileptic recordings. However, the determination of these numbers from purely experimental time series, much like the aforementioned estimations of the correlation dimension, is a complicated matter. This will be a matter discussed in sections below, but now let us only mention that, to extract Lyapunov exponents, a system of equations is absolutely needed so that the flow can be visualized in state space by numerically solving the equations, and then the spectrum of Lyapunov exponents can be precisely defined and from this the presence of sensitivity to initial conditions and the divergence of trajectories (both signs of chaos) can be clearly estimated.

It was just said that chaotic attractors are structurally stable, and yet their extreme sensitivity to initial conditions seems to indicate instability. The notion of stability of trajectories was introduced above, but it can be already appreciated that there are several manners to define stability. For example, consider the exponentially diverging trajectories on a chaotic strange attractor. These are unstable according to Lyapunov since any small perturbation (tiny variations in the initial conditions) is "amplified," meaning that the trajectory will follow a different path from the original, as shown in Figure 2.6. On the other hand, these trajectories remain in the attractor, regardless of their divergence they do not leave the bounded region of the state space that constitutes the geometric attractor, which means they are stable according to another perspective known as stability according to Poisson. This version of stability is the weakest requirement for stability, and thus it can be said that on a chaotic attractor the trajectories are unstable according to Lyapunov and stable according to Poisson. Finally, recall the structural stability notion introduced above, that deals with the whole system, this is yet another type of stability (chaotic attractors are structurally stable since their topological structure does not change). These comments are presented here to demonstrate that things depend on the viewpoint that is taken, and the application of these dynamical system notions to experimental recordings has to be done carefully and with a proper understanding of what is meant when, for instance, stability of a specific neural activity is considered.

Nevertheless, strange attractors are still structurally stable and therefore chaotic dynamics, while a source of unpredictability and thus some fun (eluding the excruciating boredom of predictability is always fun), does not allow for a full display of phase transitions (bifurcations) that can result in pattern formation, a main characteristic of natural phenomena. For this reason, most "attractors" in nature are probably of the nonhyperbolic type. These are characterized by the coexistence of chaotic and regular attracting sets, still in a bounded element of the state space that constitutes the whole nonhyperbolic attractor (Figures 2.7 and 7.3). The dynamics will wander from attractor to attractor and, in this case, as the system's parameters change bifurcations can occur.

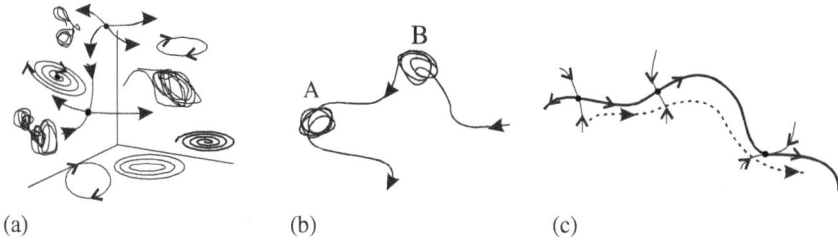

(a)   (b)   (c)

Figure 2.7. Schematic representations of three much related notions. (a), a nonhyperbolic attractor, that includes several attracting sets; (b), chaotic itinerancy, the orbit escapes from one attractor (B) and moves toward the next (A), to escape again to visit other regions in state space that will determine other dynamic patterns; (c), a stable heteroclinic channel, where the saddle nodes are connected via their unstable separatrices. The channel itself is structurally stable but the topology depends on the initial conditions.

In closing, since much previous talk has revolved around manifolds, fixed points and attractors, it should be acknowledged that the identification of these invariant sets is not a trivial matter in stochastic and nonstationary systems, as the presence of noise makes the orbits transit from one to another "invariant" subspace, hence in reality destroying the invariance. For small noise, though, studies have revealed that the dynamics can be restricted to the "almost-invariant" sets (as Billings and Schwartz (2008) call them) for relatively "long" periods of finite time. If it were really invariant it would remain there for an infinite time. Note that in the previous discussions there was no mention at all of noise and fluctuations, except perhaps for the comments on the variability of initial condition in chaotic attractors. Introducing stochasticity can be fixed by adding a noisy term to Eq. (1), operation which results in the switch from strict determinism to stochastic dynamics. Section 2.4 is devoted to the perspective from stochastic processes.

2.2.3. Are attractors relevant to neuroscience?

The title of this section is a paraphrase of Crutchfield and Kaneko's 1988 paper, where they discussed this query in the context of turbulence in fluid dynamics. They presented examples of dynamical behaviors in extended systems that are not associated with attractors, and concluded that the dominant behavior for large, extended systems is governed by long transients, not by attractors. This is in line with the notions presented in the Billings and Schwartz study mentioned in the last section. The comments in Chapter 1 on the different activity patterns found in nervous systems and those in this chapter when treating the matter of the concept of attractor, already suggests that the identification of attractors is not a trivial matter in neuroscience, even more so if one is restricted to time series alone. The question of whether attractors can be identified in experimental neuroscience is discussed in depth in other sections and in each specific case in subsequent chapters. Presently, and to start the discussion in clear terms, it has to be admitted that if nervous system activity is mainly nonstationary which means that it does not have time enough to settle onto any asymptotic state, then the concept of attractor is not too meaningful. Here, an alternative to the attractor paradigm is considered. Readers will no doubt notice that these notions about to be commented, proposed by several investigators, share a common essence: they are not focused on the existence and identification of asymptotically stable attractors, rather, on the importance of transient, not asymptotic, behavior, jumping from state to state.

The concept of chaotic itinerancy proposes that the dynamics wanders among many possible attractor states, without being trapped onto one, or if it does, it is only for a short time (Tsuda, 2001; Kaneko and Tsuda, 2003; Tsuda and Unemura, 2003; Tsuda, 2009). It describes transitory dynamical behavior that may appear nonstationary, and has been applied to neuroscience including the transitions between synchronized neural states (Tsuda *et al.*, 2004). Some efforts have been taken in finding signatures of chaotic itinerancy in neurophysiological signals (Freeman, 2003). In this framework, the state space is not represented by the aforementioned geometric attractors, rather by attractor ruins. These attractor ruins are conceptually close to the Milnor attractors (described in Section 2.2.2), recall that these are attractors with an unstable direction so that trajectories have a certain, non-zero probability of leaving the attractor. As a result, the corresponding trajectory of the dynamical system is an itineration between these attractor ruins (Figure 2.7).

Now, consider another perspective that has been used to model the processing of smell by insects, describing the continuous switching among unstable saddle points that may underlie the recorded neuronal activity in odour processing (Rabinovich *et al.*, 2001; Huerta *et al.*, 2004). This view relies not on attractor ruins but on a chain of saddle nodes, in what is termed a heteroclinic channel (Rabinovich *et al.*, 2008), as depicted in Figure 2.7(C). Recall from Section 2.1 and Figure 2.2 that a saddle has an attracting and a repelling manifold. The stable heteroclinic channel is basically a trajectory in the vicinity of the saddle chain along their unstable separatrices, and it was shown, in the above publication, to be structurally stable in a wide region of parameter space. A heteroclinic cycle is defined as a sequence of trajectories connecting a set of fixed points. There are a number of studies on the stability and robustness of these cycles, and it was shown that they behave as attractors and that, upon losing stability, different bifurcations are possible (for a review, see Krupa (1997)). A model of binocular rivalry has recently been built on this notion (Ashwin and Lavric, 2010). Note how much the heteroclinic channel resembles the previously commented chaotic itinerancy, but, according to the authors, saddle sets rather than Milnor attractors are essential for this model to show reproducibility.

Closely related to these topics, some authors have described quasi-attractors (Haken, 2007), where the attractor state exists only for a very limited time. The world of computational modeling offers several instances of networks composed of model neurons exhibiting long transients that may or may not settle onto attractors (Zillmer *et al.*, 2009). Recalling now the concept of nonhyperbolic attractors is convenient: as mentioned in the previous section, these are characterized by the coexistence of a countable set of periodic and chaotic attracting subsets in a bounded region of phase space and the system's dynamics travel between these subsets. Note again the similarity between this concept and that of chaotic itinerancy and the heteroclinic channel. But these are not the only proposals to capture transience. If the system has periodic attractors and non-attracting chaotic sets, noise-induced chaos can occur (Tél *et al.*, 2008), and indeed these authors describe that a union of saddle points (perhaps, much like the heteroclinic channel) constitutes what they term a chaotic saddle which is the main characteristic for the transients between periodic and chaotic sets in their model. Notwithstanding the technicalities involved in the differences among these frameworks, we cannot help but noticing that, in essence, they are all conveying the same message: rather than focus on identifying attractor states, it would be more useful to think in terms of transients and accept the consequent variability and nonstationarity that this implies. Because of the importance and widespread distribution of nonstationary processes, there is an increasing interest in the characterization of nonstationary systems, and efforts are being made using a variety of strategies including "snapshot attractors" (Serquina *et al.*, 2008), which may be a viable approach in some experimental situations. The concept of structural stability of a dynamical system was introduced in Section 2.1, and it may have something to add to the present discussion. For, when investigators started to develop the theory of hyperbolic dynamical systems, it was soon found that higher-dimensional manifolds (more than three in general) cannot be made structurally stable by an arbitrarily small perturbation. This means that in higher dimensions, structurally stable systems are not dense; or, in simpler words, any perturbation, regardless of how small, will have the chance to alter the whole system. Hence, this leads again to the concepts of nonhyperbolic attractors and itinerancy, because there will be transitions between different states that will appear or disappear due to the loss of structural stability.

If in reality transients are what we witness and have access to in natural phenomena and neural activity in particular, then there are some suggestions for both the experimentalist and the theoretician/computational neuroscientist. The latter, normally busy with computer simulations of neuronal models, should consider that rather than long-lasting iterations of the model that reach asymptotic behavior, short iterations that reach instabilities and update initial conditions are probably more relevant to brain function (Koerner, 1996). On the other hand, the experimentalist, engaged with the study of neurophysiological recordings, should pay careful attention to noise and fluctuations rather than disregarding them (normally these two are a source of frustration for experimentalists) and reflect on the possibilities that result from this scenario. For example, as discussed in Section 3.5.2.1 of Chapter 3, transient phenomena may underlie the current controversies on the anticipation of epileptic seizures.

All these notions capture and address a main concept in brain dynamics that was emphasized in the past chapter (Section 1.4.2), that of *metastability*: cognition depends on *transient dynamical processes in brain networks*, this was the main message in Chapter 1 after the consideration of the currently known neurophysiological events taking place during cognition. Notwithstanding some short moments of apparent stability in the neuronal recordings, variability is the rule in brain studies at all levels, from the individual neuron to the whole brain (McIntosh, 2004). It is for these reasons that the concept of metastable states is attracting the attention of neuroscientists. Metastable states have perhaps more intellectual appeal than unstable states, because the former can always become stable, at least transiently. One example of a possible metastability found in the analysis of brain recordings will be addressed in Section 2.5.2 and in the following chapter on epilepsy, where a proposal is advanced that epileptic seizures are manifestations of a transient stabilization of a metastable state in brain dynamics. Thus, the word metastability is appearing more and more often in the neuroscientific literature, with some early exponents like Kryukov *et al.* (1990). Particularly in the case of neural synchronization, or coordinated activities, metastability refers to a transient coordination reflecting the tendencies of neural tissue to synchronize, and as discussed in the previous chapter, this confers flexibility to the system and allows for the simultaneous integration and segregation of information needed for proper brain function (Kelso, 1995; Kelso and Tognoli, 2007; Fingelkurts and Fingelkurts, 2004; Oullier *et al.*, 2006). The three concepts described in Figure 2.7 can represent metastability: the transition between saddle sets, between attractor ruins, or between attracting sets in the global nonhyperbolic attractor. These ideas lead now to the study of dynamical bifurcations, and an overview of aspects that will be important in subsequent topics is presented in the next section.

2.3.   Transitions between Dynamical States: Bifurcations

The dynamic behavior of systems, obviously, changes in time. The change may be more or less abrupt and many times results in the formation of a completely different pattern from the previous one. The study of how these changes occur is the topic of bifurcation theory, an integral part of dynamic system theory. It was probably Poincaré who first introduced the notion of a bifurcation as the emergence of several stable solutions from one solution in a system of differential equations. This is what is normally represented in typical bifurcation diagrams, like that of Figure 2.9: the lines represent stable solutions of the equations describing the dynamical system. Bifurcations are also known as phase transitions, a term that comes from thermodynamics, and can be generally described as a qualitative change in the behavior of a dynamical system as a result of changes in control parameters. A more technical definition of bifurcation is the appearance of a topologically nonequivalent system under the variation of the bifurcation parameter (Balibrea *et al.*, 2009). The intuitive notion of bifurcations is easy to grasp, and the best manner to understand what bifurcations mean, why their characterization is essential, and the closely associated theme of the fundamental role of noise and fluctuations especially near bifurcations, is to look in detail at one, very striking and well-known, bifurcation. One that can be observed at home while making dinner: add oil (silicone oil works best but it is not recommended for cooking!) to a frying pan and heat it from below. In a few seconds-minutes, a geometric pattern on the oil's surface will be visible. This is the Rayleigh–Bénard instability, discovered by Henri Bénard, a paradigm to show pattern formation and emergent properties. The instability results in the appearance of a singular geometric pattern in a heated fluid (Figure 2.8). Lord Rayleigh, later on, studied it in more detail and showed that an initially motionless fluid layer becomes unstable to small flow perturbations when the temperature difference is sufficiently large. Initially, when the temperature gradient between top and bottom is not large, heat is transferred mainly by conduction, and the fluid molecules remain at rest, colliding against neighboring molecules. As the temperature gradient between top and bottom grows, heat is better transferred by convection: the coherent motion of the molecules. Thus, small convective fluctuations develop, hot fluid near the bottom thermally expands and becomes lighter than the fluid above it, rising, then cools and returns in an overturning flow. This motion is opposed by the viscosity of the fluid and the tendency of thermal diffusion to smooth out temperature gradients. Some of the fluctuations become amplified and organized into a coherent motion, originating a flow pattern that consists of rolls or hexagonal cells, this pattern depending on the details of the fluid properties. The pattern may be stable or it may become unstable and display chaos.

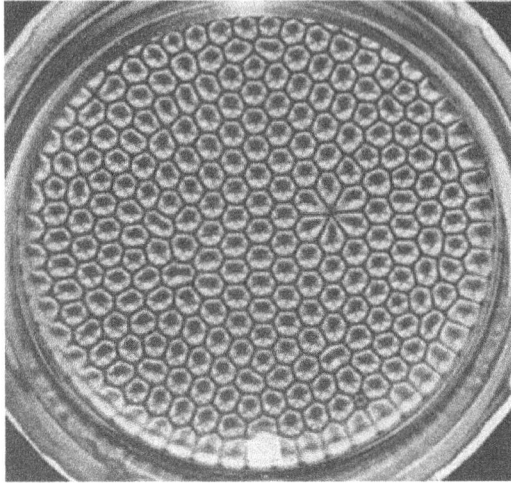

Figure 2.8.   The Rayleigh–Bénard instability, showing a typical hexagonal convection pattern.

   Thus, an important principle can be appreciated from this example: the bifurcation results from the competition between tendencies, the tendency to order due
to the molecular interactions that transfer heat by conduction, and the disordering
tendency of the external (heat in this case) and internal fluctuations (the stochastic
fluctuations in the molecules' velocities that start the convection). This competition determines the critical point, and it is a general characteristic: analysis of any
other critical state in other systems reveals a similar competitive process between
tendencies specific to each process. As an example, let us consider one possible
neurobiological phase transition: an epileptic seizure (more details will be given
in Section 2.5 and in Chapter 3). Neuronal synaptic interactions tend to order
the networks into fast and transient, more or less synchronized activity patterns,
while the fluctuations in the strength and extension of these interactions due to
internal factors and external inputs tend to disorder the activities of the interconnected cellular networks. There may be a point at which, due to changes in control
parameters like extracellular potassium concentration (or changes in inhibitory
neurotransmission), the state of the system reaches criticality and the fluctuations in synchrony are amplified to such a large extent that results in the generation of the pathological coordinated cellular activity during the seizure: a new
pattern has emerged that can be visualized in neurophysiological recordings as
high-amplitude waveforms (Figures 2.12 and 2.14 show some seizure recordings,
and more in the next chapter) due to the large numbers of cells that are synchronizing their activities (action potential firing). Coherent activity, pathological in

this case, appears as a result of the increased coupling between the cell networks that synchronizes the initially incoherent and stochastic fluctuations in synchrony. Each succeeding chapter will address possible neurophysiological or behavioral bifurcations.

The determination of bifurcations sheds light onto the understanding of the evolution of dynamical systems: qualitative behavior changes when certain parameters pass bifurcation values (Guckenheimer and Holmes, 1983; Bergé *et al.*, 1984; Hoppenstead and Izhikevich, 1997). The study of bifurcations, especially in neurosciences, offers "an entry point into the investigation of pattern formation at both the behavioral and the cerebral levels" (Oullier *et al.*, 2006). What are the relevant control parameters that can serve as bifurcation parameters? How do we know when we have one? Following the definition of bifurcation, if there is a qualitative change in the dynamics when crossing a specific value of the parameter, then it can be surmised that we have a bifurcation parameter in our hands. Section 2.7 deals with this matter, and the next chapter addresses control parameters more specifically in epileptiform activity. Subsequent chapters present other possible bifurcation parameters in a variety of behavioral transitions.

There is an advantage in studying systems near bifurcation points (also known as critical points), in that the system becomes simpler: the number of degrees of freedom is reduced, according to the "slaving principle" in synergetics (Haken, 2006). Regardless of the many degrees of freedom of the systems, near the critical point only a few order parameters become unstable and thus they describe the emerging macroscopic pattern, enslaving the myriad of microscopic degrees of freedom. The message of this tale is that in order to simplify the analysis and better capture the relevant order parameters that describe the system's dynamics, try to push the system near a bifurcation point and analyze it around that region.

The two main types of bifurcations are the supercritical and the subcritical. From our point of view in the application of these concepts to neuroscience, the most important difference between these two types is their result: if the former occurs, new stable equilibria emerge and replace the old one that exists before the bifurcation point. The supercritical bifurcation thus leads to a "soft" loss of stability, as the system's variables will remain on the newly formed attractors close to the old one. If a subcritical bifurcation takes place, the equilibria and attractors disappear, in what is termed a "hard" loss of stability and the consequences are that completely new dynamical regimes take place and the variables may grow drastically, as opposed to the former case. Figure 2.9 depicts both types, using as an example the pitchfork bifurcation. Shown in the figure are the typical bifurcation diagrams of a variable X, where the x-axis represents the evolution of a bifurcation parameter (μ).

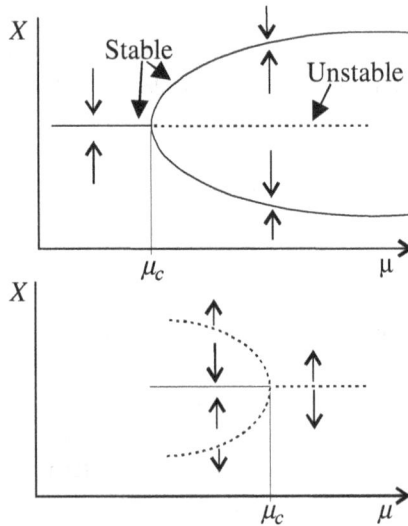

Figure 2.9. Supercritical (above) and subcritical pitchfork bifurcation. Stable equilibria of the order parameter X are plotted against the bifurcation parameter μ. Thus, to construct this type of diagram, one needs to solve the system of differential equations for different values of the bifurcation parameter and assess the stability of the solutions. Stable equilibria are plotted as continuous lines, and unstable ones as discontinuous lines. In the supercritical case, the initial equilibrium loses stability at the bifurcation point (μ_c) and two new stable equilibria emerge after the critical point. In the subcritical case, note there is no stable equilibrium after the bifurcation point, rather, one unstable equilibrium (dotted straight line) appears after the value of the parameter reaches its bifurcation value (μ_c).

From the neuroscience perspective, it is conceivable that subcritical bifurcations leading to large changes in the variables and new dynamical landscapes may play an important role, as this allows brain cells to exhibit large responses to inputs as certain parameters change. Conversely, supercritical bifurcations do not lead to great changes in the values of the variables and thus may limit the repertoire of the cell's responses (note the open arrows in the supercritical case in Figure 2.9 denoting that values of X remain close to the stable states, but since there is no stable equilibrium after the subcritical transition the parameter values can change a great deal, represented by the arrows moving away from the unstable state). One particular bifurcation that directly applies to neuronal activity is the Hopf bifurcation. It describes the onset or disappearance of periodic activity, and can be depicted using the previous figure if X were the amplitude of an oscillation: initially it is stable with value of 0 (no oscillation, neuron at rest), and after the bifurcation the two new stable states correspond to oscillations of amplitudes x_1, x_2 (periodic neuronal spiking). The monograph by Hoppensteadt and Izhikevich

(1997) contains this type of geometrical approach based on bifurcation theory to the characterization of neuronal firing patterns and offers a detailed mathematical description of, among other things, the different bifurcations in neural network models. To some extent, it can be said that this geometric approach captures the essence of these phenomena. It was shown previously in Figure 2.4(C) another illustration of the relation between geometry and physiology.

It is not only of basic but also of practical interest that there are, many times, precursory fluctuations before an impending bifurcation. Specifically, before a phase transition, a growth of the fluctuations in order parameters can be detected, more likely if the bifurcation is supercritical (Sornette, 2004). This means that, in principle, monitoring the fluctuation of an order parameter may be of value to anticipate when a bifurcation (a seizure?) may occur. Section 2.7 (Figure 2.15) discusses and presents results on the noisy precursors of a putative order parameter (phase differences) before equally putative bifurcations (epileptic seizures).

2.3.1. Local versus global considerations, once again

Section 1.7 already presented discussions on the transition from local to collective activity, and now from a more dynamical viewpoint, other aspects are considered. While these types of aforementioned dynamical analyses can be considered local, that is, they are performed near critical nonhyperbolic points, this local information near bifurcation points is useful for understanding global characteristics, namely the emergence of patterns of cellular activity in the brain. In neural network models, it has been shown that the global behavior depends on local processes taking place when fixed points become nonhyperbolic (recall once again that hyperbolic equilibria cannot display bifurcations), and these are called nonhyperbolic neural networks. To realize the interconnection between local (micro) and global (macro) levels of description is a major central theme in this book, because, as noted copiously in Chapter 1, it is fundamental for the understanding of complex systems and brain and behavior in particular. From a more theoretical viewpoint, the step from a local to a global description uses detailed neural models in the former and field models for the latter (Wright and Liley, 1996; Nunez, 2000). As with most neural models, some of the field, or global, models have been applied to reproduce one or another brain activity pattern, and their study have revealed specifics about these patterns of neural activity. Related to the idea on the possible (meta)stability of rhythmic activity that permeates all this book, the study of Rodrigues *et al.* (2007) employed a global model of thalamocortical networks to assess the global stability of limit cycles, and thus the possibility of the persistence of rhythmic behavior in the corticothalamic circuitry. This paper illustrates the difficulties in assessing global stability from the theoretical perspective. This is just an example among many other publications that have used field models.

The loss of stability of an equilibrium state is a local bifurcation, because the phase portrait does not change globally and in this sense it can be thought of as preserving most of the dynamics taking place before the bifurcation, or at least not presenting major changes. However, this thought may not be too accurate, for a local bifurcation can affect the global dynamics in a fundamental manner. In neural terms, this is the basis for important changes in the activity patterns to occur, and as the reader may suspect, it is related to the notions presented above on nonhyperbolicity and itinerancy. Just to make this point more visual and illustrative, Figure 2.10(a) depicts a saddle equilibrium point similar to that one in Figure 2.1, but notice that a branch of the stable (W_s) and unstable (W_u) manifolds have joined, forming what is termed a homoclinic loop: it starts and ends at the same equilibrium point. This situation is unstable, because perturbations can destroy the homoclinic orbit. For instance, if there were an attracting node as shown in the figure (the fixed point inside the loop), the perturbation and subsequent destruction of the loop would make the orbit fall towards the attractor as shown in the figure, thus altering in a significant manner the activity pattern resulting from that dynamical route.

Also shown in Figure 2.10(b) is another characteristic that makes a system structurally unstable, a homoclinic tangency that occurs when the stable and unstable manifolds are tangent. Both of these situations portrayed in the figure result in structural instability and are characteristics of nonhyperbolic attractors already introduced in Section 2.2.2. As these are nonhyperbolic systems, several bifurcations are associated with the situations shown in the figure, and we refer readers to other works for technical details (Anishchenko *et al.*, 2002). One aspect that is almost always presumed is that natural systems can be described by nonhyperbolic attractors because of the need for dynamical bifurcations and pattern formation in natural phenomena, however, to our knowledge, the nonhyperbolicity has not been demonstrated rigorously in neural activity (we are not talking about neural

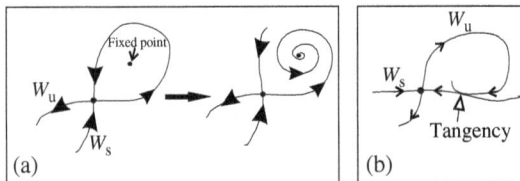

Figure 2.10. Structural unstable situations. (a), global bifurcation by perturbation of a homoclinic orbit, which is the closed loop starting and ending at the fixed point (black dot). The other fixed point (marked with an arrow) is an attracting node. (b), a homoclinic tangency.

models, but about real recordings). Indeed, in natural systems, and again to our knowledge, homoclinic tangencies derived from experiments have been inferred only in the dripping faucet experiment (Pinto *et al.*, 2001), even though there are quite a few theoretical studies showing this situation in a variety of models. There is one method that can be used to obtain a signature of a nonhyperbolic attractor, and that is the determination of the homoclinic tangencies by measuring the angles between the stable and unstable manifolds: if some angles are near zero, this indicates a situation like that reproduced in the figure and thus a nonhyperbolic saddle (this would be an indication only and not a proof because, as it turns out, some types of tangencies are persistent to perturbations so even their presence does not guarantee hyperbolicity, as detailed in Section 6.7 of Guckenheimer and Holmes, 1983). This analysis may be feasible because there are methods to determine stable and unstable manifolds from a time series, particularly this undertaking has been popular in neuroscience recently (Le van Quyen *et al.*, 1997; So *et al.*, 1998). There are algorithms as well to compute the angles between the manifolds using appropriate statistics, like the method presented in Lai *et al.* (1993). Their results, using the Hénon map, indicated that nonhyperbolic saddles are quite common. Whether this also occurs in natural systems remains to be assessed, but if this were true it may have some implications, into which we will not dwell now but only mention in passing: most results in dynamic system theory apply and have been proved for hyperbolic systems. Hence, if natural systems are not, could this pose another limitation to the applicability of the aforementioned notions to neuroscience?

2.4. Stochastic Dynamical Systems

The purely deterministic description of dynamical systems presented above is not realistic in natural systems, as these are subject to noise and fluctuations. The sources of noise are not only external but also internal; for instance, consider the well-known stochastic fluctuations in ion channel opening in the neuronal membrane (Schneidman *et al.*, 1998). Indeed, variability in neuronal activity is found at all levels of description from the molecular to the cellular and network levels. Synaptic noise has probably been the fluctuation source most studied from early times (Fatt and Katz, 1950) to the current era, where it has been observed that synaptic fluctuations (such as spontaneous neurotransmitter release at synaptic terminals) determine spike firing activity in a variety of neuronal cells (Destexhe *et al.*, 2001; Wolfart *et al.*, 2005). Due to the current recording and analysis techniques in neuroscience, it is unfeasible to separate what is termed "noise" from the "meaningful" neuronal activity. As a matter of fact, it may make no sense at all, if the notions detailed in Chapter 1 are accepted, namely, that sensory-motor processes are modulations of ongoing, background brain activity, for in this case there is little to distinguish between noise and signal. The border between noise and high-dimensional dynamics is indeed fuzzy (Kowalik, 2001). Such inherent variability in all aspects of nervous system activity invites a probabilistic interpretation of many of these phenomena.

Thus, under this perspective, the variables now are stochastic processes and require a probabilistic approach. The monograph by Anishchenko *et al.* (2002) is recommended for those interested in the basics of stochastic dynamics. Here, we only mention the very basic notions, starting with the fact that this approach is based on evolution equations not for the variables, but for the probability distributions of those variables. Hence, now the general equation reads

$$\frac{\partial p(x)}{\partial t} = Lp(x), \tag{4}$$

where L is the operator that describes the dynamics and $p(x)$ can be the probability distribution of the variable x. This is called a Fokker-Planck equation (Frank, 2004, 2005), and represents a macroscopic, collective behavior. The microscopic equations from which Fokker-Planck equations derive are known as Langevin equations, that describe the time evolution of a variable with noisy terms added (the original equation proposed by Paul Langevin was a description of the position of a molecule executing a random walk). Equation (1) of Section 2.1, plus a noisy term would be a typical Langevin equation. Once again, we encounter the shift from the microscopic description to the collective level, from the local to the global. To be sure, these formalisms based on evolution equations for transition probabilities formally apply to the description of Markovian processes (after the Russian mathematician Andrey Markov) and were originally derived for diffusion processes (the Langevin's random walker aforementioned), conditions that may or may not be realized in brain tissue. Hence, the use of this formalism in brain research is another debatable question that can join the already large number of arguable points mentioned in previous sections like the use of the concept of attractor. Nevertheless, nonlinear Fokker-Planck equations have been used in neurodynamics within the framework of coupled oscillators (Frank *et al.*, 2000), as well as in neuronal models (Galán *et al.*, 2007).

However, it is not difficult to conceive a simple Langevin approximation to nervous system activity. Consider the time evolution of the activity, A, of a cell, or a network of cells, that receives stochastic inputs, I:

$$A(t + \tau) = A(t) + I(t). \tag{5}$$

Equation (5) allows for the knowledge of the activity after a time step τ knowing the activity at time t, and it is a Langevin equation because the term $I(t)$ is stochastic; to be more precise, the inputs should be independent and identically distributed. If instead of the activity A we have the position of a random walker, and instead of the inputs we use the lengths of the steps of the walker, then we have the classical random walk problem pondered by Paul Langevin. The solution to Eq. (5) is $A(t) = I(0) + I(\tau) + \cdots + I(t - \tau)$ which corresponds to a master equation:

$$P(A, t + \tau) = \int_{-\infty}^{\infty} \prod(I) P(A - I, t) dI$$

that simply says that to obtain the activity A at time $t + \tau$ while the activity was $A - I$ at time t, the cell or network has to make the necessary "step" resulting from the received inputs which are distributed according to the probability distribution $\prod(I)$. The Fokker-Planck equation is obtained from this master formula by taking the limit when τ and I go to zero. In the context of a random walk problem, this results in the celebrated diffusion equation. In the context of neuroscience, this may be open to interpretation.

Just as linear stability analysis was described above for systems of coupled differential equations, a similar analysis can be done with a Fokker-Planck equation, namely, the determination of the stationary solutions where

$$\frac{\partial P}{\partial t} = 0$$

in order to obtain stationary probability density functions. In general, these equations can have multiple stationary solutions that define a (stationary) Markov diffusion process. Why is this apparently abstract notion revealed here? Because the evolution of a stochastic process defined by a nonlinear Fokker-Planck equation reveals an itinerant behavior defined in terms of transition probabilities and, as shown in Figure 2.11, it is reminiscent of Figure 2.7 on itinerancy. This behavior has been called a "hitchhiker" process, and it may have a deep relation to the concepts that were presented in Section 2.2.3. The details are shown in the figure legend below, but note that in this state space (the function space of probability densities), the points u^*, $u1$, and $u2$ represent initial probability distributions, and not geometric attractors.

The end result is that the stochastic process defined by a nonlinear Fokker-Planck equation evolves within the family of Markov processes with transient probability densities W defined by a linear Fokker-Planck equation. Stochastic process means that all state transitions are probabilistic. It is now the time to use the power of abstraction and consider the similarities between the transient dynamical behavior discussed in Section 2.2.3 and the hitchhiker process. In one representation the

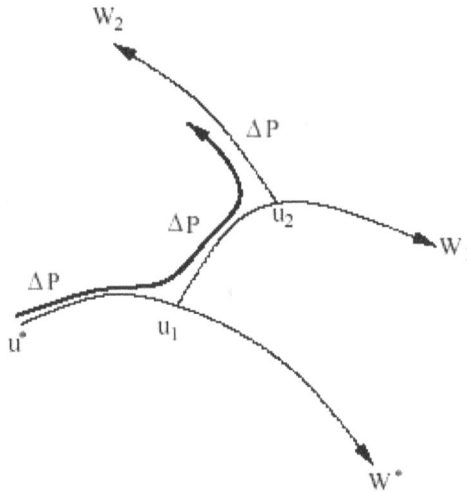

Figure 2.11. Evolution of a stochastic process defined by a nonlinear Fokker-Planck equation, representing the Markovian hitchhiker process. Each initial probability distribution u has associated a transient probability density W. Three solutions, W, W_1, and W_2, are plotted, the first determines how the system's probability distribution P evolves from the initial distribution u^*, approaching the point $u1$, which then evolves like W_1 and approaches $u2$ and so on. Reprinted with permission from Frank (2004).

space is populated by probability densities and in the other representation there are geometric objects called attractors, quasi-attractors or attractor ruins. However, recall the probabilistic interpretation of an attractor discussed in Section 2.2.2, in that there is a probability to leave the attractor, in the Milnor sense. Is it then conceivable to establish a connection between these two representations? If a probability is ascribed to leaving one attractor (the Milnor interpretation of an attractor) and approaching the neighborhood of another, perhaps these two representations are more related than a cursory consideration may indicate. Having said this, the Fokker-Planck formalism relies on Markov processes, those that do not have memory effects, basically a random process where all information about the future is contained in the present state and not on any past states, a point that may bring about the protest of the neuroscientist because neural tissue does show memory phenomena. However, the "states" may be widely defined, and there could always be a time scale, which some authors have called the Markov time scale (Ghasemi *et al.*, 2006), in which the process is Markovian. For instance, the peak of an action potential clearly is settled after the voltage threshold for spike firing is crossed, hence it is not Markovian if fractions of milliseconds are considered, but the spike does not depend at all on the membrane voltage one hour before, and thus a designation of a (admittedly wide in this simple example!) Markov time scale of about 1 h can be chosen.

To end this section, let us remark that while Fokker-Planck equations have been used in a wide variety of fields from psychology (Frank and Beek, 2003) to marketing, their application to neuroscience is still in its infancy. We believe that the value and potential of this approach in answering questions relevant to the nature of the dynamics of cognition is considerable because it emphasizes the stochastic, probabilistic approach that, we think, is very relevant to brain function and behavior.

2.5. Relevance of the Theoretical Concepts and Methodologies Directly Applied to Neurophysiology

It is now the time to take up the two main questions posed in Section 2.1, but now applied specifically to the nervous system. The queries related to the main problem of the reconstruction of the dynamical system that in our case represents the brain. The first was the *choice of variables* that may best characterize the system to define the *appropriate state space* in which the dynamics can be followed. The second task is the *reconstruction of the dynamics* in the chosen state space. Computational neuroscientists normally start by creating their own model, a system of differential equations that is supposed to capture some particular neural activity. This will not be the route followed here. Rather, we start with experimental data, neurophysiological time series recordings, and examine what can be extracted from them. In general, the reconstruction of a dynamical system from biological data, or experimental time series in general, poses no few problems and in the end it has to be understood that whatever dynamical model is used represents but a rough approximation to the real system. With this consideration in mind, some reasonable choices can be made.

2.5.1. Reconstruction of nervous system state space: State variables and control parameters

The first problem that arises when faced with a time series recording is that we have to choose a variable(s) in order to create a global model that is conveniently represented by continuous, smooth functions. What dynamical variables are best? Some thought has been devoted to this matter recently by W. Freeman (2007a,b) in the context of brain–computer interface. However this has been an old subject of thought, with early proposals about using the coordination between neural signals to reconstruct the state space (Schild, 1984), proposals that may have their continuation in the current use of phase space based on synchronization measures. There are two perspectives that will be mentioned here. First, from the biological perspective, we would like to focus on the variable that is more relevant to the particular physiological phenomena in our study, and, why not, it should be relatively easy to measure. Second, from the mathematical point of view, we may want to choose variables that will generate a global dynamical model with smooth functions (that is, differentiable functions). One study reported that two important mathematical considerations to choosing variables are the system's symmetry properties and the coupling between the variables and the arguments, ideas that revolve around the notion of "observability" of the dynamics (Letellier and Aguirre, 2002). Others have proposed choosing variables based on the analysis of the local properties of some dependencies specifically testing for single valuedness and continuity (Smirnov *et al.*, 2002), as these two aspects are fundamental in obtaining a global dynamical model with a smooth function that can then be analyzed in any manner explained in this chapter. There may be a third perspective as well, that is a sort of combination of the biological and theoretical ones, and takes into account that from a practical viewpoint, the aforementioned notion of structural stability of dynamical systems (Section 2.1) poses some questions as to the relevant physical properties that can be measured. Perhaps, the measured properties should be those that are preserved under perturbations. Which are these? Each chapter that follows will contain specific comments on this matter.

As the physiological perspective normally wins in the case of choosing state variables, voltage has been widely used to recreate phase portraits using first-return plots, as it is the most straightforward neurophysiological signal recorded in all settings, whether clinical or laboratory. Other variables that can serve as order parameters (recall the definition in Section 2.1) can be chosen from the wide neurophysiological repertoire, again based on the ease of recording them as a good-quality time series. Several studies have used the time intervals between spikes, as it was demonstrated that amplitude (voltage) time series can be converted to interspike interval time series with no loss of information (Sauer, 1994, 1995). Still others have employed the difference in the oscillating phase of the signal, as well as the instantaneous amplitude or phase of the signal extracted via the Hilbert transform (Frank *et al.*, 2000; Wallenstein *et al.*, 1995; Kelso, 1995; Kelso and Tognoli, 2007; Freeman, 2007b).

In line with the philosophy of this monograph, a clarification is in order here at the start of this discussion. Quite a few studies create a state space using the word "embedding." This term is used loosely to describe a *reconstruction* of trajectories in a state space, using, for instance, time-delay coordinates. Let us be careful here because a reconstruction is not necessarily an embedding. An embedding is a function, a relation between two spaces (or two time series), technically it is a diffeomorphism that preserves rank (preserves the original dimension of the manifold). For example, as explained in Section 2.2, a Poincaré mapping function can be found between an original time series and a reconstructed object such as the points on the Poincaré section, but this may or may not be an embedding. It is surely a mapping, but tests to assess whether a mapping is an embedding (which is a function that relates one to the other in that particular manner described above) do exist (Pecora *et al.*, 1995; Letellier *et al.*, 2008) and should be used to clarify this aspect in those particular studies where it is necessary. An embedding creates, thus, a relation between two attractors, or two time series, or in general between two objects.

However, regardless of whether the function is an embedding or not, our interest is to find a mapping from a scalar time series, voltage or interspike intervals for instance, into a reconstructed object (which will not be called attractor because, as discussed in previous sections and extensively in Section 2.6 below, we cannot be sure from the start that the geometric object reconstructed is indeed an attractor). One of the first papers that proposed a method to construct a state space from time series is due to Packard *et al.* (1980), based on time delay or time-derivative coordinates. Behind all these methods lurk the celebrated embedding theorems, the most famous being Takens' time delay theorem (Takens, 1981) that consists of using one variable that can be measured and constructing a return map with it, that is, plotting one value versus the next, which is equivalent to a Poincaré section. This theorem was subsequently generalized by others and extended to show the usefulness of the interspike interval reconstruction (Sauer, 1994), at least in those systems where spikes can be recorded. There are works with detailed explanations and applications of these "embedding" techniques (Kantz and Schreiber, 1997). In general, these works proposed that time delayed state-space portraits constructed from a single-variable time series are able to retain certain dynamical aspects of the system (such as fixed points). This does not mean that the "attractor" obtained in this simplified view (state-space) is identical to that of the original multivariate state-space, but that the new representation preserves the same topological properties that may suffice for studying essential attributes. It is relevant and completely acceptable to use this strategy, because, as explained in previous sections, when studying highly complex systems it is an established procedure to reduce the system's multidimensional continuous trajectory in state-space to a discontinuous low-dimensional projection, like the Poincaré maps presented in Section 2.2. The main problems that have to be faced when using these approaches are to find an adequate "embedding" dimension (perhaps it should be called reconstruction dimension according to the previous discussion) and the adequate time delay. Obviously, a too short time delay will result in a first-return plot along the diagonal because the values of the variable will be almost identical since the time interval is minute. On the other hand, a too long time delay may result in losing structural details that can be best resolved with an intermediate time delay. While there are proposed methods to help with these inconveniences (Pecora *et al.*, 2007), in the final analysis it is probably best to do it empirically in a sort of a trial and error process to ascertain what best fits the phenomenon in question.

After choosing the variables, the second task is the construction of the evolution operator (system of differential equations or a discrete map) that defines the temporal evolution and will allow us to study the trajectories in the state space spanned by the chosen variables. Quite a few studies have explored many methods to create models from numerical time series, some classical works have already been mentioned above, but there are many more; two other early works are Gouesbet and Maquet, 1992, and Gibson *et al.*, 1992. To make matters complicated, the spatiotemporal patterns of neuronal activity in the brain develop both in time and in space, which means that the state variables are functions of time and position; thus, an accurate model would have to consist of coupled partial differential equations with respect to time and to a spatial coordinate. Likewise, the lack of formalisms to characterize the dynamics displayed by extended dissipative systems (such as spatiotemporal chaos) does not make things easier either. Brain activity unfolds in the spatiotemporal domain, thus simultaneous anatomical and temporal description of the activity may be required to understand its dynamics. There are developments in the field of spatiotemporal dynamics (Gameiro *et al.*, 2004; Liu *et al.*, 2004) but remains to be seen whether these are applicable to nervous systems. Nevertheless, to simplify, the great majority of models assume only time dependency. As illustrations, examples where systems of coupled differential equations have been used to simulate epileptiform activity can be found in Hernandez *et al.* (1996), Friedrich and Uhl (1996), and Robinson *et al.* (2002), perhaps because the modeling of epileptiform activity is simpler than other types of brain activity. In two of these three studies, though, the equations were created not directly from neurophysiological recordings, but were rather chosen because the system could simulate the particular neural activity observed in the recordings. The Friedrich and Uhl report uses a 3D model of differential equations derived from the mode decomposition of experimental recordings to analyze spatiotemporal dynamics during epileptiform activity. Nevertheless, some insight into the dynamics can also be obtained without a system of differential equations, because a discrete dynamical system can always be used and it is as acceptable as the continuous one based on differential equations. The next paragraphs present an illustrative example of the extraction of dynamic information using first-return maps (a discrete system) obtained directly from electrophysiological time series. But before, a brief mention of one report that indicated that the most valuable parts of a time series to construct a global model in state space, particularly a discrete model, are the transient processes (Bezruchko *et al.*, 2001). This is of interest because transients could be all that is measured in nervous system activity as discussed in other sections, hence this result represents an encouraging thought in the practice of these techniques.

2.5.2. Examples of the reconstruction of the dynamics associated with specific brain activities

The illustration presented here uses the idea of the Poincaré map, or, more specifically, the first-return plot of time-delay coordinates, to explore the dynamical regimes in epileptiform activity. The following paragraphs are needed to understand and complement sections in the next chapter that is devoted to the brain dynamics in epilepsy, and specifically Section 3.4.4.3 (Figure 3.8) that gives details of a geometric interpretation of neural activity based on these concepts.

As with many studies investigating the dynamics of complex systems, we start with the "embedding" (reconstruction) of a certain scalar time series in a state space. In the present case, the time series will be the time interval between peaks (or spikes) found in brain recordings, and the resulting graph will be called an inter peak interval (IPI) first return plot from now on. Of course, as aforementioned, this procedure is not really an embedding but a reconstruction. The main idea is to obtain a 1D first-return mapping function, that is, a discrete dynamical system rather than a differential equation, to perform the typical stability analysis described in Sections 2.1 and 2.2 and uncover possible equilibrium points and their stability. In this approach we follow the advice of Kelso and Engstrøm (2006): "fixed points are the skeleton upon which dynamical systems are built. They enable one to make predictions about the system's behavior." First-return maps have been used in other areas of biology, and the work by R.M. May in 1976 on ecological models is a very early illustration of first-return maps applied to the biological scenario. Discrete models based on 2D maps have been useful to reproduce the dynamics of neuronal spiking and to analyze their synchronization regimes (Rulkov, 2002). We also build on the already discussed concepts that time-delayed phase portraits generated from experimental time series of one variable maintain the essential characteristics of the original multidimensional phase space. With the aid of the Poincaré section, a reduction of the multidimensional continuous system's activity to a first-return map can be achieved. How is the Poincaré section established? Figure 2.12 shows the original neurophysiological recordings (bottom panel) that are the starting point in

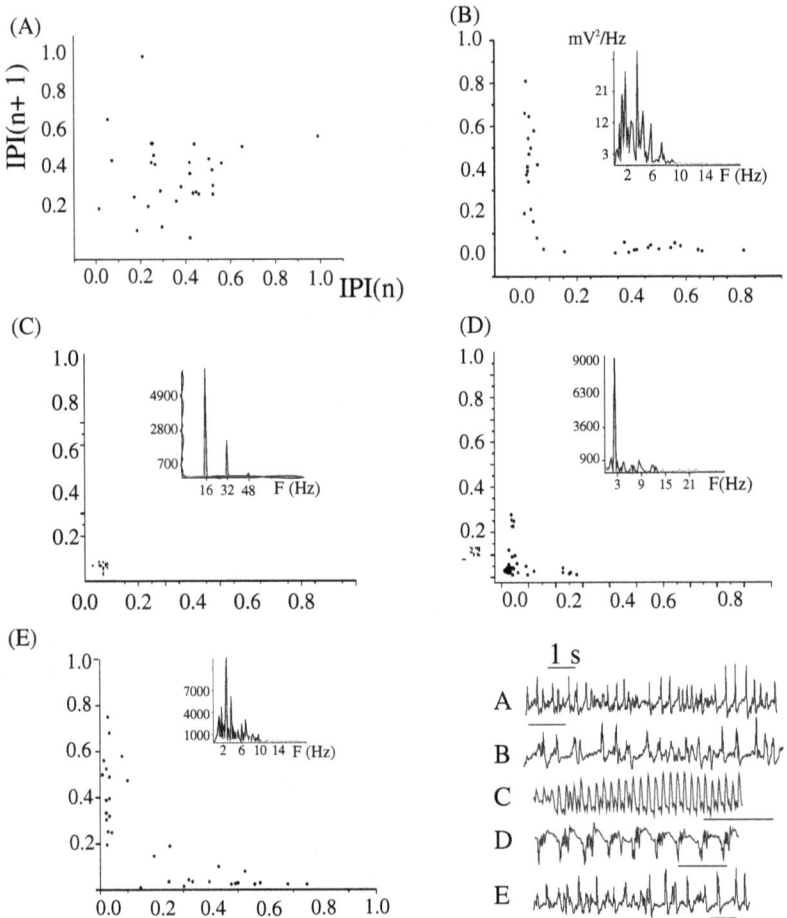

Figure 2.12. First return interpeak interval (IPI) scatter plots during interictal and ictal activity for a seizure in a patient with epilepsia partialis continua. The panel at the bottom right shows part of the five recordings used for the creation of the respective plots (time scale is 1 s for all). One IPI is plotted versus the next for 50 s during interictal activity 24 h before the seizure (A), 1 min before the seizure (B), the first 4 s of the seizure (C), 8 s before the end of the seizure (D), and 18 min after the ictus (E). The insets in B, C, D, and E represent the power spectra of the EEG recordings, that denote the more regular activity in some parts of the seizure (C and D, where a major frequency peak is identified). Reproduced with permission from Perez Velazquez *et al.* (1999).

this study. These five segments come from an intracranial recording in an epileptic patient (more details in the figure legend). Advantage was taken of the fact that the EEG in some patients showed frequent spikes during the interictal-preictal period, which, incidentally, is uncommon for most epileptic patients, nevertheless we profit from this unusual phenomenon. Note the peaks in those recordings. It is irrelevant, for our purposes, whether they are upwards or downwards. These peaks are identified and the time interval between them (the IPI or inter peak interval) calculated. Then, as is the usual method to construct a time-delay portrait, one IPI is graphed versus the next one, resulting thus in a first-return plot. So, where was the Poincarè section located? Obviously, it cannot be constructed as shown in Figures 2.4 and 2.5, since the trajectories are not available but only the raw recordings. Each time the recording crosses a threshold value, this is classified as a peak (or spike) and assigned a time point. In this sense, it is similar to using the Poincarè section crossing the torus in Figure 2.5: each time there is a revolution on the torus there is a point on the surface of the section, and the point at which the revolution crosses the section can be considered one spike of the system. Hence, looking at the trajectories on the torus or detecting the spikes represents a similar phenomenon. Recall also that the equivalence between the threshold-based spike detection and the map on the surface of a section was depicted in Figure 2.4(C). Thus, this is the practical manner with which to "construct" Poincaré sections when using time series. The rest of the panels in the figure are return plots at different times during the epileptic activity (details in the figure caption).

Note that the resulting IPI scatter plots in Figure 2.12 showed different geometries a while prior to the seizure (panel A), immediately preceding it (B), and during the seizure progression, indicating a transition from aperiodic or quasi-periodic to simpler dynamics (more periodic, as can be readily seen in the raw recordings or from the power spectra insets in those panels particularly in C and D). A characteristic L-shaped plot (panel B) was sometimes observed during the transition to the seizure, both in human EEG and *in vitro* recordings from brain slices (Perez Velazquez *et al.*, 1999, 2003; Perez Velazquez and Khosravani, 2004). This plot, naturally, is a manifestation of the nature of the peaks: look again at the bottom panel in the figure showing the raw EEG recordings and note that these peaks are, frequently, bursts composed of several spikes (see Figure 2.14 for an *in vitro* example of bursts), thus generating a long-short-long interval sequence that appears as an "*L*" in the interpeak interval first-return plot. Since there is a particular geometry in this graph, closely resembling a continuous function, this indicates that the points are distributed along an underlying curve that can be approximated by a 1D return map, and thus the sought-after Poincarè mapping function can be readily obtained:

$$IPI(n + 1) = F\{IPI(n)\}.$$

This procedure is of a very general application; it can be used in other systems using different variables that can be measured, regardless of their nature: concentrations of chemical reagents, peaks in heart activity (see Figure 3.8 panels A and B in Chapter 3), number of specimens in ecological systems etc. Interested readers can look, for example, at how the analysis of a 1D return map derived from chemical concentrations of a substrate in a biochemical reaction was used to describe the oscillatory dynamics in that biochemical system, and several other examples included in A. Goldbeter's monograph (1996).

Thus, the scatter IPI graph in panel B was approximated with a nonlinear equation which is a first return 1D map (it is 1D because it contains only one variable, the interpeak interval), as shown in Figure 2.13, so that the typical linear stability analysis to study the properties of the first-return map could be carried out. This figure shows the fitting of two first-return plots, the one in A was obtained from the neurophysiological (intracranial) recording 1 min before the transition to a generalized seizure in a patient with epilepsia partialis continua (details can be found in Perez Velazquez *et al.*, 1999). The graph in B was taken from the recording shortly after the seizure. One question that may appear now in some readers' minds is that the fitting function could have been different, for there are several functions that will approximate, with more or less accuracy, the scatter plot. The fitting procedures are standard, normally using some form of a least-squares minimization, however it is true that there can be several equations with different forms (this one shown here is an inverted polynomial with three parameters, a, b, and c) that will approximate the plot equally well. In reality, the form of the equation does not matter. This may be a surprising statement to some, but consider that the equations are approximating the graph and the graph is one and only one, while the form of the equations may vary. That is, even if two very different equations approximate a graph equally well, then the properties of the graph are included in both formulae, hence the specific form is irrelevant: upon analysis, both equations should give us equal results. The equation derived in this case is considered a global model and it has three parameters, but because of the nature of the experiments and the procedure that was followed, not much can be said about these parameters, perhaps only that they are linked to the frequencies of the peaks. A more specific interpretation of these parameters is difficult because we do not have a model that simulates the system: this is the realm of computational neuroscience, where detailed models with very specific parameters simulate particular waveforms. Here, we are not practising computational neuroscience. But, does the mapping function model this reality? One test one can always perform on the model is to assess whether it can reproduce the original, experimental observations. As an illustration, Figure 2.13(C) shows two examples resulting from the iteration of the mapping function, for different parameter values. It can be seen that the skeleton of the original plots derived from the experiments is reproduced.

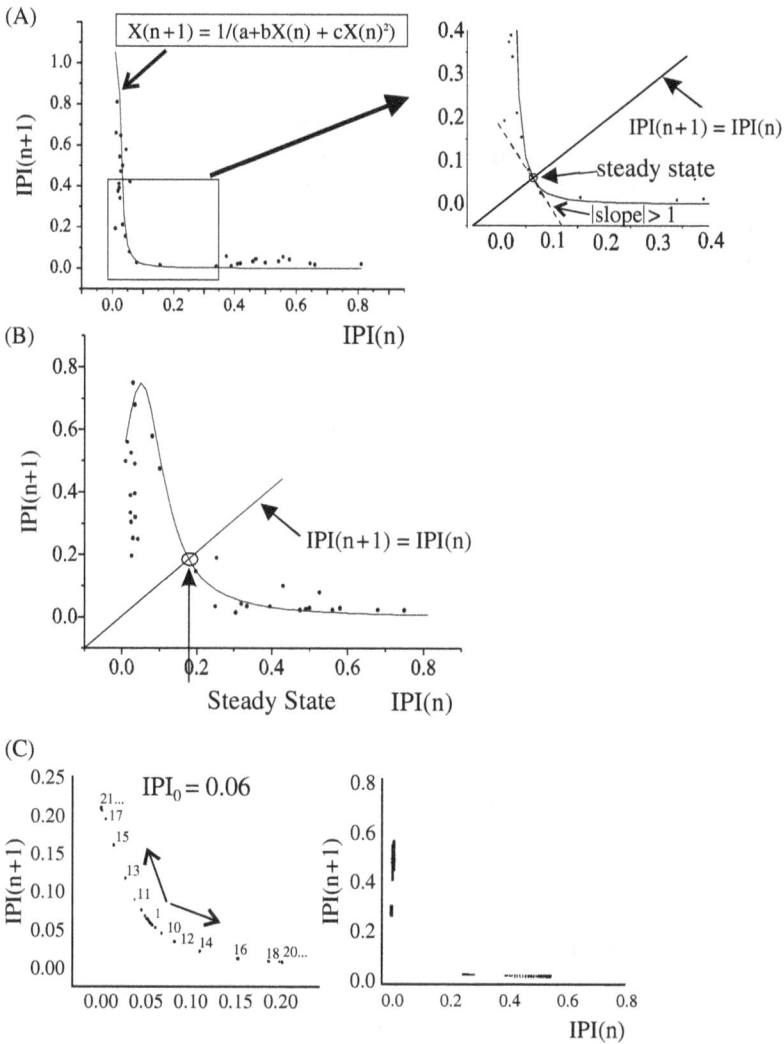

Figure 2.13. 1D first-return map for the *IPI* scatter plots 1 min before the seizure (A), and 18 min after (B). The mapping function that was found to be a best fit to the scatter plot is represented above in A and consists of an inverted polynomial. Top inset shows a magnification of a part of the graph to illustrate the steady state or fixed point at the intersection of the map with the diagonal (representing the identity map: $IPI(n + 1) = IPI(n)$), and the value of the slope at that point, greater than 1 in absolute value, as detailed in the text. In B, the same function fits the scatter graph, but with different parameter values. (C), two examples of iterates of the mapping function (each one has distinct parameter values) to ascertain that it can recreate similar patterns as the experimental ones, starting the iterations with an initial value of 0.06. The numbers shown in the left graph indicate the number of the iterations, and the arrows the path they take. A and B modified with permission from Perez Velazquez *et al.* (1999).

The mapping function is taken to represent the transition to the seizure, just because panel A was derived from the recording 1 min before the seizure. Interesting information can now be extracted from this mapping by application of the tools for the analysis of 1D maps (Collet and Eckmann, 1980; Guckenheimer and Holmes, 1983), specifically employing the linear stability analysis aforementioned in previous sections. Building on the notion that fixed points are the skeleton upon which dynamical systems are built, the identification of equilibria is of interest, and thus the first thing to do is to solve analytically the equation of the map for $IPI(n + 1) = IPI(n)$: when one point equals the next, this represents a fixed point. Geometrically, as represented in the figure, this is the intersection of the mapping function $F\{IPI(n)\}$ and the identity map which is the 45 degree line (bisectrix) where $IPI(n + 1) = IPI(n)$. In this specific case, the analysis reveals that there is one fixed point, or steady state (simple periodic behavior), at 0.063 s for the IPI (that in frequency means 1/0.063 or 16 Hz). The next aspect of interest is the stability of the fixed points. This is given by the slope of the map at the singular point. In our case it can be analytically determined by solving dF/dx at the fixed point $x = 0.063$, which results in a value of -2.4. Since, in absolute value, the slope is greater than 1, the fixed point is unstable; on the other hand, absolute values lower than 1 indicate stable equilibria. It is of interest to note that the power spectral density of the first 15 s of the ictal event reveals a main peak at 15–17 Hz (Figure 2.12(C)), perhaps indicating that the unstable fixed point becomes stable for a short time at the beginning of the seizure. Very close to the fixed point there is another point where the slope equals -1, the precise point is determined by solving $dF/dx = -1$, which yields the value for $IPI(n)$ of 0.0805 (12.5 Hz). Equilibria associated with this slope are neither stable nor unstable, they are metastable. Recall that in Section 2.2 the notion of nonhyperbolic equilibria was introduced, and what is found in the case of maps when the slope equals 1 (or -1) is precisely a nonhyperbolic fixed point: hence bifurcations are possible. Specifically, a slope of -1 is associated with a flip bifurcation (for those interested, technical details can be found in Guckenheimer and Holmes, 1983, or Bergé *et al.*, 1984) also referred to as a subharmonic bifurcation, which can give rise to two types of phenomena, the period doubling route to chaos, or type III intermittency. Section 3.4.4.2 in Chapter 3 describes that signatures of these two dynamical regimes have been found in epileptiform activity. Most work on dynamical regimes has been done at the computational, or simulation, level, and not so much in experimental research. Nevertheless, an early experimental evidence of type III intermittency associated with a subharmonic bifurcation was obtained by Dubois *et al.* (1983), in a hydrodynamical system. Chapter 3 provides evidence for these two dynamical regimes in epileptiform activity. As can be seen, then, information regarding specific dynamical regimes can be obtained from these methods.

It is explained in Chapter 3 how, within the framework of "dynamical diseases," the transition to the seizure has been conceptualized as occurring through a possible bifurcation in brain dynamics: the activity in the epileptic brain is close to a bifurcation that results in seizures, while normal brains are far from this hypothetical bifurcation point.

For the sake of completion, let us describe now briefly what these two dynamical regimes resulting form the flip bifurcation, period doubling and type III intermittency, may indicate about brain activity. A more in-depth consideration of this and other brain dynamics is presented in Sections 3.4.4.2 and 3.4.4.3 in Chapter 3 (also in Perez Velazquez and Wennberg, 2004). Intermittency takes place in the neighborhood of a fixed point, in the case shown above the singular point was determined at 0.063 s, as already mentioned. The classification of intermittencies into three types is due to Pomeau and Manneville (1980), who proposed it according to different modes of instability of fixed points. The proposal that intermittency could underlie epileptic phenomena was advanced some time ago, perhaps more explicitly in Iasemidis and Sackellares (1996). Once again, we find ourselves applying a precise physico-mathematical concept, intermittency, to empirical brief time series recordings. Neuroscientists do not have the advantage of the physicist studying, for instance, the dynamics of turbulence in fluids where many of these mathematical concepts have been so clearly applied. In empirical neuroscience, experimentalists cannot easily control parameters of brain activity that can switch one oscillation into another, as for instance the variation in the Rayleigh number does in fluid mechanics. This is a dimensionless number that describes the relation between diffusive and buoyant forces acting on a fluid, and can be changed by, for instance, changing the temperature gradient in the fluid. By controlling this parameter, physicists are able to let the system run into one rhythmic state and then alter it to determine how oscillatory phenomena change in the fluid. These experiments led to accurate empirical observations of intermittency and period doubling in the classical studies in the field of fluid dynamics, clearly detailed in Bergé *et al.* (1984). On the other hand, in brain activity, the parameters are changing themselves without external control. Even in the few exceptions when we naively think we have isolated a part of the system and have it under control, as in the case of *in vitro* experiments using brain slices, the activity still depends on so many parameters that we can cast a serious doubt as to our real "control" of the experimental situation. Nevertheless, these aspects should not deter us from trying to find some signatures of specific dynamical regimes, understanding that it is only indications that can be found, and not clear proofs, unless, of course, a computational model is constructed from which precise mathematical statements can then be derived. Type III intermittency makes sense in brain dynamics as this dynamical regime is characterized by short, rather that long, epochs of periodic activity. Other intermittency regimes, type I for instance, is characterized by a predominance of long episodes of periodic activity. Short, transient periods of rhythmic activity is what is normally observed in brain neurophysiological recordings in healthy conditions, whereas long-lasting periodic rhythms are not as common and many times associated with pathologies like seizures or Parkinsonian tremor, that will be addressed in Chapter 4. On-off intermittency has also been reported in human behavior (Cabrera and Milton, 2002). On-off intermittency corresponds to the Pomeau-Manneville type III intermittency (Čenys *et al.*, 1997). Intermittency is a dynamical state unique to nonlinear systems where the transitions between periodic, quasiperiodic, and aperiodic (possibly chaotic) states occur without the need for external stimuli, reminiscent of the itinerancy concepts introduced in Section 2.2.3. Hence, because it reflects the switching between synchronous (laminar) and nonsynchronous regimes, intermittency seems adequate to describe nervous systems. As to period doubling, this is the typical route to chaos from periodic activity (Feigenbaum, 1983), and, to our knowledge, it has been reported only once in brain recordings, specifically during seizures (Figure 3.7 of Chapter 3, also Perez Velazquez *et al.*, 2003). However, it has been noted a number of times in theoretical/computational studies that demonstrated model systems going through period doubling bifurcations during the changing neural firing patterns. Because of the alternation in the almost rhythmic activities observed in neurophysiological recordings, it is conceivable that period doubling occurs at several levels. Chapter 3 presents the possible consequence of this dynamical regime during the progression of an epileptic seizure.

To recapitulate, then, within this framework it is conceivable to hypothesize that seizures start as a materialization of a metastable bifurcation point (the one with slope of -1 aforementioned), corresponding to high frequency, more or less synchronous, activity that is commonly observed at seizure onset; then type III intermittency or period doubling ensues, which results in the destabilization of the cellular synchronized firing patterns and brings about the termination of the seizure. These matters are engaged again in Chapter 3 on epilepsy.

2.6. Attractors in Brain Activity?

The geometry observed in the first-return graphs presented in the figures above, specially the appearance of the L-shaped formation indicating that this could be the "skeleton" of a continuous function that describes the dynamics, begs the question of whether or not this graph represents a possible attractor in brain activity. The search for attractors in neural activity has had a long history, from the classical computational works of J. Hopfield (1982) on neural nets and the search for persistent brain states (Little, 1974), to more neurobiological oriented studies (Mandell and Selz, 1993). Now, it is time to recapitulate the notions presented in Section 2.2, and to fundamentally consider the comments in Section 2.2.3 regarding the relevance of the attractor concept to neuroscience. A first step to clarify this query is to find out whether we are looking at asymptotic or transient dynamics. The formal answer is relatively clear: transients are all we can measure, since we cannot perform infinitely long recordings that would be needed to ascertain the asymptotic quality of an attractor. However, a more real answer closer to the experimental condition is not easy to retrieve. One argument that can suggest that in fact attractors may exist in brain activity is that a main characteristic of dissipative systems is the large number of constituents giving rise to coherent activity whose long-term behavior may fall into specific regimes that define attractors. Thus, the brain, being an open, dissipative system made up of millions of interacting cells, could be considered to be governed by attractor dynamics, either high or low dimensional (low dimensional during specific rhythmic patterns like alpha oscillations or spike and wave seizures, for example, and high dynamics if quasi-periodic attractors are considered). Early studies using first-return maps suggested that low-dimensional chaos was present in epileptic activity (Babloyantz and Destexhe, 1986). This observation implies the existence of an attractor, a hyperbolic chaotic attractor as described in Sections 2.2.1 and 2.2.2. However, this was based on the extraction of geometrical properties (correlation dimension) of the attractor underlying the observed activity. Here, once again we face the question we posed previously in the introductory paragraphs to this chapter, does the "cloud" of points representing the first-return plot obtained from the data represent an attractor? It seems sometimes to be routinely assumed that any sort of first-return map of neuronal activity represents an attractor. A few considerations are in order here, as the presumption of attractors in brain dynamics underlies heated controversies in neuroscience like the old-time dispute about chaotic dynamics in brain (whole books are devoted to this topic, a representative example that readers may want to explore is "Chaos in Brain?" edited by Lehnertz *et al.*, 2000). This matter was treated previously in depth in other publications (Perez Velazquez, 2005); here, only the principal considerations will be pointed out.

A main consideration is that, as explained in the sections above, measures like correlation dimension or Lyapunov numbers have meaning only as descriptors of attractors: recall that the correlation dimension is a geometric measure of an attractor, but if there is no attractor, then, what is that dimension evaluating? Current methods of nonlinear time series analysis (Kantz and Schreiber, 1997; Lehnertz, 2008) allow the extraction of correlation dimension or Lyapunov exponents from almost any time series one can put his hands on, but the underlying assumption that the time series represents the system's activity within an attractor may not be correct. Thus, here lies the risk: dimensions can be easily computed and if non-integer values are obtained then a claim is made that a strange chaotic attractor has been found, which may or may not be true. This line of research has become enormously popular particularly in epileptology because of some clinical implications, which will be detailed in the next chapter.

The term attractor has been repeatedly mentioned in applications in neuro-science in manners that can be considered too loose. A few instances of the use of the attractor terminology defining some neural activity patterns are memories stored in neural networks identified as fixed point attractors; the observed coordinated activity in a large neuronal ensemble using calcium imaging (Cossart *et al.*, 2003); or the maximum or sustained high-frequency spike firing in single neurons (Amit, 1989; Griniasty *et al.*, 1999). In general, if some cellular activity detected by any measurement, whether calcium imaging of spike firing, is relatively persistent, then the tendency is to consider this as an attractor. These assumptions about neuronal activity are adequate if the purpose is to model and study artificial neural networks (Hopfield, 1982), indeed there exists the fertile world of attractor neural networks (Amit, 1989). However, what is normally sought in these computer simulations is the asymptotic, long-term behavior of the dynamics of the network under consideration, finding adequate energy functions that can be investigated in terms of local and global minima and maxima. However, in spatially extended, dissipative systems displaying long transients, "asymptotic" behavior may not exist at all, which thus prevents identification of attractors.

What, then, could be a sign of a "brain attractor"? Perhaps, some functional brain states that could be relatively persistent, measured as electrical activity that persists over time as proposed several decades ago (Little, 1974; Mandell and Selz, 1993). But here again the term "brain state" is encountered, which augments the problematic nature of this question (recall the few comments at the end of Chapter 1 regarding the nature of the presumed brain states and the difficulty in designating them). Nevertheless, persistent activity is known to occur in brain networks, especially those involved in the maintenance of working memory like prefrontal cortices (Curtis and D'Esposito, 2003). In this regard, recordings from individual neurons in monkeys have demonstrated that some cells fire spikes tonically (not bursting) while specific memory items have to be maintained (the monograph by J. M. Fuster [1995] provides an account of many of these findings). Conceivably, it is in simpler systems where persistent neural activity can be found more consistently, due to the fact that, as thoroughly explained in Chapter 1, there is an extraordinary tendency of nervous system to synchronize the activity; it was also mentioned that the simpler the nervous system the higher the probability to synchronize persistently. In this regard, persistent activity patterns have been found in neuronal cultures and have been subjected to study since some of these were, remarkably, dissociated cell cultures (Wagenaar *et al.*, 2006). It is remarkable because the connectivity in dissociated neuronal cultures is random and, sometimes, not very dense (although this depends on the plating density of cells). In other culture types, like organotypic brain slices, persistent rhythmic activity is common but not as surprising as these cultured slices retain most of the original connectivity found *in vivo*. If the focus is on brain states, then we encounter the difficulty of defining a functional brain state (Nunez, 2000). Perhaps, it was Lehmann *et al.* (1984, 1987, 1989) who attempted the first measurements of what they termed "brain microstates," while studying the alpha rhythm in human EEG. These functional microstates were defined according to the stability of the recorded waveforms, whatever the period of time the spatial distribution of the brain's electrical activity measured by EEG remained relatively stable. These time segments were characterized by what they called quasi-stable (electrical) potential distributions, words that resonate with the already mentioned concept of metastable brain dynamics. In one study, the authors claimed that two mental states, or reported experiences (abstract thought and visual imagery), were associated with the different microstates in EEG segmentation (Lehmann *et al.*, 1998). Considering that, in the final analysis, mental states are associated with electrical activity patterns, this finding is not as surprising as it may sound, even though the recording method used in these studies (scalp EEG) seems too crude to obtain this type of information about specific mental states. The more disparate the mental states, the better the chances of distinguishing them by analysis of scalp electrical activity, as these will recruit different cortical areas. Thus, these studies represented a pioneering effort to "read the brain." Based on these studies, Wackermann (1994) introduced a method to segment EEG recordings, using trajectories in a high-dimensional space derived from the empirical data, and to evaluate similarities of recurrent field patterns. Other investigators have devised other methods to define mental macrostates from EEG recordings based on the notion that metastable states are "almost invariant sets" and can be extracted from electrical recordings (Allefeld *et al.*, 2009). Once more, the notion of metastability is encountered. Their algorithm assumes that the sequence of microstates is a realization of a stochastic process (a Markov process, recall the comments in Section 2.4). The application of this algorithm to an EEG from an epileptic patient was successful in identifying two "mental states": the seizure and the no-seizure. While it may not be very impressive to classify these two states from the recordings (in fact, any seizure-detecting algorithm can achieve this feat), this study has to be taken as an initial approximation, along with others (Freeman and Barrie, 1994), to more clearly specify states in nervous system activity. In general, the brain states found with these methods were brief, ranging from 75 to 500 milliseconds in duration. Of course, there are other notions of brain or mental state that may not require electrical recordings to be identified, for instance if sensory-motor or other cognitive states are considered, which may be based on more behavioral aspects (but behavior can be measured too, nevertheless). This discussion is related to the ideas of the emergence of mental states from neurodynamics, and for the moment we shall leave it at this point. In summary, there have been different proposals for definitions of brain states, but what almost all of these have in common is that the "states," or at least the elements of the state space, consists of activity present in a subset of nervous cells. As described above, the classification of Lehman *et al.* depends on the spatiotemporal pattern of electrical activities in cortical areas, and the study by Allefeld *et al.* considered electrical waveforms from the recordings, while others, more theoretical models, start with a space of excited cells in a space of neural ensembles in the identification of invariant measures in brain dynamics. Particularly in this regard, the proposal by Boyasrky and Góra (2006) is that invariant measures can be detected in brain activity based on the concept that long-time averages of the activities of ensembles reveal a time-independent invariant measure in the space of cell ensembles. Regardless of the method or concept, it seems that it is then in the activity patterns that attractors, if there are any, may be found. Building on the consideration introduced in Chapter 1 on the characterization of coordinated nervous system activity based on phase differences (phase locking patterns, or phase synchrony), it is ventured the idea that a more proper state space can be constructed from these observables, in that phase differences represent a measure of cellular interactions. Thus, rather than using the raw electric signal, namely amplitude of waveforms and frequency components, a more adequate representation is endowed by the relations between the activities in distinct cells or networks. The considerations about phase difference as an order parameter of the nervous system will be treated in Section 2.7.

But there is, in principle, an empirical way to demonstrate that the geometric object obtained in a return graph or other similar reconstruction is indeed an attracting set: perturb the activity and see if it returns to the "attractor." This sounds simple enough, but is not trivial, for the perturbation cannot be too large, nor too small since the activity should be displaced from the "attractor" but should not destroy it. Previous sections presented theoretical considerations on attracting sets and their basins of attraction. If the perturbations are too large, the dynamics leaves the basin of attraction and may never come back. Could this represent what occurs in some brain activity, like epileptiform activity? This will be treated in the next chapter, but here let us just say, as a manner of introduction for what will come in that chapter on epilepsy, that small electrical perturbations stop very transiently rhythmic (limit cycle-like) seizure activity (Figure 3.12 of Chapter 3), which already suggests a sort of "epileptic attractor" because the cessation is very ephemeral. On the other hand, relatively large perturbations (more of chemical nature, as described in Section 3.5) can stop seizure activity for a prolonged time, indicating perhaps that the neural dynamics has been much displaced from the basin of the "epileptic attractor" by these more robust perturbations. However, these routes to seizures and out of seizures based on the notion of hyperbolic attractors can also be conceptualized in terms of itinerancy (Figure 2.7) or stochastic dynamics, as aforementioned. The experimental example described in the previous Figures 2.12 and 2.13 provides one possible indication of attractor-like dynamics. That is, can the persistence of the first-return map in the case of the transition to seizure shown in those graphs be demonstrated? In an attempt to find answers to these queries, direct electrical perturbations were applied to alter the spontaneous epileptiform activity observed in an *in vitro* seizure-like model. These experiments were performed with the ultimate goal of using electrical stimulation to halt the transition from interictal activity to seizures. As described in Khosravani *et al.* (2003), perturbations applied as extracellular current pulses to brain slices that were made "epileptic" resulted in the temporary displacement of the data points away from the L-shaped geometric object, only to return to it in a few seconds. Figure 2.14 depicts the experiment. In this manner, seizure-like activity could not be aborted by many of these stimulations, unless these were timed to a specific pattern of the slice's activity. In more neurophysiological terms, this means that the extracellular stimulation briefly altered the spontaneous bursting activity of the brain slice before the seizure-like event but reverses back to the pre-seizure activity relatively fast and the ictal event materialized regardless. Can this be considered as an empirical proof that, under some circumstances, the activity of neuronal networks manifests attractors? To be sure, these were *in vitro* experiments that may have little to do with reality... but research is a rough approximation to reality, or, more eloquently expressed in the words of Vladimir Nabokov (in *Ultima Thule*): "When a hypothesis enters a scientist's mind, he checks it by calculation and experiment, that is, by the mimicry and the pantomime of truth."

These *in vitro* experiments were chosen because, among other factors, the dynamics of the brain slices seem to be simpler than that of the whole brain. In fact, under these conditions that make slices present spontaneous seizure-like events (the term "seizure-like" is used in these artificial epilepsy models as a respect for clinicians, because seizures are defined based on behavioral alterations in the patient), it was noted that about 50% of the slices would have no seizure-like events, rather the activity was reminiscent of the interictal type in some patients in whom bursts are evident: these slices had a continuous spontaneous and quite periodic

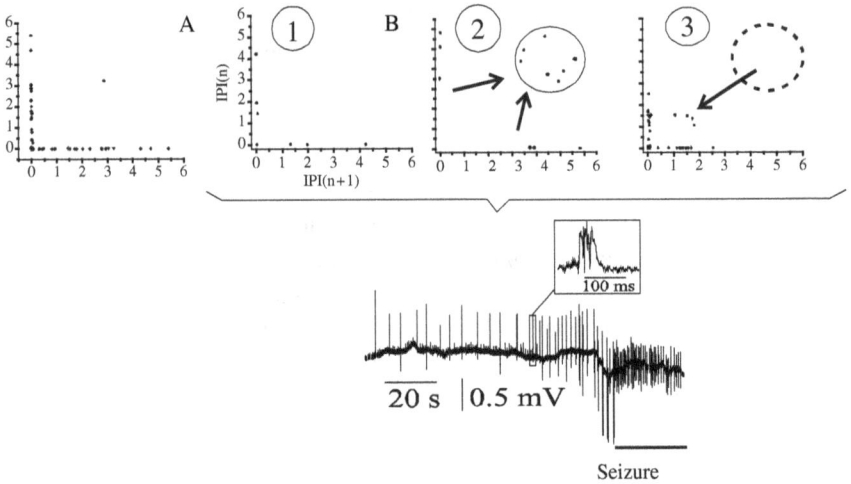

Figure 2.14. Attractor dynamics in epilepsy? (A) and (B), first-return IPI plots constructed with the peaks found in field potential recordings during spontaneous activity in hippocampal slices *in vitro*. One recording (corresponding to the data in panel B) is shown in the bottom, representing the activity before and during the seizure (the seizure is the high-frequency activity at the end of the recording), and the inset shows a detailed burst. Plot in A depicts a typical IPI first-return graph obtained in a slice during the transition to the seizure event, similar as those in Figure 2.12. The plots in B are constructed from the spontaneous bursting activity in another brain slice, for short time intervals lasting 10 to 35 s, before (1), right after a 20-s 0.5 Hz electrical perturbation (2), and a few seconds later (3), showing that the activity was displaced from the L-shaped skeleton to the diagonal (circle) after the periodic stimulation (plot 2), but returned shortly afterwards (plot 3) and a seizure-like event developed, as depicted in the electrophysiological recording below. Modified with permission from Perez Velazquez (2005).

low-frequency bursting that would continue for hours, indeed for the duration of the recordings. Hence, it looks like hippocampal slices maintained under these conditions of low magnesium and high potassium to induce hyperexcitability (more details in the original articles by Khosravani *et al.* (2003) and Perez Velazquez and Khosravani (2004)) present only two dynamical regimes: low-frequency periodic bursting or periodic activity that generates seizures from time to time. Perhaps, the former could be considered a stable limit cycle attractor, since that activity lasted for the whole duration of the experiment, which normally was a 1 h recording. In the successful perturbations that prevented the generation of seizure-like events in the experiments described above, it was noted that instead of developing the seizure (as shown in Figure 2.14), the periodic low-frequency bursting activity remained stable after the electrical perturbation (pulses at 0.5 Hz) which was applied for a relatively short time (10–20 s), as if the limit cycle had been stabilized by the stimulation and thereby the seizure was not produced (the bifurcation did not take place).

Similarly, regarding the seemingly everlasting unresolved question of chaotic attractors in the brain, there is an empirical manner to uncover chaotic dynamics, as explained in previous sections: sensitivity to initial conditions and short-term predictability are two characteristics of chaos amenable to experimentation. Noise, on the other hand, is not predictable at any time scale. These themes have been discussed in detail in other publications (Perez Velazquez, 2005). Many words have been devoted to the issue of chaos in brain activity, of deterministic versus stochastic brain operations, but for our purposes this is not essential. Perhaps, only to mention the title of Lopes da Silva *et al.* (2000) book chapter, because it expresses the main message: "Rhythms of the brain: between randomness and determinism"; this is where brain activity most likely resides. Readers can find a very extensive literature on this topic, which we cannot cite here because it would add a few hundred more pages to this monograph. The appeal of chaotic brain dynamics is entirely reasonable in terms of the information generation, adaptability, and flexibility that chaos offers, as these are characteristics of brain function as well, but the determination of chaotic dynamics has not even been achieved in the relatively simpler brain slice *in vitro* preparation (Aitken *et al.*, 1995), as the demonstration of chaos in any physiological (therefore noisy) time series is not a trivial matter (Kantz and Schreiber, 1997; Hunt *et al.*, 2003). As brain dynamics can be expected to be high dimensional, this already restricts the applicability of the theory of low- dimensional chaos (chaos is low dimensional since it is deterministic, otherwise it would be noise). Of course, some specific neural phenomena may be low dimensional, but perhaps not chaotic.

Pondering about these queries on attractors in neural activity may be facilitated by considering primitive nervous systems, or simpler systems like the aforementioned *in vitro* slice preparation, where the dynamical regimes seem to be simplified. The central pattern generators (CPGs) in primitive systems (also in brain stem and spinal cord of mammals) seem to display, much like the brain slice, stable rhythmic patterns as discussed in Section 1.2 of Chapter 1. Del Negro *et al.* (2002) provide an illustration of the periodic behavior of the *in vitro* brain stem preparation. Computational studies have revealed the great stability of the activity in these networks, for instance Prinz *et al.* (2004) showed that their model of the neuronal ganglia network is seemingly able to perform in an almost identical manner to a very wide variety of inputs or changes in the model variables. Studies like this indicate that altering substantially a variety of parameters in invertebrate nervous systems does not modify the responses (normally rhythmic). In some nervous systems, then, a sort of attractor that determines a periodic limit cycle-like behavior may be desirable, apparently needed for the invertebrate to crawl along its world or for us to breathe normally. It is, therefore, conceivable that these simple systems/preparations offer better chances for empirically finding attractors. On the other hand, the emergence of advanced nervous systems, particularly those of the vertebrates, requires a more advanced flexible processing of information, and thus the anatomy and physiology of these newer systems allow for a state space in which neuronal dynamics is not governed by regular, periodic activity but by stochastic, metastable dynamics, as emphasized in Chapter 1. Neuronal ensembles and single neurons operate at the edge of stability (Makarov *et al.*, 2001). As a result, any input, either external or internal, can alter the brain dynamics substantially and therefore be processed with greater efficiency. Along these lines, it is noteworthy that the resting state of the brain, that is, slow-wave sleep (we are not talking about the resting state of the brain default network which is another matter), is characterized by more pronounced rhythmic activity, perhaps vestige from past evolution. Hence, it can be conceived that while the resting activity of brain cell networks may be characterized by relatively periodic, synchronized dynamics, external or internal inputs destabilize the long-range synchronization resulting in metastable states so that efficient information processing takes place.

In closing, as proposed in other works, rather than introducing flexibility in the precise mathematical definition of attractors, perhaps we should use the term "attractor-like" state for experimental purposes, as much as seizures *in vitro* are termed seizure-like activity because "seizure" denotes not only a paroxysmal electrical recording but also a behavioral manifestation which, obviously, is not present *in vitro*. These attractor-like states could then be specific activity patterns that occur for some time interval. What these activities are will depend on the level of description, whether cellular, network, or behavioral responses are evaluated. From this perspective, and based on the previous discussions and observations presented in these sections, there seems to be evidence for the presence of these "attractor-like states".

2.7.   On Neurophysiological Order and Control Parameters

The choice of a variable as an order or control parameter is not arbitrary, but it is not quantifiable and only indications as to whether a particular variable is one or the other can be obtained. Along the comments in Sections 2.1 and 2.3 on this topic, the most straightforward manner to demonstrate that the parameter is a control parameter is to see whether the dynamics of the system changes as the parameter smoothly changes. On the other hand, there are other indications that a parameter can serve as an order parameter. First of all, its measurement serves as a decent descriptor of the system; this is a sort of subjective reason. A more objective indication is that if the parameter increases its fluctuations on approaching a bifurcation, then we probably have an order parameter. Subsequent chapters will introduce different possible control parameters for each specific case. Here, we would like to propose that the phase difference of neurophysiological oscillations, the relative phase, is a convenient order parameter. By "neurophysiological oscillations" we refer not only to those electrophysiologically recorded, but also to behavioral rhythms. The notion that the difference in the oscillating phase extracted from neural signals or from specific oscillatory behaviors is an adequate order parameter has been proposed by some investigators (Kelso, 1995; Breakspear *et al.*, 2004). We tend to favor this concept because phase differences represent a measure of coordinated activity and, as frequently mentioned in this book, it is in the coordinated activity patterns where a most meaningful description of a complex system can be reached. In general, any bivariate measure that can capture the interactions between components would be a preferred order parameter.

How then does one go about finding indications that phase difference is an order parameter? A first indication that a parameter may serve as an order parameter is given by the fact that it is useful in capturing the system's dynamics. In this regard, synchrony studies based on phase differences, particularly seeking phase locking, have been used numerous times to characterize the brain coordinated activity patterns, as was described in Chapter 1. Another more "quantitative" way to find a sign for an order parameter (to our knowledge, there is no quantitative proof that can demonstrate the existence of an order parameter) is to consider that, within the framework of complex systems, an order parameter should increase its fluctuations before a critical point (Sornette, 2004). Hence, one can ask whether the magnitude of fluctuations in phase differences increase before a bifurcation in brain dynamics. In a previous study we reported the observation of an absolute minimum in the values of a phase synchronization index immediately before epileptic seizures (Perez Velazquez *et al.*, 2007c). This phenomenon is already indicative of the amplification of fluctuations in phase differences, because, recalling (Chapter 1, Section 1.4.3) that measuring synchronization involves calculation of a synchrony index from the phase locking, the appearance of more fluctuations as jumps in phase and phase slips will decrease the synchrony index value. One problem here is the clear identification of the possible bifurcations in brain dynamics that can be used to assess the hypothesis of the pre-bifurcation increased noise in the relative phase. As noted above and in greater detail in Chapter 3, epileptic seizures can be considered bifurcations. Hence, it is then conceivable to test two hypotheses at once: the critical point hypothesis (that the seizure represents a bifurcation, critical points represent bifurcations in complex system theory) and that phase difference is an order parameter, by monitoring the growth of fluctuations in the phase

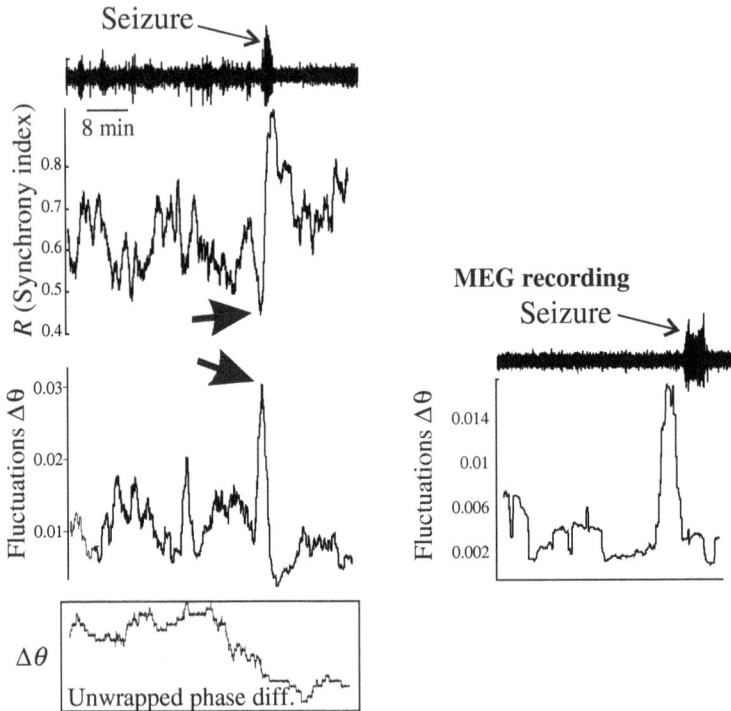

Figure 2.15. Pre-seizure growth of the fluctuations in the phase difference ($\Delta\theta$) between two intracerebral electrodes located in the left hippocampus of a patient suffering from temporal lobe epilepsy (left panels), and between two magnetoencephalographic sensors (one MEG recording shown on the right-hand side) in a patient with absence epilepsy. The upper trace on the left-hand side is the recording from one of the hippocampal electrodes, lasting 54 min, with the seizure visible as the large amplitude signal located at minute 36 and lasting $\sim$55 s. The MEG recording shows the signal in one of the two sensors used to compute $\Delta\theta$, lasting 2 min, and the seizure occurring near the end. In the first case, the phase synchrony index R, evaluated at 30 ± 2 Hz, between the two signals is plotted (upper graph) for the duration of the recording: note the increase in R during the seizure and the drop immediately before the seizure, the latter coinciding with the growth of the fluctuations in phase difference (lower graph) seen just before the seizure (arrows). The rate of fluctuations in $\Delta\theta$ was computed as described in the text. The bottom panel shows the time evolution of the unwrapped phase difference between the two signals, exhibiting phase jumps and the relatively "flat" region coinciding with the seizure. Time scale (x-axes) is the same for all panels. For the MEG case, the phase difference was evaluated at 5 Hz because of the nature of the seizure in this patient, with spike and wave at 3–4 Hz.

difference extracted from brain signals. Figure 2.15 shows the result of testing these two complementary premises in a couple of recordings from two patients (Perez Velazquez *et al.*, 2011). Note that the growth of fluctuations in phase difference occurs just before the ictal event, in parallel with a minimum in the synchrony index. The computation of the fluctuations in phase difference ($\Delta\theta$) was done by taking its time derivative absolute value: $|d(\Delta\theta)/dt|$.

Similar results obtained from several patients seem to indicate that the phase difference in brain signals can serve as an order parameter, and that the ictal event represents a critical point in brain dynamics. This empirical procedure has been used to address parallel questions in, for example, seismic studies, to test whether earthquakes represent critical phase transitions. Because criticality entails large susceptibility to external influences and long-range correlation between the constituents of the system, it represents a cooperative phenomenon, and the sudden and seemingly unpredictable occurrence of seizures, as will be treated in detail in Chapter 3, can be considered a natural consequence of these critical phenomena.

The increase in the amplitude of the fluctuations in the relative phase in brain activity has also been reported in sensorimotor coordination, where the phase difference was measured between EEG signals and auditory stimuli presented to individuals performing a sensorimotor coordination task. More specifically, the growth of the "noise" in this relative phase was detected before the behavioral bifurcation occurred (Wallenstein *et al.*, 1995). Experiments on sensorimotor coordination dynamics are described in Chapter 4, Section 4.2.1.

Like order parameters, control parameters are fundamental in the characterization of dynamical systems. Parameters that seem reasonable to be chosen as control parameters are those involved in synaptic function, which is probably a safe start (Wheal *et al.*, 1998). Mandell and Selz (1993) proposed noise from the brain stem as a control parameter since they claim that it modulates global brain activity. Nevertheless, it is not easy to perform these experiments in the biological field because of the manipulations required to assess the quality of the parameters, like continuously changing the parameter values and looking at the resultant activity; this is more easily done with computational models. Some factors, however, have been manipulated in experiments *in vivo* or *in vitro* that suggest some control parameters. For example, the well-known fact that diminishing inhibition results in seizures, which is a considerable change in the dynamics, plus all the comments in Section 3.4.4.1 of Chapter 3 regarding the continuous variation in GABAergic transmission during epileptiform events, indicates that inhibitory neurotransmission is a control parameter. This one may be trivial, but other factors that have been thought to be manipulated in empirical science and have altered brain dynamics are not so easy to interpret, due to the fact that the experimenter can never be completely sure that it is the change in that specific parameter of interest that brings about the different dynamical states, due to the interrelatedness in the actions of the many parameters in nervous system function. Manipulations of gap junctional communication represent an illustration of this complication in the interpretation of results: many studies have shown that pharmacological alterations of gap junctional coupling result in either more or fewer epileptiform events, depending on whether the gap junctional conductance is enhanced or decreased, respectively (Perez Velazquez *et al.*, 1994; Perez Velazquez and Carlen, 2000; Proulx *et al.*, 2006). Recently, a change in gap junctional coupling among horizontal cells in the retina was associated with a behavioral switch — the switch in visual integration time that occurs from daylight to nightlight conditions because during the day, and due to the abundance of photons, the signal-to-noise ratio is high and the visual system tends toward short integration times, while the opposite occurs during night (Pandarinath *et al.*, 2010). The closing and opening of gap junctions shifted the retinal circuit behavior, and consequently that of the animal, from one to another state, hence these experiments provide direct evidence for a control parameter, gap junctional coupling, in behavioral dynamics. In addition to possible and almost always unknown side effects of the drugs used or the experimental manipulations performed, measuring the extent of the alteration in gap junctional coupling is very difficult *in situ*, even though some efforts have been devoted to assess how much the coupling is changed by experimental manipulations of different concentrations of the drugs that alter gap junctional coupling. In the final analysis, these alterations in neuronal and glial networks result in a myriad of changes so that the whole concept of "specificity" loses its meaning, as expounded in the next section.

A variety of neural bifurcation parameters can be found in the literature, a common one in computer simulation studies is the current density (or magnitude of ionic currents moving through synapses or ion channels) that cells receive, but from the perspective of collective activity some others are more relevant at this level of description. In this regard, an interesting bifurcation parameter was considered by Diesmann *et al.* (1999) to estimate the conditions for a stable propagation of synchronous activity in model networks, specifically a model of integrate-and-fire units. The group size of the cell assembly (number of cells) was their control parameter, and the evidence presented indicated the necessity of a minimum number of cells to maintain synchronous activity and stable propagation of inputs from network to network. Because of the great importance of synchronization phenomena in nervous system, that will appear constantly under different aspects in all subsequent chapters, this study is enlightening since it suggests that to reduce synchronous activity, which may be desirable in circumstances like epilepsy or Parkinson's disease as explained in Chapters 3 and 4 respectively, it is enough to reduce the cell number. This reduction can be achieved by distinct manipulations, and is not meant only by killing cells; methods to reduce the effective number of active cells in specific networks that are involved in some pathologies will be discussed in the following chapters. Remaining within the synchronization viewpoint, another bifurcation parameter that has been studied empirically and in theoretical works is the concentration of extracellular potassium, $[K^+]_o$. A rise in $[K^+]_o$ depolarizes cells and thus increases their excitability, that brings about the tendency of some brain regions to exhibit synchronized bursting in high-potassium conditions. *In vitro* evidence indicated that rising $[K^+]_o$ increased the functional relations between neurons such that they became more functionally coupled within the network (Netoff *et al.*, 2004b), observation that has implications for a variety of pathological conditions that result from synchronous cellular activity. Computational models have presented evidence for the important role of this parameter in the emergence of different stages of ictal activity and spreading depression (Florence *et al.*, 2009). The profound effects of rising $[K^+]_o$ on the emergence of synchronization can be clearly seen by recording extracellular field potentials from *in vitro* hippocampal slices submerged in artificial cerebrospinal fluid where the potassium concentration can be increased from the considered normal, 3 mM, to, say, 5 mM or so. Specific comments on possible bifurcation parameters will be presented in each chapter. Nonetheless, these need not be too specific; bear in mind, for instance, that temperature, a relatively nonspecific bifurcation parameter, is one control parameter in the Rayleigh-Bénard convection (Section 2.3).

2.8.   On the Notion of Specificity

The observations on control parameters above invite a few comments on the notion of specificity in biological research, for it is almost a fundamental principle in research: the search for specific effects and actions of a drug, of a gene, of a mutation in a protein, of a brain malformation. The comments that follow on the concept of "specificity" will apply to the rest of the chapters. Consider the scheme of Figure 2.16. It illustrates that a gene A produces protein B that is involved in a couple of processes C and D, which in turn act on other metabolic pathways E, F, G, and so on. Imagine a mutation in gene A, which causes a misfunction of protein B. This is very specific, granted. However, if the phenomenon of interest lies in the downward processes E, F, G ... specificity becomes blurred and basically impossible to discern. If we are happy to concede that whatever pathology is the result of the specific mutation in gene A, this is fine, but then we have specificity restricted to the first reaction: the translation from gene A to protein B. Equally specific are the small microdomains of the particular reactions acting on short time scales between C and E or F, etc. But all this reveals little about the reasons why a certain pathology that involves E and G, for example, derives from a mutation in A. Along the same lines, a neurotransmitter has very specific effects when it acts on a receptor

Chapter 2

Microdomains of specificity

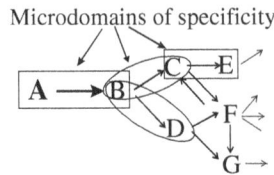

Figure 2.16. Scheme to exemplify the restriction of the concept of specificity. The effects of a mutation in gene A are shown. The mutation causes a dysfunction of its product, protein B, which in turn causes a deficiency in the rest of associated metabolic pathways C, D, etc. In the end, processes E, F, G will all be altered resulting from the specific mutation in A, but for those investigators looking at the level of E, F, G, and beyond, this "specificity" has lost significance because now there are various simultaneous alterations.

(the microdomain in space) in a very short time scale, but its actions in a large mass of tissue and in longer time scales may be better understood if considered as a nonspecific control parameter (Kelso and Engstrøm, 2006) that changes the dynamical state space of the brain networks. As Alwyn Scott remarked: "Nonlinearity is less convenient for the analyst because multiple causes interact among themselves, allowing possibilities for many more outcomes, obscuring relations between cause and effect and confounding the constructionist." (A. Scott, 'Reductionism Revisited', *Journal of Consciousness Studies* 11(2)). In natural phenomena, the realm of specificity is restricted to a microdomain in space and a short time scale.

2.9.   Concluding Remarks

This chapter has introduced concepts and methods that will be used in the following chapters to describe brain activity in diverse conditions. One aspect that deserves to be stressed is that dynamical system theory provides conceptual frameworks to understand nervous system activity, and not only numerical techniques to analyze neurophysiological recordings. It is important to emphasize this point because many physiologists see only the methodological aspect of the theory and remain stuck on performing time series analysis to generate numbers such as correlation dimensions and Lyapunov exponents, although the theory was never meant to constitute an analytical arsenal for time series analysis, but rather a conceptual body of work to comprehend the nature of dynamic systems. It is in this sense that we will make more use in the following. We are interested in understanding brain and behavior, not in simulating their activities. Understanding is not the same as simulating (Laurent, 2000; Stevens, 2000).

To sum up, the main methodology that has been described throughout this chapter, the general procedure in dynamic system theory (if one starts with a set of equations that describes the system) is thus: (1) locate the invariant manifolds (fixed points) of your (unperturbed) system, normally by solving for the zeros of the system of coupled algebraic equations; (2) linearize the system about the invariant manifold; and (3) study the stability of the linearized system, which will provide information about the dimensionality of the stable and unstable manifolds found in step 1. If no equations are known at the start and the only thing at our disposal is time series from some variables, then the procedure is to try to obtain the underlying structure of the dynamics in a reasonable state space using time series analysis methods and perhaps fit this structure to some mapping function, and then, from this function, the estimation of the fixed points and their stability can proceed following the above steps 2 and 3. With regards to finding order and control parameters for the nervous system, it is advised that order parameters should be chosen those that can be measured with some accuracy and have descriptive value, and control parameters should be those that alter the system's activity, and of these there are myriad.

Several limitations on the applications of dynamical system theory have been exposed here, but our opinion is that, in spite of these constraints, these investigations including time series analysis of nervous system activity are still worth pursuing. Rather than obstacles, these limitations can be seen as challenges to the intellect to achieve more complete understandings of the nature of mental function. Metaphors can be stretched to some extent, determined by what is measured and described.

In these final words, it is worth emphasizing the approach of stochastic dynamical systems (Section 2.4) to neuroscience, as the language of probabilities is best suited to capture the nature of brain and behavior. Paraphrasing the words of Alwyn Scott (in *Reductionism revisited, Journal of Consciousness Studies* 11(2), 2004), as much as no two candles burn alike in spite of the common process that generates it, namely the energy supplied by the wax, in people's behavior what remains (more or less) constant is the higher-level processes (such as the personality or individual dispositions) but a probabilistic interpretation for specific behaviors is necessary. Unless we are much mistaken, a fertile field in theoretical and also practical brain research awaits for those interested in focusing on the importance of characterizing phenomena not by averages (or any other central tendency) and standard errors, as is the common scientific well-honored tradition, but by the consideration of full probability distributions in the space of realizations of the processes. The study of dynamics, noise, variability, and fluctuations are penetrating disciplines.

And finally, after all the technical discussions and what seemingly are very specific quantifications that have been presented in this chapter, we would not like to leave the reader with the impression that this type of studies needs to be precisely quantitative. The statement of A. Winfree in this regard, already presented in Chapter 1, is worth reconsidering and having it always present in the mind, for what we are really seeking is a qualitative understanding of brain and mental phenomena. To achieve this purpose, as emphasized in the previous chapter and in those to come, integration between levels of description is essential. A global view may widen the horizons for the undertaking of the creation of theoretical and empirical frameworks to study mental dispositions and their relation to behavior. The main purpose of this monograph is to emphasize that brain and behavior should be viewed not as separate phenomena but as manifestations of an underlying common reality, and that dynamic system theory can offer theoretical frameworks to understand the intricate relation between brain function, behavior, and the environment.

CHAPTER 3

Epilepsy: The Transition to Seizures

This chapter is devoted to epileptic seizures, also called ictal events, not only because the dynamics of the brain cell networks during and between seizures has been studied in extraordinary depth during the past few decades, but also because the collective coordinated activities of cellular networks in seizures represent an informative depiction of the transition among different activity patterns as the molecular and cellular events dynamically progress in time from interictal to ictal phenomena. Indeed, epilepsy has been termed a "dynamical disease" (Milton and Black, 1995; Belair *et al.*, 1995; Lopes da Silva *et al.*, 2003), and a profusion of theoretical and empirical evidences, overviewed in sections below, has revealed dynamical paths of the transition to seizure from normal brain activity. In line with the comments in Chapter 2 regarding the application of mathematical frameworks to neurophysiological recordings, it is fair to say that indications, or signatures, of specific dynamical regimes during the studied epileptiform activities have been identified, rather than asserting that clear dynamical paths that make brains develop seizures have been indisputably proven.

Epileptiform events can be considered as a guide to understanding brain function in general, because these are manifestations of certain tendencies in brain cell activities explored in Chapter 1, particularly the propensity of nervous system networks to synchronize their activity. The focus is on the relevant molecular, cellular, and network phenomena illustrating the continuum in brain activity that results in epileptic phenomena, with an effort to integrate those levels of description. Much like cancer, which represents a (not too surprising) deviation from the normal division properties of cells (hence, cancer is really a natural extension of normal life processes), epileptic phenomena, too, represent a probable extension of the fundamental characteristics of nervous systems, specifically synchronization. For this reason, and before addressing in subsequent chapters other brain pathologies, it is instructive first to understand how deviations in molecular, cellular, and brain network properties result in the transition from the normal to the pathological coordinated activity typical of epileptic phenomena. In this regard, it is conceivable that, at the collective level of description, epilepsy shares common aspects with those of other deviations such as schizophrenia or depression.

In order not to offend the clinician, we note now that we have lumped together all types of seizures, absence and limbic, observations in humans or animals and even *in vitro* experiments are described in same sections, with some crucial differences mainly pertaining to the network dynamics remarked. Our monograph focuses on common biophysical aspects of diseases and is not intended to precisely describe and differentiate the specific clinical aspects, for which there are excellent textbooks that interested readers can consult.

3.1.   General Concepts about Epileptic Syndromes

The paroxysmal brain events called seizures represent a useful illustration of the ideas discussed in Chapter 2. Epileptic seizures are transient neurological states characterized by abnormal synchronization of the brain cellular networks, which may involve focal, circumscribed brain regions, or, alternatively, the majority of the neuronal population of both brain hemispheres. At a global, collective level of description, and within the conceptual framework examined in Chapter 2, it is not surprising that there are many molecular and cellular routes to the seizures: disparate microscopic events lead, as the end result, to similar macroscopic patterns characterized by the apparently organized form of synchronized cellular activity typical of seizures. This may be, perchance, another biological manifestation of the central limit theorem previously discussed (end of Section 1.3 of Chapter 1).

The fundamental characteristic of epileptic seizures that makes them ideal for the study of dynamical brain function is that they start — and stop — spontaneously, with the transient ictal episodes typically lasting from a few seconds to a few minutes. In adults, Jenssen *et al.* (2006) reported median durations between 78 and 130 s, while in children they can last up to 5–10 min (Shinnar *et al.*, 2001). Certain seizure types can be triggered by specific external stimuli, such as flickering visual photic stimulation (Träff *et al.*, 2000; Parra *et al.*, 2003), although this is relatively uncommon. Also relatively uncommon are the instances when a given seizure does not stop with some external input (typically administration of some form of antiepileptic or anesthetic medication); this situation is known as status epilepticus, and fortunately is rare. Readers who have gone through Chapter 2 will probably be now asking themselves what is meant by the "spontaneous" start and end of seizures. If any small and unmeasurable input to the brain can affect largely the dynamics, to what extent can the term "spontaneous" be used here? This consideration is very much related to the self-organizing phenomena discussed in Chapter 2, where sometimes "invisible" inputs give rise to complex pattern formation. To be more precise, the term "spontaneous" is used in this chapter with the connotation of phenomena that develop without any apparent, measurable external intervention.

Epilepsy is the name given to the condition characterized by recurrent spontaneous seizures. It should be viewed as a dynamic process characterized by the occurrence of seizures due to abnormal alterations in brain tissue. Seizures, also termed ictal events, are transient paroxysmal discharges that cause a behavioral change (Gastaut and Broughton, 1972). In a broad sense, there are two main types of epilepsies (Bruni, 1995). One type is referred to as generalized epilepsy, and the clinical seizures in this condition manifest as either brief "absence," or "petit mal," episodes or as generalized tonic–clonic or "grand mal" events. The other main type of epilepsy is focal, or partial, epilepsy. In this situation, seizures arise in one part of the brain, where they may remain confined or from where they may spread to involve greater neighboring brain areas. The classification of seizures, based on clinical and electroencephalographic (EEG) criteria, includes partial, generalized, and unclassified seizures. Consciousness is impaired in complex partial seizures, while during simple partial seizures patients remain conscious. The loss of consciousness during complex partial seizures is a consequence of the aberrant brain activity during the ictal event (Blumenfeld and Taylor, 2003). Importantly, the coordinated activity of cerebral networks is significantly altered due to an abnormal synchronization, as detailed in sections below, with the consequent impairment of information processing, and thus unconsciousness results. The aforementioned generalized seizures may or may not be convulsive: absence seizures are nonconvulsive, for instance. While prior to the 1940s seizures were considered by most as cortical phenomena, current knowledge indicates that subcortical structures have an important role in distinct types of seizures, and, in general, that different seizures will involve distinct brain structures not only restricted to the neocortex (Norden and Blumenfeld, 2002).

Seizures have distinct EEG signatures, reflecting the different activities of the underlying brain cellular networks. For example, the EEG hallmark of absence seizures is the widespread 3 Hz "spike-and-wave discharge," first described by Berger in 1933, where the high-amplitude epileptiform activity appears abruptly and ends just as abruptly, with near instantaneous return to the cerebral background activity (Gloor and Fariello, 1988). Figure 3.1 shows representative examples of EEG recordings of seizures. On the other hand, the tonic–clonic seizures show a faster rhythmic ictal EEG pattern that occurs in association with major motor manifestations of these seizures, normally convulsions. The recordings depicted in the figure illustrate the similarities between the epileptiform activity recorded in

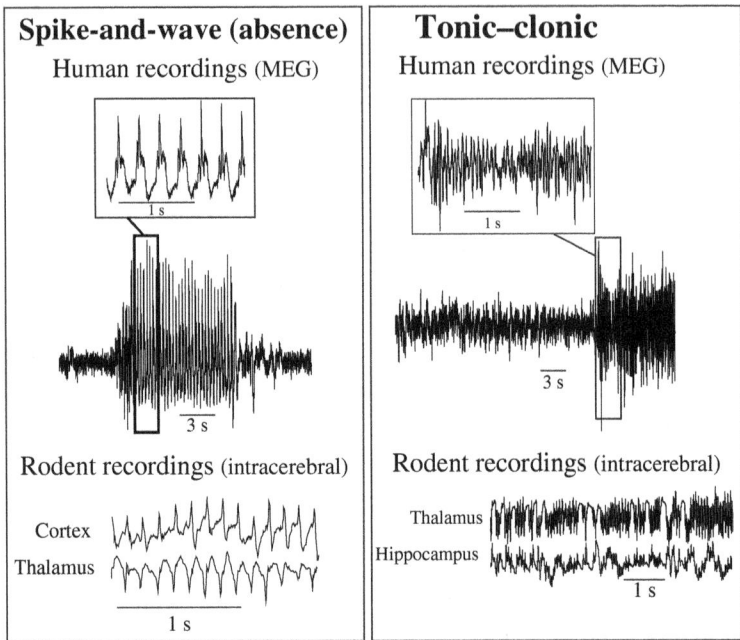

Figure 3.1. Brain electrophysiological recordings from human patients and rat models of seizures, showing absence seizures with the typical spike-and-wave discharge (left-hand side) and tonic–clonic seizures (right-hand side). Insets include a magnification of some parts of the traces. Note the sudden start of the seizures and, in the case of the rodent recordings, that similar activity is simultaneously recorded from different brain areas: the thalamus and the hippocampus during the limbic seizure and the cortex and thalamus during the spike-and-wave discharge. Note that in this case, the absence seizure in the rat has a frequency of ~7 Hz, while in the patient it is ~3 Hz. Spike-and-wave activity is a common phenomenon in rodents and have frequencies ranging from 6 to 11 Hz (Gloor and Fariello, 1988). Some rodent models of absence seizures that employ drugs such as gamma hydroxybutyric acid (GHB) can make the spike-and-wave frequency resemble that of the human absence condition (~3–4 Hz).

patients and in animal models of seizures, which are widely used to gain insight into epileptogenesis (Sarkisian, 2001; Coulter *et al.*, 2002; Engel *et al.*, 2003). These recordings also exemplify that the waveforms are very similarly recorded with distinct methods: the human recordings were taken using magnetoencephalography (MEG), and the rat recordings with intracerebral electrodes.

The recordings in Figure 3.1 represent collective activities of brain cellular networks, recorded basically as variations of local field potentials, as described in Chapter 1. There should be a correlation between local field potential activity and neuronal spike firing, for after all these recordings are representations of activity at two levels of description: network and cellular. Figure 3.2 is an illustration of this correlation, showing simultaneous recordings from individual neurons and

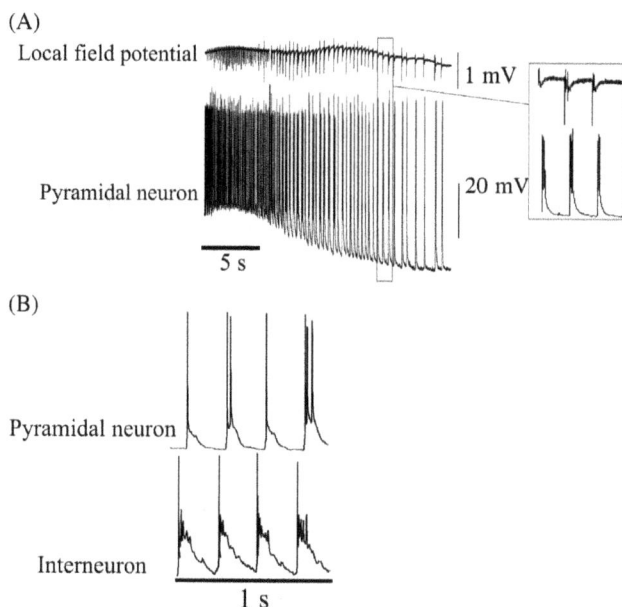

Figure 3.2. Simultaneous recordings of a local field potential and the activity of an individual pyramidal cell (A), and two neurons (B), during seizure-like activity in the *in vitro* hippocampal slice. The neuronal recordings were obtained in the whole-cell configuration of the patch-clamp method, while the local field potential in panel A was obtained through an extracellular recording electrode placed in the CA1 pyramidal layer of the hippocampal slice. Seizure-like activity was promoted by bathing hippocampal slices in a medium containing low magnesium and high potassium (see Khosravani *et al.* (2003), for experimental details regarding this *in vitro* seizure-like model). Note the correlation between field activity and neuronal bursting, in panel A, and the synchronous firing of a pyramidal neuron and an interneuron in panel B during a seizure-like event.

field potentials during paroxysms *in vitro*. Note the synchronous firing pattern in different types of neurons during the seizure-like event, for example a pyramidal cell and an inhibitory interneuron in Figure 3.2 (B). These are two cell types that differ not only in biochemical aspects but also in the action potential firing patterns. Yet, during seizures, their activity patterns seem homogeneous. It could be said that the collective behavior has been "simplified" by the paroxysmal activity. Changes in complexity of the signals may thus be of interest to determine; an account on the spatiotemporal complexity studies during epileptiform activity is presented in a section below.

In focal epilepsies, seizures seem to originate in one restricted area and then may progress to recruit other connected regions into the paroxysmal discharge. The end result of a focal onset seizure that has spread to involve the entirety of both hemispheres will be a (now secondarily generalized) tonic–clonic seizure, indistinguishable in its final clinical manifestations from a generalized tonic–clonic seizure as seen in the generalized epilepsies. Thus, those interested in investigating the mechanisms of spread of paroxysmal activity can take advantage of this form of epileptiform activity. The elucidation of the network mechanisms of the spread of seizures, or how cellular populations become recruited in a global abnormal synchronized activity, is an important matter not only in epilepsy research but also for what it has to reveal about the normal and physiological spread of synchrony associated with cognition. Below, empirical observations obtained *in vitro* are discussed, demonstrating how a "normal" brain region can sustain seizure-like activity if connected to an "epileptic" piece of brain tissue (Kahlilov *et al.*, 2003). Related to this, studies on the spread of seizures are examined in Section 3.4.3. Sometimes, the brain region responsible for generating the initially local seizures is identifiable using noninvasive (meaning that electrodes do not penetrate the brain tissue) scalp EEG recordings. However, in a large number of patients, multifocal and bilateral independent epileptiform abnormalities exist that necessitate further invasive intracranial EEG recordings to identify the area of brain most responsible for the generation of the seizures, an area that can then be targeted for surgical resection. From such investigations, performed either with depth electrodes inserted into the brain or with subdural electrodes placed along the cerebral cortex, a vast amount of electrophysiological information has been gathered, which has facilitated the understanding of normal and abnormal brain processes in general. Although focal seizures may arise from anywhere in the brain, a most common location occurs in the temporal lobe, a very epileptogenic area. A summary of the clinical and experimental features of temporal lobe epilepsy, from the molecular to the behavioral level, can be found in Gloor (1997).

3.2. Fundamentals of Epileptiform Activity: Tendencies to Increase Excitability and Synchrony

In this section, the basic and main features that lead to epileptiform phenomena are introduced. These are hyperexcitability and hypersynchronization of cellular activity. Of note, the term "hyper" before "synchronization," usually denoting a massive synchrony encompassing the whole cortex, is not too accurate, perhaps "more synchronization than normal" reflects better the consensus from the studies performed so far, as discussed in the sections on synchronization (Sections 3.4.1 and 3.4.2). Recall the evidence presented in Chapter 1 stressing the propensity of nerve cell networks to synchronize their actions. Depending on the extent and magnitude of this synchronization, different degrees of paroxysms can be expected. Is this true?

Consider the interictal spike, for instance. Interictal activity is a brief (less than a second) abnormal activity observed in scalp EEG or other type of brain recordings, which occurs, as its name indicates, between ictal events (Gastaut and Broughton, 1972). Interictal events are not associated with behavioral manifestations of the patients. These abnormal waveforms are normally either spikes or short bursts, and the intracellular manifestation of the extracellularly recorded interictal EEG spike is a burst of action potentials on a depolarizing envelope, as has been revealed in paired intracellular and field potential recordings (Ayala *et al.*, 1973; Jefferys, 1990; Staley *et al.*, 2005). This transient neuronal depolarization that triggers the burst of action potentials is termed paroxysmal depolarizing shift (PDS), and occurs synchronously in many cells during the interictal spike. Interestingly, the neuronal PDS during interictal activity looks very much like the neuronal firing patterns toward the end of a tonic–clonic seizure (in the "clonic" part of the seizure), as can be inspected in the inset of Figure 3.2 (A). This may not be a coincidence: it has been proposed that interictal discharges are related to inhibitory mechanisms that limit the spread of seizures (Binnie and Stefan, 1999). Anticipating the discussion in Section 3.4.4.1, inhibitory neurotransmission is altered at seizure onset and returns to normal in the later stages of the ictus (inhibitory postsynaptic potentials (PSPs) tend to become excitatory at the beginning of the seizure as explained below). If interictal discharges represent inhibitory mechanisms, could there be some correlation between interictal and ictal events? There has been debate as to whether interictal activity promotes or diminishes seizures (Staley *et al.*, 2005), and early evidence already indicated that the more interictal activity, the fewer the number of seizures (Swartzwelder *et al.*, 1987), a view with which most posterior accumulated evidence agrees (Barbarosie and Avoli, 1997; Avoli, 2001; Librizzi and de Curtis, 2003; Benini *et al.*, 2003). The cellular mechanism of the PDS were controversial in the 1960s and 1970s, and a dichotomy appeared (no surprise!): discussion arose as to whether or not this was a manifestation of the existence of the "epileptic neuron" as opposed to the abnormal synchronization in a population, the "epileptic aggregate" (Dichter and Spencer, 1968). Intracellular recordings demonstrated synaptic currents as major components of the PDS (Johnston and Brown, 1981), thus concluding that the PDS represents a large synaptic input to the cell as well as contributions of some intrinsic ionic conductances in the cell membrane (calcium currents particularly). Current epilepsy research has gone beyond, mercifully, the search for the "epileptic neuron." It is, however, useful, or entertaining at least, to contemplate the dichotomy behind the search for epileptic cells or aggregates, consequence of the major trend in scientific research: divide and disintegrate, the everlasting and omnipresent reductionism paradigm. For, if it is granted that seizures represent an abnormal collective synchronization in many cells, does not this presume that individual cellular events will be necessary to enhance synchronization? Either synaptic potentials from connected brain areas or intrinsic ionic conductances will necessarily be the reasons for the tendency to a higher than normal synchronous activity. The PDS caused by a large excitatory potential already implies a synchronous arrival of excitatory inputs to the cell, in turn derived from an enhanced excitability (spike firing and other factors) in those cells providing the inputs; thus the "epileptic neuron" originates from the "epileptic aggregate" and vice versa. A diversity of molecular/cellular factors, that can be considered control parameters if dynamical system formalisms are to be employed, will cause alterations in synaptic input and/or intrinsic biophysical cellular properties (such as ionic currents) that can result in enhanced spike firing and thus increasing the probability for a higher-than-normal synchronization in local neighborhoods, the end result being a paroxysm. The next section describes a few of these factors. Excitability and synchronization cannot be easily separated in the nervous system, as already stressed in Chapter 1, these are two aspects of a similar phenomenon.

From the perspective of dynamic system theory, the interictal spike represents an interesting phenomenon since it seems to be a manifestation of a tendency of neurons to synchronously depolarize and discharge, very briefly in this case of interictal bursts, as if indicating the instability of this state. It is estimated that the extension of the cellular activity that is detected in scalp EEG as an interictal spike comprises an area of about $6\,cm^2$ in the cortex (Cooper *et al.*, 1965). Could it be that seizures represent a more stable state in the repertoire of possible synchronized states in network activity? Is this a matter of mass action, the larger the mass the more extended the paroxysm and the more stable the state? Empirical evidence is presented in Section 3.4.1 indicating that different seizure types originate depending on the number of cells that present abnormal coordinated activity, and dynamic considerations are addressed as well in later sections. But to start analyzing and describing possible dynamical regimes in brain activity and epilepsy in particular, it helps to take a close look at factors that can be conceived as control parameters, as well as variables (measurements) that could serve as collective variables, using the dynamical system theory jargon developed in Chapter 2.

3.3. Molecular and Cellular Factors that Contribute to Seizures. Control (Bifurcation) Parameters in Epileptiform Activity

A vast investigation into the molecular and cellular aspects of epileptiform activities has revealed that a wide variety of microscopic factors can lead to a common macroscopic behavior in terms of enhanced cellular synchronization. This section introduces an overview of the main facts at the cellular and biochemical levels. An extensive literature exists, and for those interested in details about particular aspects, the reviews by Prince and Connors (1986), Jefferys (1990), Glass and Dragunow (1995), and McCormick and Contreras (2001) are recommended. While much talk is on neurons, it should never be forgotten that glial cells contribute in a fundamental manner to neuronal functioning, and that the interplay between glial and neuronal activities (Grisar, 1986; Amzica and Steriade, 2000; Oberheim *et al.*, 2008) will result in distinct collective patterns of coordinated activity.

Knowing that the main characteristics that generate (or increase the probability of) seizures is enhanced excitability and synchronization, it is not difficult to envisage some principal molecular mechanisms that can cause ictal events. Anything that promotes neuronal depolarization in large neighborhoods of cells, thereby increasing spike firing, will increase the probability of the synchronization among those cells and the development of a paroxysm. Figure 3.3 describes some factors that contribute to triggering paroxysmal discharges, and to their end. For, as important

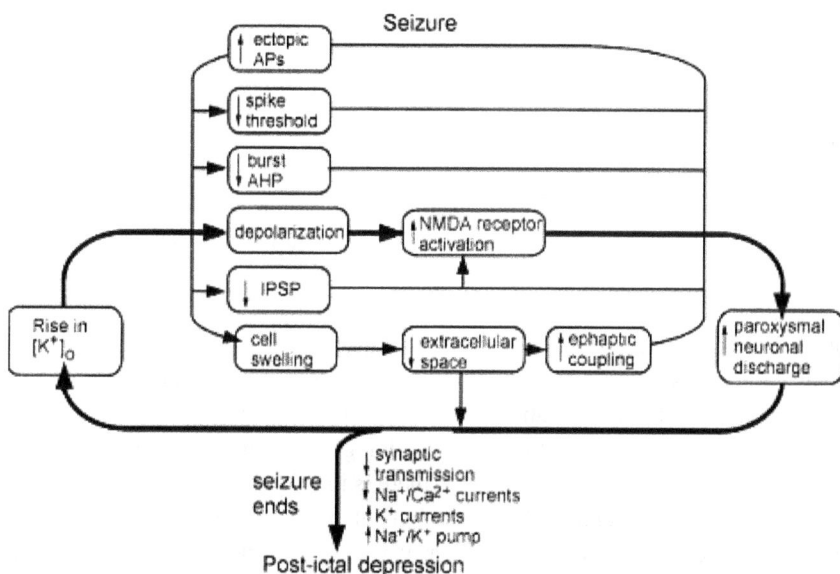

Figure 3.3. Factors that contribute to the onset and termination of seizures. Postictal depression is a common occurrence after seizures, and probably represents the exhaustion of synapses and the massive accumulation of external potassium $[K^+]_o$ that causes spreading depression. Reprinted with permission from the *Annual Review of Physiology*, Vol. 63, (www.annualreviews.org), McCormick and Contreras (2001).

as the study of mechanisms of seizure onset, is the elucidation of factors that end paroxysms, at least with regards to the possibilities to control seizure activity, as discussed below. As aforementioned, any factor that contributes to enhancing cellular excitability will contribute to increase the probability of a paroxysmal discharge, and any factor that contributes to a decrease in excitability will play a role in its termination. A rise in extracellular potassium is all that is needed to generate seizure-like activity in hippocampal slices *in vitro* (Schiff *et al.*, 1994; Jensen and Yaari, 1997) like those shown in Figure 3.2. Rising potassium was found to enhance the functional coupling between neurons (if only because cells depolarize and fire more action potentials which results in stronger synaptic inputs to receiving cells) and, therefore, favors the tendency toward more synchronous activity than normal (Netoff *et al.*, 2004b). Some brain areas are more prone to synchronize their activities than others, and the temporal lobe regions are very epileptogenic. This lobe includes the hippocampus, a great favorite for *in vitro* experiments. One reason for this is that the cell networks of the hippocampi are remarkably prone to synchronize: synchronous seizure-like discharges can be easily recorded even in the absence of synaptic transmission blocked by reducing extracellular calcium (Taylor and Dudek, 1984; Shuai *et al.*, 2003)! Figure 3.4 shows some recordings in this situation. This phenomenon of synchrony without chemical synapses may seem surprising, however, factors contributing to the spontaneous synchronization in the absence of synaptic transmission include field electrical coupling (known as ephaptic coupling) and gap junctional communication (reviewed in Perez Velazquez and Carlen (2000)), because the hippocampal cell layers are very densely populated and the effects of direct electrical interactions are enhanced (Costalat and Chauvet, 2008). Not everything in nervous system function relies on synaptic interactions.

The significance of the schematic Figure 3.3 is that it illustrates the diversity of processes leading to paroxysms: microscopic parameters (potassium ions and channels), mesoscopic (cell activities and synaptic interactions), and macroscopic (synchronization and coordination of cellular activities). The serious student of the brain should always consider that, when studying the possible roles of any of these factors on seizures or other phenomena, there is interaction at multiple levels, and one factor (almost) never acts alone. Just think of the osmotic effects that are known to contribute to synchronization particularly in the hippocampus (again, an interesting synchronizing phenomenon occurs in the hippocampus, and due to the same reasons as detailed above). Turns out these effects on excitability by osmotic changes are due to increasing electrical field coupling or ephaptic transmission (Ballyk *et al.*, 1991). Thus, a microscopic parameter (changes in osmolarity or the ionic composition of the cerebrospinal fluid) brings about a mesoscopic factor (enhanced

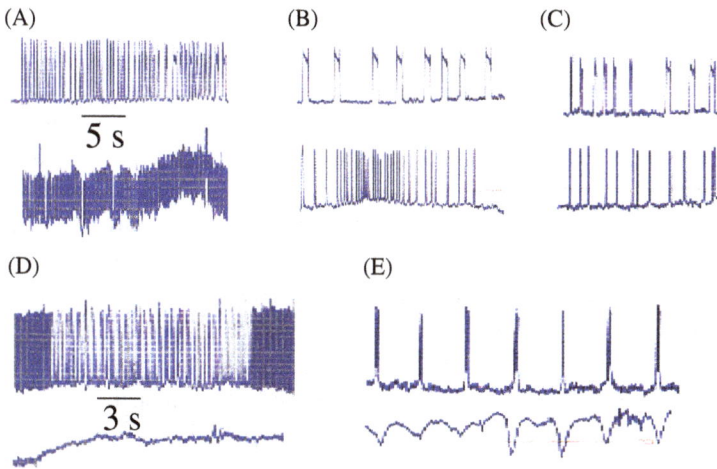

Figure 3.4. Spontaneous activity in the CA1 layer of the hippocampal slice immersed in calcium-free solution. (A), (B), and (C), simultaneous whole-cell recordings from two pyramidal cells at three time points, at the start of the perfusion (A) showing the asynchronous tonic high-frequency spike firing, and a few minutes after, depicting the transition to a more synchronized activity in panel C (about 10 minutes after the recording in panel A). (D) and (E), the simultaneous whole-cell and field potential recording in the cell body layer demonstrates that, right at the start of the perfusion (D), there is no high-amplitude field potential activity and the cell fires tonically, which becomes, a few minutes later (E), bursting and now synchronized to the high-amplitude field activity.

field coupling) resulting in a macroscopic observable: synchronization measured as local field potential. Some of these seizure-precipitating factors have their origins in the genes, or in mutations of proteins. Readers interested in this topic are recommended to look at the reviews by Rogawski (2000), Noebels (2003), Steinlein (2004), and other works (Baulac *et al.*, 2001), for more specifics about ion channel and neurotransmitter receptor mutations in epilepsy. In principle, then, each of these factors in the figure can be considered as a control parameter that will switch the dynamics of cells and networks. Normally, one parameter change will lead to other parameters' alterations, making it difficult to reason in terms of cause and effect (recall the comments on causality and specificity in Chapter 2, Section 2.8). While thinking in terms of cause and effect is the standard, the necessity to switch between the levels of description when studying complex phenomena such as epilepsy makes it highly advisable to restrain this type of either/or thought (Kelso and Engstrøm, 2006).

What ends a seizure? A variety of factors, too. Figure 3.3 shows a few. From depletion of synapses to excessive depolarization of neurons (which results in the impossibility to fire action potentials until the resting membrane potential recovers to relatively hyperpolarized values around −50 mV), their joint action will contribute to the ending of the paroxysm. Some factors are quite nonspecific and general, such as energy failure (recently reviewed in Lado and Moshé (2008)). Depending on the brain area where seizures are taking place, one parameter will be more important than other, because, as explained in Chapter 1 when discussing thalamocortical activity, the combination of the cellular circuitry and the intrinsic biophysical characteristics of the constituent cells in those networks results in one factor having completely different effects on the paroxysmal activity in distinct networks, as for instance inhibitory transmission promoting absence seizures in the thalamocortical networks and reducing limbic ictal episodes in other brain regions. The description in that chapter of the thalamocortical activity is pertinent here to understand how specific paroxysmal (or synchronized) activity depends on the networks and its constituents.

3.4.   The Perspective from the Network/Collective Level

The consequence of those aforementioned molecular/cellular mechanisms is to enhance activity in the neuronal networks, which leads to an intensification of the synchronization of neuronal spike firing. It can be stated that these two phenomena — hyperexcitability and synchronized activities — are necessary and sufficient for the development of ictal events (but bear in mind that synchrony can also occur in absence of hyperexcitability).

3.4.1.   A variety of synchronous activities leading to and within the paroxysms

There is wide empirical evidence for both of these phenomena, hyperexcitability and higher than normal synchronization, during seizure activity. However, with regards to the precise measurements of synchronization, there have been different reports that will be commented upon in Section 3.4.2. Most of these observations have been derived from a variety of electrophysiological recordings: scalp EEG, subdural grids, intracerebral recordings, and MEG. A point worth emphasizing again is that the majority of these methods, as detailed in Chapter 1, record the *collective activity* of cellular ensembles, *rather than the individual spike firing* in neurons. The comments on these recording methodologies in the previous chapters drew attention to some considerations in the physiological interpretation of the recordings. It is then natural to ask whether or not, by knowing the collective activity through these measurements that mostly reflects potential differences generated by PSPs received in neurons, it can be surmised that neurons fire in a synchronized manner during seizures; and whether or not it can be established with some degree of accuracy how many neurons are recruited into this presumably synchronized firing. An early study addressing spatiotemporal correlations between neuronal spike firing and local field potentials (Babb *et al.*, 1987) tried to answer the question of how many neurons increase their spike firing rates associated with different epileptiform activities. Babb and colleagues used chronically implanted microelectrodes into areas of the limbic system in patients to determine the individual spike firing (unit activity) simultaneously with the local field potentials. Their results indicate *that there is a population of focal neurons that depolarize at the same time and are more prone to discharge synchronously, and that the number of these neurons that increase their firing rates determine the extent of the seizure propagation.* For example, they estimated that about 7% of the recorded neurons increased the firing rates during sub-clinical seizures (these are seizures that do not alter consciousness and have no behavioral manifestations, thus the term "sub-clinical," however these are recorded electrophysiologically), while about 14% of neurons augment the firing rates during auras and 36% of neurons during clinical seizures. This is of interest because there is a gradual transition in the consciousness levels from auras (which precede the epileptic seizure) to the full development of the seizure. Auras, considered mild forms of clinical seizures, represent an altered perception by patients, while during the clinical, proper ictal event, consciousness may be completely impaired (Blumenfeld and Taylor, 2003; Johanson *et al.*, 2003). Hence, the more neurons increase synchronously the firing rates, the more impaired consciousness is. Note that these investigators could not determine how many cells are firing during the paroxysms, because the methods do not allow the estimation of this number that would need the evaluation of the firing pattern of many cells at the same time. However, they obtained relative changes in firing frequency by recording from specific cells for long times. The conclusions of Babb *et al.* are today substantiated with intracellular recordings from neurons during seizures, particularly *in vitro* situations during seizure-like activity in brain slices. Note, as an illustration, the depolarization observed in the pyramidal cell recording in Figure 3.2 (A) at the start of the seizure-like event recorded as a local field potential in the upper trace. This depolarization occurs in most neurons, at least in this *in vitro* situation (Jensen and Yaari, 1997), thus supporting Babb and colleagues' concept of synchronous depolarization of many cells in local areas. Signs of increased cortical excitability have been reported for a long time, even 24 hours, preceding a seizure (Badawy *et al.*, 2009). The increased excitability that is at the root of the emergence of seizures is the reason why acute cerebral trauma often leads to paroxysms, since traumatic or ischemic brain injuries cause an almost immediate cell depolarization, due to several reasons such as calcium influx into the cytoplasm, and thus hyperexcitability develops and the consequent trauma-induced epileptogenesis (reviewed in Timofeev *et al.* (2010)).

When addressing the topic of synchronization in epilepsy, this is usually understood as spike firing synchrony. Which, of course, will lead to synchronous volleys of synaptic potentials arriving to connected cell networks, and, therefore, the aforementioned synchronous depolarization of many cells can be expected. Are there other synchronous events that may be taking place, besides synchronous firing? Glial cells also have activities and have been observed to develop synchronous depolarizations of their membrane potentials during spike-and-wave paroxysms (Amzica and Steriade, 2000), depolarizations that were, in addition, synchronous to the extracellular potassium concentration, an expected result since glial cells avidly buffer external potassium. The ionic composition of the extracellular fluid also changes in "synchrony" with other cellular and network alterations: pH shifts in epileptiform events have been studied (de Curtis *et al.*, 1998), which in turn modify neuronal excitability (Jarolimek *et al.*, 1989). There are, then, several synchronous events during paroxysmal activity, most of these so interconnected that trying to separate one from another is not the best strategy to understand the system's behavior. In principle, any of these observables can be used to construct a state space where the brain dynamics can be scrutinized, some being easier to record than others. This will be one topic discussed in Section 3.4.4. Evaluation of the synchrony between the phases of the brain rhythms is a common favorite to estimate coordinated activities, and next section describes phase synchrony studies in epilepsy. It can be thus understood why so much focus on synchrony exists in this field, since the presence of synchronicity at several levels seems to be a most fundamental characteristic of epileptogenesis.

To summarize, these observations indicate a tendency for cells to depolarize, fire spikes and thus synchronize their firing in local neighborhoods at or near seizure onset. This tendency is brought about by control parameters (Figure 3.3) that are themselves continuously changing during brain activity. If the dynamics that govern that activity is characterized by metastability (as discussed in Chapters 1 and 2), then at some point in time, and depending on what region of the state space the dynamics are visiting (the trajectories that define the system's evolution), these changes in control parameters will lead to possible dynamical bifurcations (discussed in Section 3.4.4.2) with different outcomes. One of these can result in the trajectory visiting an attractor-like state in the form of a synchronization manifold, thus favoring synchrony and the recruitment of more local neighborhoods into the paroxysm resulting in sub-clinical seizures, auras, or any other seizure type. Is this recruitment by synchrony a major mechanism for the spread of seizures through the brain? Before considering this query, a look at the evaluation of synchrony in epilepsy will facilitate the discussion on the spread of seizures in Section 3.4.3.

3.4.2. Synchronization studies in epilepsy. Phase difference as a collective variable

The comments in Chapter 2 on phase difference as a possible collective variable of interest is illustrated in the following discussion about experiments and analysis of brain recordings to address the network mechanisms of epileptiform events. Because of the high-amplitude waveforms observed in neurophysiological recordings during seizures, synchronization has always been a favorite topic in epilepsy research. Studies of synchronization (not necessarily phase synchrony) in epileptic activity have had a long history (Cohn and Leader, 1967; Petsche and Brazier, 1972) and have used a variety of methods to estimate correlations in cellular activity, from cross-correlations to the analytic signal approach in a variety of *in vitro* and *in vivo* situations (Gotman, 1981, 1983; Lachaux *et al.*, 1999; Mormann *et al.*, 2000, 2003; Le van Quyen *et al.*, 2001; Netoff and Schiff, 2002; Quiroga *et al.*, 2002; Chavez *et al.*, 2005; García Dominguez *et al.*, 2005; Perez Velazquez *et al.*, 2007c, d). Interestingly, rather than supporting the concept of global hypersynchrony during seizures, many of these studies have cast doubt on it. The general inference that emerges from a wide variety of synchronization studies on epileptiform activity is that, while the raw electrophysiological recordings look hypersynchronous throughout the entire brain neocortex (to the eyes of the clinician electroencephalographer), upon a more sophisticated synchrony analysis this seizure hypersynchrony tends to become very unclear: there is synchrony but it is not as widespread as appears to the eye. Local synchrony, in neighboring cortical areas, is enhanced during seizures, but the widespread and long-range synchrony that the electroencephalographer would swear is apparent from the recordings has rarely been found in any study.

It was already known from early studies on penicillin-evoked seizures in rabbits that seizures begin locally and that the estimated coherence in oscillation frequency was inversely proportional to the area of cortical tissue involved (Petsche *et al.*, 1984). This early experimental evidence provides already an indication of what may be occurring at seizure onset: paroxysms start locally and then spread to encompass connected brain areas by a synchronization mechanism. This scenario complements the ideas discussed in the previous section. Increased local synchronization in seizures was reported from EEG, MEG, or intracerebral recordings (van Putten, 2003; Mormann *et al.*, 2000, 2003; García Dominguez *et al.*, 2005;

Schevon *et al.*, 2007; Amor *et al.*, 2009). *In vitro* experiments offer a better opportunity to study in detail cellular and collective activity, and again failed to support the concept of extensive hypersynchrony (Netoff and Schiff, 2002). However, it may be argued that the fact that large amplitude field potentials are recorded during seizures is already a demonstration of strong local synchronization, either from cell spike firing or because the area around the recording electrode is receiving a strong synchronized synaptic input from other distant areas, which again indicates synchrony at some level. In line with the comments in Chapter 1, the physiological interpretation of synchronization as measured from these field potential-like recordings (EEG, MEG) is not trivial since field potentials represent a summation of synaptic inputs to the area where the electrode is located, in addition to spike firing. Another complication relates to the analytical methods: referential montages such as those commonly used in EEG recordings are prone to artifactual synchronization due to the reference electrode (Fein *et al.*, 1988; Schiff, 2005; Guevara *et al.*, 2005; Trujillo *et al.*, 2005).

As can be seen, then, most of these studies of synchronization reflect mixed levels of description: coordination among individual cellular output activities (spikes) and those derived from collective postsynaptic inputs and glial cell contributions. Then, it is conceivable that there could be differences in coordinated activity at distinct levels. For instance, desynchronization in the spike firing of individual neurons (this would be at short time scales, or high frequencies) and synchronization at the slower time scale (low frequencies) of bursting activity and synaptic inputs, could be found. To most properly address these queries, *in vitro* systems offer advantages. Some studies have addressed the time scales at which synchronization occurs during epileptic events (Netoff and Schiff, 2002; Lasztóczi *et al.*, 2004), indicating that there are transient periods of synchronization at different time scales: synchrony at higher frequencies was commonly observed between local neighborhoods during seizure onset (Chavez *et al.*, 2005). These results are consistent with the observation that seizures usually start by high-frequency activities (Fisher *et al.*, 1992; Wendling *et al.*, 2002; Lasztóczi *et al.*, 2004), and prompt the question of whether or not more complex synchrony during seizures can be expected, like multi-frequency synchronization. The majority of synchrony analyses study the 1:1 case, but it is known in natural phenomena the existence of higher ratios; particularly, in cardiac activity and in cardiorespiratory synchronization, multi-frequency locking ratios have been reported (Castellanos *et al.*, 1995; Schäffer *et al.*, 1998), as well as in muscle–brain synchronization during Parkinsonian tremor (Tass *et al.*, 1998). A recent study reported the observation of complex phase synchronization (devil's staircase) only during seizures (Perez Velazquez *et al.*, 2007d). This observation could be of importance in the explanation of re-entrant mechanisms in the generation and/or maintenance of seizures, a topic widely discussed in the epilepsy literature (Bertram *et al.*, 2001; Norden and Blumenfeld, 2002; Amor *et al.*, 2009). Multi-frequency locking has been seen in pathological cardiac re-entrant arrhythmias, specifically atrial flutter (Castellanos *et al.*, 1995) and, in the heart, re-entrant arrhythmias are dependent on the presence of asymmetrical conduction within cellular networks (Spach and Josephson, 1994). Since cardiac arrhythmias and epileptic seizures share certain characteristics as dynamical disorders of human physiology, the common finding of complex phase synchronization in both conditions raises the possibility of re-entrant activity as a mechanism for seizure stabilization, perhaps one of several parameters that stabilize a possible metastable seizure state that will be commented in Section 3.4.4. Computer simulation studies using neuronal networks have addressed these concepts of re-entrant activity in coordination dynamics (Tsuda *et al.*, 2004; Izhikevich *et al.*, 2004), indicating the relative importance of this recurrent network topology for the expression and maintenance of coordinated activity.

As will be discussed below, an area of great interest relates to the possible prediction, or anticipation, of ictal events. Not surprisingly, then, trends in synchronization have been searched for as possible indicators of approaching ictal periods. Phase differences have been put to use to gain some knowledge into these particular topics. However, a variable number of results can be found in the literature. Some studies have described a decrease in phase synchronization during the preictal period just before the seizure (Mormann *et al.*, 2000, 2003; Le van Quyen *et al.*, 2001, 2003), while other studies reported an enhancement in local synchrony (van Putten, 2003), and still others found no clear trend in synchrony patterns (Netoff and Schiff, 2002; Jouny *et al.*, 2005; García Dominguez *et al.*, 2005), or found trends in spatiotemporal characteristics that were unique for each patient hence not reproducible from case to case (Schevon *et al.*, 2007). To those adepts of nonlinear dynamics and complex systems, the notion of irreproducibility is not that revolting. In any case, because the frequency bands (time scales aforementioned) at which phase synchrony was studied differed among these studies, and the recordings methods also varied (scalp EEG, subdural grids, intracerebral depth electrodes and others), these results in different works may not be comparable and require careful interpretations. The most comparable and reproducible aspect of all these studies, indeed, is the variability in the patterns of synchronization observed in different patients, and even within same subjects in different periods of their epileptiform activities. It could be surprising, though, to find this high degree of variability in synchronization during seizures, as these events represent robust higher-than-normal correlated activity, at least in local areas. However, such is the nature of *relative coordination* (Kelso, 1995) as already presented in

Chapter 1 in the context of nonpathological nervous system activity, now extending to the activity we label pathological. Although absolute coordination tends to be found during seizures, indications of relative coordination during these epileptiform events derive from the spread of the distribution of phase differences (Perez Velazquez *et al.*, 1999). Fluctuations in phase locking patterns were also apparent any time during seizures, depending on the cortical area studied (to be more accurate, and because the MEG sensor space was used in this study rather than the source space, the term "cortical area" is not too accurate, better to say "on the signals studied," details of this work in Perez Velazquez *et al.*, 2007c). Recall also Figure 2.15 of Chapter 2, where important decreases in phase synchrony between some MEG signals were observed just before, and sometimes also after, seizures. This may support the aforementioned findings of decreases in synchronization before paroxysms. In any case, fluctuations in an order parameter certainly can be expected before critical bifurcations. In neurophysiological terms, these observations suggest that, before developing higher synchronization within seizures, cell networks receive a diversity of inputs that can be very locally synchronous but globally asynchronous, resulting in a transient period with the formation of small cell clusters with synchronized activity and thus the apparent decrease in a more extended synchrony (Le van Quyen *et al.*, 2003b).

In summary, the evidence obtained from the assessment of brain coordinated activity in epilepsy, while variable (as expected from the study of a complex system), strongly suggests that distinct spatiotemporal synchrony patterns develop during epileptiform activity, and provides conceivable scenarios for seizure initiation, propagation, and termination. Anticipating some aspects of the discussion on seizure spread in the next section, it can be envisaged that as the seizure onset approaches there is a tendency for local cellular neighborhoods to synchronize their activities, which in turn will create synchronous inputs to connected networks thus favoring the spread of excitation, as well as a propensity to synchronize their activities even further. As the seizure progresses, there is a fluctuating pattern of synchrony in different networks, not necessarily phase-locked, but sustained by the abundant recurrent circuitries of the brain, and, all the time, by the control parameters (Figure 3.3) continuously changing; until, finally, some combination of these parameters result in the apparent abrupt termination of the paroxysm. The usual onset and termination of ictal events seem abrupt to us because we normally judge (measure) the collective level, as activity seen in EEG, field potentials, or MEG. However, if we were to assess the cellular level and perform more detailed analysis of synchronization as discussed above, we could already perceive some indications of the incoming paroxysm or of its termination: the transition to and out of the seizure seems more subtle at these levels of description. Next section describes subtle changes in phase synchrony that may participate in the recruitment of cell networks into paroxysms.

3.4.3. On the spread of paroxysms: Tendency to abnormal coordinated activity

The observations described above indicating the tendency to increase phase locking in epileptiform activity prompt the use of phase differences to address a very important matter in the field, that of seizure spread. A few words on the mechanisms of the spread of seizures are in order, as this is a question of major importance not only to the clinician but also to the basic scientist interested in the mechanisms of spread of synchronized activity. Moreover, this section will provide another illustration of the usefulness of considering the oscillation phase and phase differences. After the previous section's review of the main observations in the analyses of coordinated activity in epilepsy, the discussion that follows concentrates on the results that attempt to explain how seizures spread throughout the brain and how different seizures remain circumscribed to particular regions.

The topic of the distinct involvement of brain structures in different seizure types has been continuously debated in the epilepsy literature, particularly amongst clinicians. For instance, as an illustrative example, take the case of typical and atypical absence seizures. The former results in spike-and-wave discharges (Figure 3.1) restricted to the thalamocortical circuitry, but in the latter, these paroxysms spread over other limbic areas (Gastaut and Zifkin, 1988). Why this difference? Animal models have been used to address this question. Here, we will not go into the discussion of whether these animal models mimic the human condition in the patient with absence epilepsy or whether the spike-and-wave discharges in the rats deserve to be called seizures, a topic that almost always arises when confronted with clinical discussions. Whether we decide to call these periods of high-amplitude synchronous activity in rat brains absence seizures or not, is only a matter of definitions and, as we all know, definitions tend to impose limitations on the taking of broader perspectives. The important matter for our purposes is that these animal models display rhythmic and synchronized spike-and-wave activity (Figure 3.1) and that the study of these synchronized activities, as natural phenomena, is important in itself; whether or not they represent a very close approximation to human absence seizures is another issue. The two rat models in question have been described in the literature, and for the present purposes the crucial aspect is that, while the specific molecular mechanisms by which the drugs induce these seizures remain obscure, the GHB-induced seizures are typical, restricted to the thalamus and cortex, while the AY9944-induced spike-and-wave discharges are atypical, invading other brain structures besides thalamus and cortex (GHB and AY9944 are the name of the drugs used in these pharmacological models, detailed in Smith and Bierkamper (1990), Williams *et al.* (1995), and Snead and Gibson (2005)). Figure 1.2 of Chapter 1 shows the simultaneous spike-and-wave discharge in three brain areas of the AY9944-treated rat: amygdala, cortex, and thalamus, thus this activity is considered to represent atypical absence seizures (in the GHB-treated rat, the amygdala recording would be "flat"). Hence, a conventional and totally accepted manner of expression would be to state that limbic structures (like the amygdala) are "involved" in atypical absence seizures. But perhaps the nature of "involvement" should be considered. Normally, this means involvement as detected in electrophysiological recordings as relatively high-amplitude waveforms. Thus, the electroencephalographer will insist that those three areas in Figure 1.2 of Chapter 1 are "involved" in the paroxysm. But since these intracerebral recordings mostly represent a summation of postsynaptic activities, does not this mean that the high-amplitude waves received in that area where the electrode is located are consequence of synchronous inputs from other connected brain areas? Many other areas could be thus "involved," all is needed is that a few of them send a synchronous synaptic input to the one recorded, regardless of whether or not these other areas show high-amplitude waveforms. It is described in the following discussion that the term "involvement" may have a wider connotation than just the concept based on the amplitudes of the recordings, if the phases of the oscillations are considered.

For all that is known about the neuroanatomical circuitries, it is clear that in the brain, all areas are connected to all, directly or indirectly. So, why typical absence seizures are restricted to the thalamocortical network? Why do some seizures remain local and others spread to become generalized? Rather than concentrating in waveform amplitudes, perhaps it is worth to assess the oscillation phase, particularly analyzing phase synchronization, to see whether or not something that is not apparent to the eye can be disclosed. Hence, examination of the phase synchronization between two areas (the hippocampi, a limbic region) that do not display spike-and-wave discharges in the GHB-treated rats (exhibiting typical absence seizures) revealed that there was an increase in synchronization between the two hippocampi when spike-and-wave discharges occurred in the cortex and thalamus (Perez Velazquez *et al.*, 2007e). These observations indicate that the seizure activity in the thalamocortical circuitry enhances the propensity of limbic areas to synchronize, but it does not seem to be sufficient to drive hippocampal circuitry into a full paroxysmal discharge, hence no high-amplitude waves are observed. This increase in phase synchrony between hippocampi occurred in frequency ranges between 25 and 36 Hz, which may be surprising because the spike-and-wave discharge has a frequency, in these GHB-treated animals, around 3–5 Hz. However, this is consistent with other observations aforementioned: Chavez *et al.* (2005) who found an increase in phase synchrony in the fast time-scales (high frequencies) at seizure onset in human patients, and Parra *et al.* (2003) who reported enhancement of synchrony in the gamma frequency ranges preceding photic-induced seizures; and with theoretical studies using simulations of neural networks that have demonstrated that input frequencies of low frequencies do not induce a coherent and synchronized output, while at higher frequencies, large scale oscillations emerge (Menendez de la Prida and Sanchez-Andres, 1999). This computational result, in the experimental context here addressed, suggests that high-frequency inputs to the two hippocampi of these rats are already started to become synchronized, however, due to some other factors, inhibition a likely one, there is no development of a hippocampal paroxysm. Having the advantage of the existence of another animal model where the spike-and-wave paroxysms invade limbic areas (atypical absence seizures), in the aforementioned rats treated with AY9944, then the question asked was whether or not an alteration in inhibitory processes in hippocampi in atypical absence seizures may be related to the spread of seizure activity beyond the thalamocortical networks to the hippocampus. Lower inhibition in the dentate gyrus area of the hippocampi of rats that displayed atypical absence seizures was indeed found, supporting the view of diminished inhibitory transmission in animals that display a widespread distribution of synchronous spike-and-wave paroxysms. In sum, all these results thus suggest that neuronal circuitries in brain areas that do not display apparent seizure activity become synchronized as seizures occur within other connected regions (such as the thalamocortical circuitry in this case), and that a weakened inhibition may predispose to develop an initial and modest synchronization into full paroxysms. The reasons for the decreased inhibition in the dentate gyrus of AY9944-treated rats remain unclear, but this could be one of the effects of the drug. This result is supported by the known evidence that blocking $GABA_A$-mediated inhibition in the thalamocortical slice preparation causes a normal oscillation in that system (spindle waves) to become spike-and-wave discharges (von Krosigk *et al.*, 1993; Lee *et al.*, 2005). Furthermore, histological and neuroimaging studies correlated with intracranial EEG recordings in patients have indicated decreases in $GABA_A$ receptor binding in brain regions remote from the primary epileptic focus, areas that are rapidly involved in the propagation of the seizure (Juhász *et al.*, 2009), suggesting a similar correlation between lower inhibitory mechanism and seizure expansion.

But the changes in inhibitory transmission do not need to be as drastic as changes at the receptor level as found in the Juhász *et al.* study. More transient and subtle alterations that occur very fast diminish the inhibition in networks undergoing intense activity, and in these cases, there are no molecular structure alterations of the receptor or channel. It is described in Section 3.4.4.1, how these fast-occurring changes in GABAergic inhibitory transmission occur during seizures. For now, the important aspect is that alterations in excitation/inhibition will contribute to the spread of the paroxysmal discharges (Lopes da Silva *et al.*, 1994). Phase synchrony analysis during generalized seizures indicated that these events propagate by progressive recruitment of neighboring areas (García Dominguez *et al.*, 2005; Amor *et al.*, 2009). In dynamical system theory terms, the spread of the paroxysm throughout brain networks can be conceptualized as the progressive incorporation of those networks into a relatively stable synchronization manifold. More on this is described in Section 3.4.4.

These studies may have implications for a better understanding of the spread of seizure activity through diverse brain areas in general. Neuronal circuitries in brain areas that do not display apparent seizure activity could begin to develop, or receive, synchronous synaptic potentials in particular frequency ranges, at the time when seizures are occurring in other connected areas, and a weakened inhibition, among other factors, may predispose these brain circuits for recruitment into the seizure network and, therefore, be responsible for the development of this early synchronization into full-blown paroxysms in regions of the brain that were unaffected originally by distant seizure activity. Going back to the discussion of the meaning of "involvement" of brain structures in seizures, these results indicate that, even though not apparent to the eye since there are no high-amplitude waveforms, networks may already be involved as revealed by a more subtle synchronization analysis of phase differences. Thus, there seem to be different types of involvement. It is still early to say what this means to the neurosurgeon that attempts to remove particular brain areas to relieve the patient from seizures; perhaps that, in addition to looking at the amplitudes in the recordings, it is also worth considering the phases of the oscillation. The more general moral of these tales is that the recruitment of connected regions into paroxysms depends on the network characteristics at the time of the seizure. The experimental work on these two animal models of paroxysms described above illustrates the continuum and gradation of the transition toward highly synchronous activity — a slight elevation of synchrony between nets may predispose these to develop or be recruited into a paroxysm.

A picture that emerges from the discussion in the past few sections can be summarized as follows. Any factor (control parameter) that promotes neuronal firing in local neighborhoods (that may, or may not, be widely distributed throughout the brain) will enhance the tendency for the phase locking of the firing in those neighborhoods and thus the propensity for a more global synchronization that will maintain the state of increased phase locking via re-entrant circuits, thus creating the possibility of not only 1:1 frequency locking but other more complex ratios during the seizure (Perez Velazquez *et al.*, 2007d). This scenario can in fact be witnessed *in vitro*. The interested reader, technology permitting, can perform the following recordings using the calcium-free model of seizure-like events (SLE) in the hippocampal slice to witness how an initial asynchronous cellular hyperexcitability progresses in a short time toward synchronous bursting (Taylor and Dudek, 1984): record simultaneously the activity of one CA1 pyramidal neuron (a whole-cell recording of the patch-clamp configuration will do) and the extracellular local field potential in the CA1 cell body layer; upon perfusion with calcium-free artificial cerebrospinal fluid, it will be noticed that the cell starts to depolarize and fire action potentials tonically at high frequencies, while the field potential recording still remains flat. In a few minutes, the field potential will start to show activity, mostly as bursts at low frequencies, coinciding with the appearance of bursts in the cell, synchronous to the field activity (Figure 3.4 (D and E)). If two cells were recorded simultaneously (Figure 3.4 (A–C)), the bursting activity in both would at this point be synchronous: neurons have synchronized their firing patterns after a few minutes of relatively unsynchronized high-frequency firing (there will be some synchrony in this period, though, because cells are firing tonically and just by chance some spikes will synchronize, however, not enough to be detected as local field potential activity). Cells have thus undergone a depolarization that caused a period of tonic firing and then switched to a synchronous bursting pattern in the whole CA1 area. Hence, in a matter of a few minutes, the plausible scenario of seizure development can be witnessed in this, of course, very artificial *in vitro* model, and yet very revealing. Never underestimate the usefulness of simplified model systems.

In this state of affairs, does it matter where a seizure is initiated? Perhaps in some cases, but most likely, in the majority of cases, the starting point is irrelevant (well, of course, it is relevant to the neurosurgeon who has to remove a focal brain area, but irrelevant in that the activity will spread regardless of where it starts). The classical question so much debated in absence epilepsy regarding where absence seizures start, whether in the thalamus or in the cortex, may not be worth pursuing, because wherever it starts, it will be maintained by the thalamocortical loop. This is demonstrated by the fact that *one single* intracerebral electrical stimulation either to the thalamus or to the cortex is able to generate a spike-and-wave discharge in the rat pharmacological models aforementioned (Proulx *et al.*, 2006), indicating the propensity of these brains to seize. Extending this scenario to a patient with absence epilepsy, we can conclude that in the hyperexcitable brain, any extra activity in thalamus or cortex will be amplified resulting in an absence paroxysm through these two interconnected networks. For other seizure types, for example limbic seizures, the reverberating circuits between hippocampus and other cortical areas will support the paroxysmal discharge (Bertram *et al.*, 2001;

Norden and Blumenfeld, 2002). The demonstration, *in vitro*, that one hippocampus made "epileptic" with a convulsive agent was able to make the other (nontreated) connected hippocampus "epileptic" as well (Khalilov *et al.*, 2003), draws attention to the point that, first, it is not difficult to make a brain area seize, and, second, that the spread of synchronous activity requires some hyperexcitability and synchrony in at least one network.

3.4.4. Dynamical regimes of epileptiform activity. Large-scale insights into the transition from normal neurophysiology to paroxysmal discharges

The view from dynamic system theory is presented in this section, for which the reading of the previous chapter is encouraged as the main concepts were treated there. This dynamical perspective allows one to take a global glance at the brain's activity, regardless of the microscopic factors that may be acting, always with the possibility to then concentrate on the major microscopic factors that are responsible for the system's behavior, for instance those parameters that lead to dynamical bifurcations.

Within the framework of "dynamical diseases" (Milton and Black, 1995; Belair *et al.*, 1995), the transition from interictal activity to the seizure event has been conceptualized as occurring through a dynamical bifurcation in the brain dynamics. Not only the preictal, but also the postictal states (or the transition from the seizure to normal EEG activity) have been conjectured to represent a system close to a dynamical bifurcation (Breakspear *et al.*, 2006). The occurrence of bifurcations implies the switching between dynamical regimes. In general, there are indications of distinct dynamical states, evolving in time, as seizures approach and develop (Jing and Takigawa, 2000; Schiff *et al.*, 2005). There have also been indications of criticality in epileptiform recordings, showing that seizures are heralded by fluctuations in the space-time evolution of the signals using the correlation integral over moving time windows as the descriptor of the evolution (Cerf *et al.*, 2004), as well as energy fluctuations (Worrell *et al.*, 2002). Section 2.7 of Chapter 2 presented the growth of the amplitude of the fluctuation in phase difference just ahead of the seizure (in Figure 2.15 of Chapter 2). The magnification of critical fluctuations in collective variables, as explained in the previous chapter, is indicative of an approaching bifurcation. All these observations indicate the occurrence of bifurcations and distinct dynamical states in epileptiform activity. But before going into details, some important matters have to be considered. The introduction to dynamical and complex systems theory presented in Chapter 2 emphasized the contentious topic of mapping brain physiology and experimental recordings to the theoretically well-established schemes and frameworks of dynamic system theory. Reiterating the point presented in Chapter 2, time series analysis of nervous system activity (and epilepsy in particular) is still worth pursuing; rather than obstacles, these limitations can be seen as intellectual challenges to achieve a more complete understanding of the nature of mental function. Thus, having in mind those limitations (or challenges!), the reader is now introduced to a wide variety of results that have provided indications for dynamical regimes in epileptiform activity and thus opened the possibility to implement control methods to alleviate these pathological events.

To apply these analyses, the first thing is to reconstruct a state space where brain dynamics can be followed and studied to understand the transition from normal to epileptic activity, which brings up the first point in this discussion: what variables can be used to reconstruct the phase space? The usefulness of phase difference as a collective variable was discussed in Chapter 2, but there has been only one study, to our knowledge, that has used the relative phase directly derived from experimental recordings in the context of epilepsy; we are not talking about the phase synchronization analysis discussed in Section 3.4.2, that, even though these are still dynamical studies, do not allow for the determination of features such as stability of fixed points. These are the themes treated in this section. The study that used phase differences to construct a state space and perform a linear stability analysis of a Kuramoto-type coupled oscillator model extracted from experimental recordings in epileptic rats, revealed the existence of various stable phase differences suggesting the phenomenon of multistability discussed in Chapter 2 (Perez Velazquez *et al.*, 2007b).

In general, the most popular method to build the state space, not only for epilepsy studies but for many others, has been the time-delay method, derived from Takens' embedding theorem, already discussed in Chapter 2, Section 2.5.1 (Takens, 1981). The collective variables that have been used to apply the time delay method are constrained by the experimental time series; hence, voltage (the simplest since it is the raw recording) and inter-spike intervals or inter-burst intervals have been mostly employed. A recent review by K. Lehnertz is recommended for those interested in specifics (Lehnertz, 2008). A few studies have used more sophisticated variables to create multi-dimensional models; for instance, the amplitudes extracted from the modes after some suitable mode decomposition of the recordings (Friedrich and Uhl, 1996). Computational neuroscientists that work with models can use a variety of other variables, such as ionic currents, but the experimentalist recording brain time series is severely limited in this regard. From the reconstructed phase spaces, which are equivalent to Poincaré sections in case of these time delay methods, a host of measures have been derived to assess aspects such as complexity in the time series, degrees of freedom, nonlinearity, and chaoticity. Working with computer simulations rather than with experimental time series offers some advantages if precision is required. Specifically for epilepsy studies, one advantage of the modeling of epileptiform activity is that it is simpler than other types of brain activity: take a look at the periodic absence spike-and-wave seizure in Figure 3.1. Conceivably, almost a sine wave can serve as its model! Thus, it is no surprise that absence seizures have been modeled as limit cycles using systems of coupled differential equations (Hernandez *et al.*, 1996; Friedrich and Uhl, 1996; Robinson *et al.*, 2002). The interest now is to understand, at this more global level of description, how brain networks develop a relatively highly synchronized activity that turns into paroxysms, and what control parameters are more important and can be used to alter this transition from the normal to the pathological. This global view to comprehend brain dynamics may be fundamental due to the unfeasibility of evaluating most of the molecular and cellular events in the activity of nervous systems, and also, to recall one of last chapter's main points, that complex systems are not completely understood by just analyzing the individual components. So, equipped now with a state space reconstructed using any of the methods mentioned above, the task is now to uncover dynamical regimes that could be implicated in the transition from interictal to ictal activity, or from the normal to the epileptic brain. Dynamical states found in epilepsy will be discussed in Section 3.4.4.2, however, as an introduction to the flavour of the matters that will be treated in more technical terms in that section, first some comments are devoted to a general principle in brain function applied to epilepsy.

3.4.4.1. *Same factor, diverse effects: Dynamical evolution of cellular events in seizures*

From the discussions in the preceding sections, it seems apparent that what is needed to stop epileptiform events is to decrease the propensity for neuronal synchronization, and thereby the probability to trigger seizures will be lower. Hence, the usefulness of drugs that promote inhibitory actions in brain, as will be detailed in the section on therapies. But, this is an incomplete story. As it turns out, if inhibitory drive is increased at the "wrong" time, seizure activity is enhanced. This is but a reflection of the dynamic nature of cellular function: what is inhibition at one time point, can become excitation at another. To see this clear, a short digression is illustrative: a look at what happens to inhibitory transmission mediated by $GABA_A$ receptors provides a clear demonstration of the fact that, depending on where the dynamics of the system lies, one "cause" will result in a different "effect" from the one expected. This underscores the importance of constructing an adequate state space to follow the system's activity.

The action of the neurotransmitter GABA acting on $GABA_A$ receptors causes hyperpolarization of the cell by activating a chloride conductance: chloride ions negatively charged (Cl^-) flow down their electrochemical gradient from the outside into the cell's cytoplasm, thereby hyperpolarizing (usually inhibiting) the cell. That is GABAergic normal action. Then, it is conceivable that when a seizure starts, a shot of GABA (or GABAergic-promoting inhibitory substances) in the cell network experiencing the paroxysm will tend to diminish it, while on the other hand, blocking GABAergic conductances should increase the seizure activity, right? Wrong. Empirical evidence in animal and human epileptic tissues indicated that the gradient for the chloride current is greatly reduced or even reverses during seizures, so what was supposed to be an inhibition (negative charges flowing into the cell) actually becomes excitation: negative charges flowing out the cell because of the reversed gradient which means, from the cell perspective, a depolarization. During the seizures, thus, the excitation produced by the glutamatergic excitatory transmission adds to the one produced by the now reversed flux of chloride ions. The reason for this phenomenon is that the $GABA_A$ channel is permeable to chloride and, to a much less extent, to bicarbonate ions (HCO_3^-), and the bicarbonate electrochemical gradient is such that it flows from the inside to the outside of the cell. At seizure onset, inhibitory interneurons fire for long durations at high frequencies, and such massive firing causes a very strong GABAergic input to the pyramidal (or "principal") cells. The result of so much activation of $GABA_A$ channels is that large amounts of chloride flows into the pyramidal cell accumulating to such an extent as to diminish or even reverse the gradient for chloride, but the gradient for bicarbonate is still maintained and now the bicarbonate outflow is predominant over the chloride inflow, thus, negative charges flowing out result in a depolarization; and upon more continuous activation even chloride may exit the cell instead of flowing into it. Experimental facts of these molecular/cellular events contributing to the network paroxysm can be found in numerous publications, both in animal (Higashima *et al.*, 1996; Lopantsev and Avoli, 1998; Perez Velazquez and Carlen, 1999; Uusisaari *et al.*, 2002; Perez Velazquez, 2003; Cossart *et al.*, 2005) and in human tissues (to be precise, the human tissue study addressed reversal of GABAergic conductances in interictal, not ictal, activity (Cohen *et al.*, 2002)). Then, perturbing the system, in this case blocking $GABA_A$ receptors, will have different effect depending on where the dynamics of the system is in state space. For instance, adding the $GABA_A$ blocker bicuculline to brain tissue in "normal" conditions will promote paroxysmal activity, as expected, but adding it near the start of tetanus-evoked seizures will shorten these paroxysms (as shown in Figure 6 in Perez Velazquez and Carlen (1999)) because the excitatory contribution of GABA is eliminated.

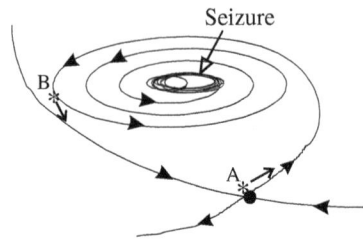

Figure 3.5. Idealized state space reconstruction of a particular brain dynamics that can lead to seizures, shown as the stable (arrows pointing toward the saddle point) and unstable (arrows pointing away from it) manifolds of a saddle point (black dot). See text for details.

A very simple depiction of the importance of the action of a parameter depending on where in phase space the system's activity is located, is shown in Figure 3.5, where the unstable manifold of the saddle point leads to a seizure state, so perturbing the system near point A (close to the saddle) may shift the dynamics so that it moves along the unstable manifold toward the seizure. However, perturbing it near point B may shift it toward the stable manifold, so that now the seizure does not occur and the dynamics will settle onto the fixed point. This very simplistic example illustrates a feature that will be developed in sections below when discussing seizure control: it is of importance to know the brain's dynamics so that a perturbation can result in the desired effect (such as aborting the paroxysm). This is fundamental not only in the case of brain dynamics but also in many other systems. As commented below in the section on seizure control, there have been successful and unsuccessful studies about the control of cardiac arrhythmias using perturbations to heart tissue (Christini *et al.*, 2001; Gauthier *et al.*, 2002), and interestingly, the study of Gauthier and colleagues reported that one possible reason for the ineffectiveness of their method was that some stimuli delivered to the heart did not induce an activation, because the tissue was refractory after the previous activity: thus, the results of the applied perturbations depend on where the system's dynamics lies.

To end this section, it is noted that, as an alternative to the section's title, it could be said too that different factors originate the same result, in that seizures may have many distinct precipitating factors but the end result is the paroxysm. This other variant is discussed at the end of next section.

3.4.4.2. *Dynamical bifurcations and metastability/multistability in epileptiform activity*

The dynamics depicted in Figure 3.5 are reminiscent of models described in the literature, that conceptualize the activity in the epileptic brain as close to a dynamical

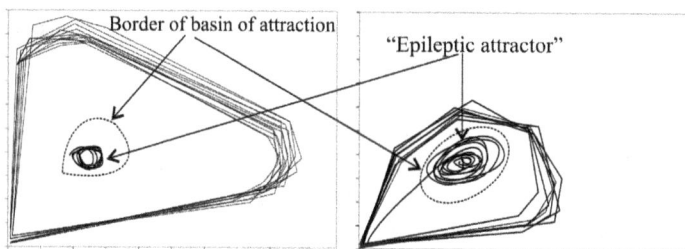

Figure 3.6. Scheme showing a dynamic representation of the normal brain (left) and the epileptic brain (right). It is assumed that there is an "epileptic attractor" in all brains, and that in the normal brain the basin of attraction (dotted line) of the epileptic attractor is far from the other attractors representing nonpathological brain activity. On the other hand, these two attractors are closely associated in the "epileptic brain" (right-hand side panel), thus the transition form normal to pathological activity is easier than in the first case. The shift between these attractor-like states can be caused by, for instance, noise, that is, fluctuations in the control parameters. Adapted from Lopes da Silva *et al.* (2003).

bifurcation that results in seizures, while normal brains are very far from this hypothetical bifurcation point (Lopes da Silva *et al.*, 2003; Suffczynski *et al.*, 2004). For instance, Figure 3.6 is an adaptation of Figure 2 in Lopes da Silva *et al.* (2003) proposing a plausible scenario that assumes that all brains can be epileptic (Lopes da Silva *et al.*, 1994), a most likely assumption (recall previous discussions above). There are other possible dynamical scenarios that can account for these transitions, like Figure 1 in Prasad *et al.* (2003) for example. Consequent with the comments in Chapter 2 about the importance of dynamical bifurcations, these have been sought in pathological brain activity, and, specifically in epilepsy quite a few studies have appeared indicating the presence of bifurcations, that, by analogy with other physical systems, have also been termed "phase transitions" (Iasemidis and Sackellares, 1996).

The exact nature of this dynamical bifurcation is hard to identify particularly for those studies using experimental electrophysiological recordings. However, the characterization is feasible from computational models, with the convenience of a system of equations ready to be used for a formal bifurcation analysis. As already discussed in Chapter 2, whether or not the investigator is content with the use of a computational model rather than the real physiological time series measured by some method, this is an individual matter.

Having said this, and always keeping in mind that finding clear evidence for dynamical regimes and bifurcations in experimental time series is not always possible, some progress has been made from electrophysiological recordings providing evidence for specific bifurcations in epileptiform activity (reviewed in Perez Velazquez and Wennberg, 2004). The analyses of the experimental recordings focused on finding equilibrium points and studying their stability, consequent with the idea that "fixed points are the skeleton upon which dynamical systems are built" (Kelso and Engstrøm, 2006). Thus, using as state space the aforementioned time-delayed plots, specifically first-return plots using inter-spike intervals as the variable (Section 2.5.2), and deriving a one-dimensional first-return mapping function (analogous to the Poincaré maps detailed in Chapter 2) from intracerebral recordings in patients and rodents (Perez Velazquez *et al.*, 1999, 2003) and from *in vitro* models (Perez Velazquez and Khosravani, 2004), indications of a subcritical (also termed subharmonic) flip bifurcation in the transition to seizure were found, observation which complements computational studies based on a mean field model of brain dynamics, showing the presence of a subcritical bifurcation associated with tonic–clonic seizures (Breakspear *et al.*, 2006). Along the same lines, Suffczynski *et al.* (2004) reported that nonconvulsive absence seizures represent bifurcations occurring in the activity of their model neuronal network. Furthermore, linear stability analysis of another one-dimensional map, derived from a small-world network, revealed a qualitative scenario for the transition from bursting fire patterns to seizures, which includes the loss of stability of a fixed point via a flip bifurcation (Netoff *et al.*, 2004a). The flip bifurcation represents a *metastable fixed point* (Guckenheimer and Holmes, 1983; Hoppensteadt and Izhikevich, 1997). What is the significance of the possible occurrence of a subharmonic flip bifurcation in epileptogenesis? A subharmonic bifurcation can lead to two dynamical regimes: intermittency (specifically of the type III) or period doubling (Bergé *et al.*, 1984). Consistent with the possibility that this flip bifurcation in epileptiform activity takes place, signs of type III intermittency and period doubling were found in some cases. For intermittency, these signs include the distribution of the duration of the periodic epochs during seizures, the walk of the data points along the identity map (or diagonal of the phase space) in first-return plots, and, perhaps more clearly and in less abstract terms, the intermittent bursting observed in the electrophysiological seizure recordings like those in Figures 3.1 and 3.2. Certainly, not much mathematics is needed to perceive an intermittent pattern of activity in those recordings! On–off intermittency as a mechanism to generate bursting behavior was proposed and studied in mathematical detail by several groups (Platt *et al.*, 1993; Venkataramani *et al.*, 1996). Experimentally, however, not much has been done. Cabrera and Milton (2002) reported on–off intermittency in human behavior, specifically in a balancing task where individuals had to balance a stick at their fingertips, and the method used included the measurement of the vertical displacement angle of the stick. Studying a hydrodynamical system, early experimental evidence of type III intermittency associated with a subharmonic bifurcation was obtained (Dubois *et al.*, 1983), where the investigators used similar analytical methodologies to those here described, but in their case based on optical measurements of light intensity to construct return maps and to study the distribution of laminar periods in the fluid. These examples demonstrate the global value in the application of these analytical and conceptual methodologies, regardless of the nature of the systems under scrutiny.

As to period doubling, the other regime that results from the flip, Figure 3.7 depicts an example of period doubling in a seizure from a patient with temporal lobe epilepsy (Perez Velazquez et al., 2003). The length of one period (defined as a basic pattern that is repeated) "almost" reproduces itself every period and the period length doubles after some time, which is the classical description of period doubling (Feigenbaum, 1983; Bergé et al., 1984); the word "almost" indicates that we are dealing with natural recordings and not with computer simulations; hence, the omnipresent noise and fluctuations preclude us form observing a "perfect" period doubling. This is one definition of period doubling; however, there is another denotation: the split-up of one period into two as observed in typical bifurcation diagrams (for comments on this topic see Perez Velazquez et al., 2003). As the seizure progresses, the successive doubling of the periods results in very complex activity (right-hand side trace in Figure 3.7 (A)), perhaps a manifestation of the famous scenario described by Landau, Feigenbaum, and others: the period doubling route to chaos. For comparison with other, disparate systems, and to illustrate the universality of these phenomena, period doubling in a thermoconvection experiment is also shown in Figure 3.7 (B). In this experiment, the investigators were able to change a control parameter of that system and observe the subharmonic cascade with successive period doublings. In the case of the neurophysiological recording, obviously, access to control parameters was not feasible, nevertheless parameters are continuously changing during the seizure (like the inhibitory transmission discussed in the past section) and thus results in a "spontaneous" subharmonic cascade.

Returning to intermittency, signs of type III intermittency were noted in the paroxysmal burst-suppression pattern in patients (Rae-Grant and Kim, 1994), in analyses of limbic and absence seizures in patients and rats (Perez Velazquez et al., 1999, 2003), and in recordings from rats undergoing absence seizures (Hramov et al., 2006). This latter study in rats examined signatures of on–off intermittency. Both are equivalent: on–off intermittency corresponds to the Pomeau–Manneville type III intermittency (Pomeau and Manneville, 1980) with a driven bifurcation parameter (Čenys et al., 1997). Intermittency, a dynamical regime that fits well many observations in natural phenomena, has been studied by many investigators. The classification into three different types, I, II and III, is quite technical, based on the linear instabilities of the periodic trajectories according to Floquet theory; in other words, Pomeau and Manneville proposed their classification of

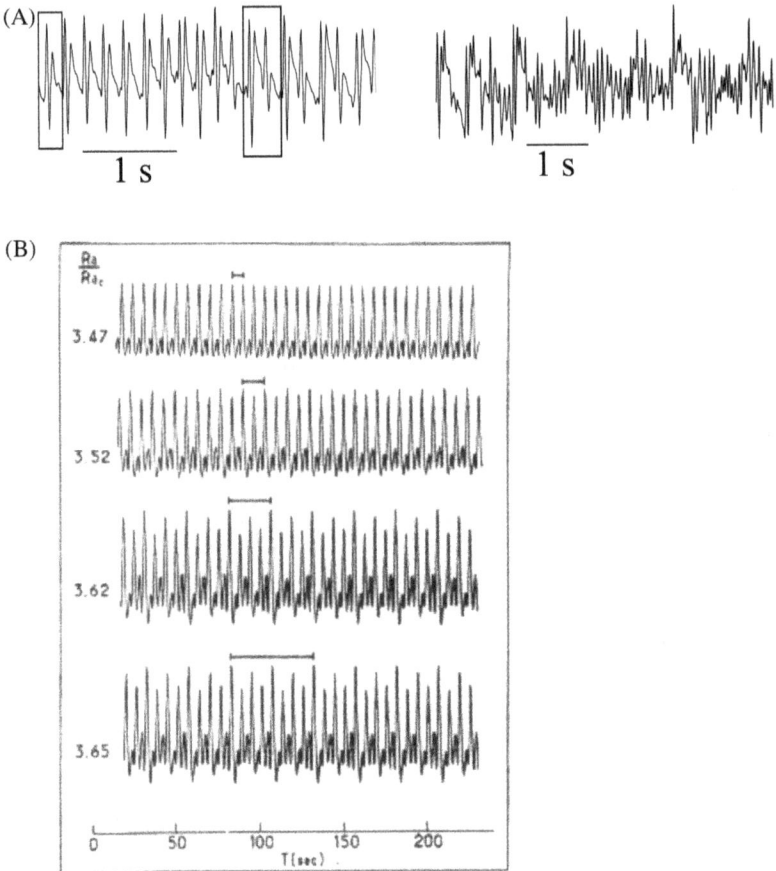

Figure 3.7. Period doubling in seizures and in thermoconvection. (A), intracranial recording from a patient suffering temporal lobe seizures. The left-hand side recordings show how the length of the period, marked by the rectangles, doubles during the seizure, while the right-hand side recording corresponds to a posterior time toward the end of the seizure where more complex aperiodic activity is apparent due to successive period doublings. (B), to illustrate how period doubling was defined in classic studies and the similarity with the intracerebral recording, doubling of the periods in a thermoconvection experiment is shown. The time series is the temperature of a fluid, note the subharmonic cascade as the control parameter of the system changes. The line segments indicate the duration of one period. Panel A is reprinted with permission from Perez Velazquez *et al.* (2003) and panel B from Bergé *et al.* (1984).

intermittencies according to different modes of instability of fixed points (the classic monograph by Bergé *et al.* (1984) is recommended for those interested in these matters). A universal characteristic of type III is the distribution of the lengths of the laminar (periodic) epochs: a power law with a certain exponent, -1.5 for moderate laminar lengths, however, the precise value depends on a relation that involves a control parameter (Bergé *et al.*, 1984; Čenys *et al.* 1997), thus hard to determine from experimental time series. This distribution means that the main feature of this intermittent regime is short epochs of periodic activity, sometimes also called "laminar phases" as this was the original term used when physicists studied intermittency in fluid dynamics; the aforementioned monograph provides an account on the origins of these studies and analyses. As commented in Section 2.5.2 of Chapter 2, because brain recordings are characterized by short, rather than long, epochs of periodic activity, this form of intermittency seems adequate to describe some aspects of brain dynamics.

Recall from Chapter 2 that bifurcations may lead to a soft or a hard loss of stability, and subcritical bifurcations, like that described above in epileptiform activity, imply the latter. As commented in Hoppensteadt and Izhikevich (1997), this bifurcation is of interest in terms of brain function, as it allows cellular networks to exhibit large responses for relatively small parameter changes. In the case of epileptiform activity, this suggests that small changes in certain parameters (those commented upon in Section 3.3) in a vulnerable brain can lead to a profound alteration of the activity, which poses a problem and a hope. A problem, because any small, subtle change in a factor may result in a seizure, and a hope since small perturbations could, in principle, control epileptiform events. Efforts to achieve this hope will be related in Section 3.5.2.

An interesting study has derived a three-dimensional dynamical system model from experimental recordings to study the spatiotemporal dynamics during absence seizures (Friedrich and Uhl, 1996), and has produced observations indicating that the activity in absence seizures has a Šil'nikov-type behavior, named after the investigator who described this dynamics (Šil'nikov, 1965). This dynamical regime is characteristic of systems exhibiting homoclinic orbits. Recall from Section 2.3.1 (Figure 2.10) of Chapter 2 that the homoclinic orbit is an unstable dynamical situation. The aforementioned study found that, for some data set of absence seizures, Šil'nikov conditions were fulfilled indicating the existence of an infinite number of *unstable periodic orbits* (UPOs). Related to the previous comments about intermittency in brain dynamics, the Šil'nikov-type chaos indicates that the dynamics are intermittent, never settling onto a stable fixed point but always evolving in the vicinities of saddle points, scenario that is reminiscent of the heteroclinic channel of Figure 2.7 (C) of Chapter 2. Perhaps, the dynamics of the epileptic brain is such that it has a propensity to settle onto fixed points, albeit transiently. This may apply to a variety of neuropsychiatric conditions, as will be detailed in the next chapters. Let us remark that Šil'nikov-type dynamics exhibit UPOs, which relates to the quest for the identification of these in brain recordings that has been a main approach in the field. The importance given to their identification is based on the notion that chaotic attractors contain an infinity of UPOs; thus, they make up the skeleton of the under-lying attractor, although finding them is not trivial because, for a start, UPOs are unstable, as the name indicates. Algorithms to find them do exist and Figure 3.11 shows one example. UPOs have been found in recordings from *in vitro* seizure models and epileptic patients (So *et al.*, 1998). In one of these studies, the UPOs, interestingly, were evoked by perceptual tasks in the patients (Le van Quyen *et al.*, 1997). The findings of the UPOs in epileptiform activity has been put to use to try to control the dynamics, implementing chaos control techniques in rat brain slices to "pace" the spontaneous neuronal firing activity in hippocampal slices made hyper-excitable by the elevation of extracellular potassium (Schiff *et al.*, 1994), which is described in Section 3.5.2.1 (see Figure 3.11 for a representation of UPOs in neural activity and their use in chaos control). However, finding these orbits is not an easy task, as Slutzky and colleagues remark in their study using the interburst intervals calculated from the spontaneous bursting activity of hippocampal slices made "epileptic" (Slutzky *et al.*, 2003). These authors implemented chaos control techniques and noted that the estimation of the stable manifold was challenging due to the noise in the system. The method relies on fitting local neighboring points near the fixed point using least-squares algorithms, the pictorial representation of the method is shown in Figure 3.11, but the linear approximation in these methods could be accurate only for very local neighborhoods around the fixed point because these manifolds need not be linear.

As described in Chapter 2, bifurcations represent a qualitative change in the system's dynamics as control parameters change. After all these comments on possible bifurcations in epileptiform activities, there has not been much discussion on what could the most relevant control parameters be that may be responsible for the transitions, among those shown in Figure 3.3. Inhibitory neurotransmission, that has been mentioned in Section 3.4.4.1, could be an essential parameter that can be added to the long list, but it is not easy to find one or another that may be more fundamental because of the inter-relatedness of all these factors: take as an example the increase in external potassium that will enhance cell firing which in turn will increase ephaptic coupling and excitatory input that will promote synchrony. It is up to the individual and his/her level of description (or biases and preferences) which of these four factors in the chain that precede the onset of synchrony will be chosen as the bifurcation parameter. They are all equally valid; however, some may be easier to handle than others, for example monitoring potassium with potassium-sensitive electrodes is easier than measuring field effects. Hence, perhaps a useful manner to investigate possible control parameters is to choose those that are easy to measure and are very general. In this regard, Lopes da Silva and colleagues proposed to use the balance between excitation and inhibition as a possible control parameter, which is easily evaluated using a paired-pulse paradigm, a common methodology in *in vitro* and *in vivo* electrophysiology. In their study, this paradigm was performed on kindled rats and revealed a shift from paired-pulse depression to facilitation as the kindling process advanced (it takes a few days to develop seizures in these rats that receive powerful electrical stimulation to specific brain areas on a daily basis, the amygdala and hippocampus being preferred areas to elicit kindling), thus suggesting this depression/facilitation ratio as a relevant control parameter to explore (Lopes da Silva *et al.*, 1994). *In vitro* recordings from "epileptic" hippocampal slices, that can be performed at shorter time scales (minutes–hours) than *in vivo* kindling that takes days, demonstrate, too, an important enhancement of excitation as the spontaneous epileptiform events emerge, as shown in Figure 3.8.

Of course, there is the need, then, to demonstrate that the parameter is indeed a control parameter. As extensively detailed in the previous chapter, the dynamics of the system should change with changes in the values of the particular chosen parameter. This could be done, in principle, by measuring at successive time points, for example, the ratio of excitation/inhibition or the potentiation of the evoked responses shown in Figure 3.8, and plotting them against the collective activity

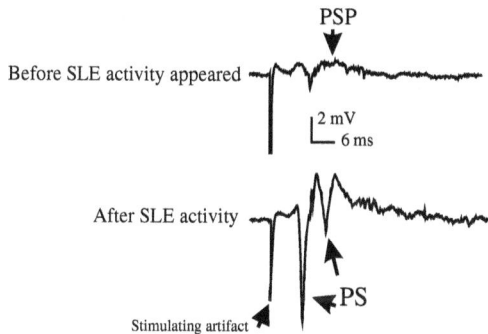

Figure 3.8. A possible bifurcation parameter in epilepsy: potentiation of evoked responses during spontaneous epileptiform activity in the hippocampal slice preparation. The local field potential recordings in the CA1 area of the slice depict the progressive increase in the evoked responses as the slices developed spontaneous epileptiform activity (termed seizure-like events, or SLEs) after being submerged in high-potassium/low-magnesium solution. The stimulation was done in the mossy fibers, that is, two synapses away from CA1 (mossy fibers are the synaptic input from the dentate gyrus to the CA3 field in the hippocampal slice). Note that the amplitudes of the population spike (PS) and the PSP are larger after SLEs have fully developed in the slice, which normally occurs after a few minutes in the high-potassium solution. There is a 482±150% average potentiation in the PS and 177±44% potentiation of the PSP. Before SLEs appeared (upper trace), only a minimal PS was recorded, while a large PS and even a second one were evoked when the slice was already "epileptic." Note also that the evoked PS appears earlier (average time was 8.9±0.5 ms poststimulus), after the onset of spontaneous SLEs, as compared with 10.3±0.4 ms before SLE activity fully developed, which is another sign of amplified excitability.

recorded as field potentials in the slice. In any event, choosing some type of synaptic function as a control parameter is a safe start (Wheal *et al.*, 1998).

The discussion above has been basically centered on findings about possible dynamical regimes and bifurcations derived directly from *experimental* neurophysiological recordings. The findings that have resulted from computational models that, one way or another, simulate epileptiform activity, are much more numerous and a few of them have been mentioned above. More formal dynamical analysis can be done with systems of equations, and the insights that have been obtained from these modeling efforts complement those from experimental time series analyses. A more complete review of these computational works is beyond the scope of this monograph.

After this overview of possible dynamical regimes in the epilepsies, a question that arises is whether or not the dynamics are different in each seizure type. Evidence presented above suggests that all seizures, regardless of cause or waveform, have commonalities in their dynamics. Some studies have addressed this query using principal component analysis and nonlinear autoregressive analysis of the extracted components, showing evidence for a common dynamics in, at least, absence and temporal lobe seizures (Schiff *et al.*, 1999). In general, it can be concluded that specific alterations in any of the microscopic factors aforementioned can lead to the same dynamical result: the development of paroxysms. Further evidence suggesting that different routes may be taken to seizure generation comes from the observation that, within the same patient, identical clinical seizures may have different cerebral localizations of ictal onsets, for example occasionally starting in the amygdala while at other times in the parahippocampal gyrus or, occasionally, beginning synchronously in these structures (Wennberg *et al.*, 2002). There are many routes to seizures, and numerous causes end up having the same result, perhaps because, as mentioned in the introduction to this chapter, epileptic activity is a manifestation of the strong tendency of nervous system networks to synchronize their activities. This is in apparent contrast to the title of Section 3.4.4.1 where it was discussed that the same parameter can have a diversity of effects, so we go from "numerous causes, same result," to *"same cause, different results."* But such is the nature of complex dissipative (nonlinear) systems, as illustrated in the words of Didier Sornette (2004): "… the richness of out-of-equilibrium systems lies in the multiplicity of mechanisms generating similar behaviors."

3.4.4.3. *The geometry of epilepsy*

Notwithstanding the limitations, already mentioned several times in previous sections, of time series analysis in the finding of specific dynamical regimes, let us consider for a moment that these two aforementioned dynamical regimes, intermittency and period doubling, may in fact be present in brain activity, and let us ponder about some consequences derived from those regimes and their dynamical bifurcations, specially in the understanding of seizure onset and termination.

Period doubling is a typical route to chaos (Feigenbaum, 1983), normally from periodic activity to complex and aperiodic activity (possibly chaotic, but we are never sure of this until proven, recall from the previous chapter the stated problems about the clear determination of chaos). This scenario can be translated, albeit speculatively, to the epileptic case as a route to desynchronization of cellular activity that favors the termination of the ictal event. Period doubling cascades during the seizure (Figure 3.7 (A)) may lead the cellular networks to more complex collective activities resulting in progressive desynchronization.

On the other hand, the presence of intermittency makes sense in brain dynamics as this dynamical regime is characterized by short, rather that long, epochs of periodic activity. Other intermittency regimes, type I for instance, is characterized by a predominance of long episodes of periodic activity. Short and transient periods of synchronous rhythmic activity (a manifestation of metastability or relative coordination) is in fact what is normally seen in brain neurophysiological recordings, at least in healthy conditions, whereas long-lasting rhythms are not as common and many times associated with pathologies like the activity during Parkinsonian tremor that will be addressed in Chapter 4. Intermittency is a dynamical state unique to nonlinear systems where the transitions among periodic, quasiperiodic, and aperiodic (possibly chaotic) states occur without the need for external stimuli. Here again, the concept of an "external" input is encountered, and same considerations as aforementioned apply: there could be unmeasurable and undetected external inputs that could qualify as "internal" because nobody can perceive these. Another feature specific for type III intermittency is that nonlinearities present in the system tend to destabilize the activity (Bergé *et al.*, 1984). Thus, this destabilization can plausibly favor desynchronization and may determine the advance toward the cessation of the seizure. One difference in the mechanism between on–off and type III (the Pomeau–Manneville type) intermittencies is that the onset of the former is due to the loss of stability of a chaotic attractor, while in the latter it is a periodic orbit that loses stability (Pomeau and Manneville, 1980; Venkataramani *et al.*, 1996). Theoretical studies have demonstrated that transitions from turbulence to periodic behavior are often realized via intermittency (Chaté and Manneville, 1987; Kiss and Hudson, 2001), and has been seen in experimental settings (Jeffries and Perez, 1982); hence, it can be argued that this dynamical regime may take place in the transition from irregular brain activity to the more periodic activity during seizures.

Within this framework, then, and summarizing the previous findings, it can be hypothesized that the seizure starts as a materialization of a metastable bifurcation point (the flip bifurcation), which in terms of neurophysiological activity is observed as a high-frequency activity commonly detected at seizure onset, and that the ensuing dynamical regimes of type III intermittency or period doubling, with their special characteristics, can result in the destabilization of the cellular synchronized firing patterns and bring about the termination of the seizure. One conceivable consequence of this framework is that seizures should be short lasting, as control parameters continuously change within seizures (potassium concentration, neurotransmission, any of those shown in Figure 3.3). Indeed, as discussed in previous sections, cases of long-lasting seizure activity, as in status epilepticus, are not common; the most common seizure event is one that lasts a few seconds only. Seizures, thus, are not much unlike most of the other brain states, all having a brief existence.

Part of the results that have been discussed here were obtained using what can be called a geometric approach, as the typical linear stability analysis in state space and similar techniques are geometric approximations to a system's dynamics, as explained in Chapter 2. The generality and usefulness of the geometric approach described in these sections to find dynamical regimes should be appreciated, as a similar approach has been used for describing the behavior of disparate systems: biochemical (Decroly and Goldbeter, 1987), physical (Bergé et al., 1984), chemical (Roux, 1983), and cardiac (Christini and Collins, 1997; Hall et al., 1997) systems. For instance, in the case of the biochemical system, the "spikes" measured to construct time delay plots were those of a concentration of a chemical in the reaction; in the case of the cardiac system, the "spikes" used to make the first-return plot were the cardiac beats (to be precise, the atrio-ventricular conduction times, Figure 3.9 (A, B)). To illustrate the remarkable similarity, at this dynamic level of description, in the activity of cardiac and brain systems, Figure 3.9 shows the Poincaré (first-return) maps and associated stability analysis of the fixed points extracted from the cardiac activity and from the transition to seizure. To understand this figure, the explanation in Chapter 2 about the relation between dynamical trajectories and one-dimensional maps (Section 2.2 and 2.5) should be recalled. Additionally, it is assumed that readers know how the iterations proceed in such graphs, termed web diagrams or cobweb plots (see http://mathworld.wolfram.com/WebDiagram.html for a pictorial description). In the cardiac study, the investigators used a geometric approach to stabilize an unstable target cardiac rhythm to control arrhythmias. Details can be found in the figure legend. This point emphasizes the comments in the previous chapter on the generality of these frameworks to study complex systems: at this level, there is no need to consider the system's constituents, *similar dynamics develop in disparate systems due to the interactions among the components*. Focusing on interactions

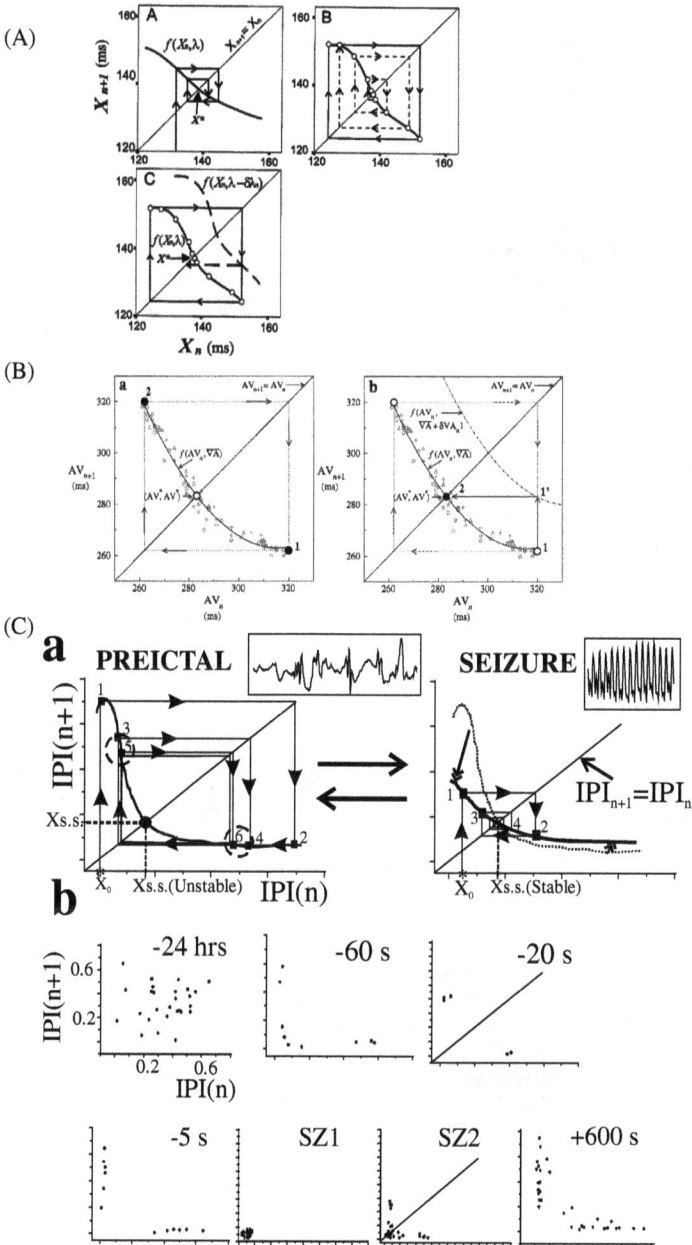

Figure 3.9. (*Continued*)

Figure 3.9. (*Continued*). Common geometric approaches to the determination of cardiac and brain dynamics. (A), state space reconstructed using a time delay (first-return) plot of atrioventricular (AV) nodal conduction intervals measured in the heart. The three graphs in this section show the schematic of the dynamics and control mechanism of a cardiac alternans. The mapping (difference equation $X_{n+1} = f(X_n, .)$) relates successive AV nodal conduction times, and, in graph A, these converge to a stable fixed point X^* since the slope of the map at that point is less than 1 in absolute value (as explained in Chapter 2 this slope implies the stability of the equilibrium). This is the normal cardiac rhythm dynamics, but it can develop into an alternans (graph B) when the mapping function changes: now the previously stable fixed point is unstable (slope is greater than 1), resulting in an alternation between the two extremes. Graph C shows the effect of a perturbation to the tissue to control the activity. This stimulation, which alters the mapping function, can be more clearly seen in section (B). Here, graph *a* represents the AV dynamics without control and graph *b* with a nonlinear-dynamical feedback control. The distribution of data points defines a mapping function (f (AV_n, VA)) approximated by a quadratic curve fit, and the intersection with the diagonal indicates the period-1, rhythmic steady state (white circle, AV^*) that, due to the slope of the map at the point, is unstable. Without control (*a*), there is an alternation between points 1 and 2 (dotted lines), this being a stable period-2 cycle (the pathological alternans) but with the nonlinear control (*b*), the perturbations shift the function to the dash-dot curve so that point 1 now becomes point 1′, and when the function returns to the previous position at the next beat, the AV interval falls near point 2 (black circle). Thus, the stable and pathological period-2 cycle is avoided by stabilization of the period-1 cycle. Section (C), geometric representation of the possible transition from preictal to ictal activity. The first-return plot in this case uses the inter-peak intervals (IPI) taken from brain recordings (as in Figures 2.12 and 2.13 of Chapter 2). Left panel in part **a** represents activity in preictal states, with the dynamics starting at an initial IPI(n) value of X_0 (asterisk), which then follows the points 1, 2, 3 . . . according to the mapping function represented by the continuous curve. The dynamics, finally, settle as a period-2 cycle switching between points 5 and 6 (circles), which corresponds to long-short and short-long inter-peak intervals, characteristic of the interictal activity as the recording trace in the inset shows. The value Xs.s represents the unstable fixed point. At seizure onset (right-hand side), the mapping function changes shape from the original (dotted line) to the new one (continuous line) due to alterations in neurophysiologic parameters, and as a result, now the IPI corresponding to the fixed point (Xs.s) is stable: notice how starting from any initial condition (X_0) the dynamics settles onto Xs.s. This IPI represents a period-1 activity at higher frequencies than the previous period-2, which is recorded as the typical high-frequency rhythmic seizure onset (inset). As parameters continue to change, the shape of the function can return to the original and thus the transition from ictal-interictal-ictal forms a sort of continuum. Below (**b**), the IPI scatter plots are shown from an experimental recording 24 hours prior to the ictus, and 60, 20, and 5 seconds before the seizure, then during the seizure (SZ1 and SZ2), and 600 seconds postictal (+600s). Note that the experimental data points 20 seconds before the ictus show a period-2 cycle, while during the seizure (SZ1) are concentrated at short IPI values (high frequencies), and then return to the L-shaped curve after the seizure (+600 s). The distribution of data points follows what can be a continuous transformation of a mapping function as described in part **a** (readers interested to see this directly may want to try the iteration of points using successive changes in the mapping function as depicted above). Panel A, reprinted with permission from Hall *et al.* (1997), Panel B from Christini *et al.* (2001) (copyright 2001 National Academy of Sciences, USA), and Panel C from Perez Velazquez *et al.* (2003).

using the view from collective dynamics, should always complement (the much more common) studies that disintegrate the system into its components.

The description of the dynamics depicted in Figure 3.9 (C) suggests that there could exist several steady states, or fixed points, in brain dynamics, but two are most prominent during epileptifom activity. One represents a period-2 limit cycle during preictal or interictal periods and another represents a period-1 cycle of higher frequencies at seizure onset. The transient stabilization of these fixed points determines some of the observed activity in recordings. Multistability, or the presence of several stable fixed points, has also been derived from the linear stability analysis of a Kuramoto-type coupled oscillator model extracted from experimental recordings in epileptic rats (Perez Velazquez *et al.*, 2007b), where the values of the stable phase differences were evaluated and compared with the experimentally recorded.

In summary, it is conceivable that a critical point may be defined as the brain activity evolves toward the seizure event, and that, through specific dynamical bifurcations, the brain activity enters a relatively stable synchronized regime (the seizure) that destabilizes progressively, due to changes in internal and external constraints until finally the ictal event halts. Recall that critical points have also been revealed in nonpathological conditions, associated with cognitive processing and sensorimotor coordination (Kelso, 1995) as will be detailed in Chapter 4. Hopefully, these studies may help to define the network dynamics in the transition from normal to ictal activity, and complement other investigations using different methods.

3.4.5. On complexity measures in epileptiform activity

Due to the contemporary interest in quantifying complexity in nervous systems, a brief account of some complexity measures in epileptic recordings is in order. This has implications not only for the global comprehension of this pathological activity, but also for practical purposes, such as the anticipation of seizures (Section 3.5.2.1 is partially devoted to these matters). A cursory look at any seizure recording (Figures 2.1 and 2.2 of Chapter 2) instantly suggests that the brain activity becomes more regular than in most of nonpathological situations. It was also mentioned above how distinct cells become similar in their firing patterns during paroxysms (Figure 2.2 B of Chapter 2). Normally, these regularities are understood as the activity becoming simpler with fewer degrees of freedom, for almost a sine-wave can model an absence seizure (for sure a limit cycle can do it, see Hernandez *et al.* (1996)). Hence, a decrease in complexity could be conceivable during paroxysmal activity. Some studies have indeed reported on a variety of measurements indicating that this may be true. These analyses have used correlation dimension or similar measures performed on EEG recordings (Lerner, 1996; Elger and Lehnertz, 1998; Jing and Takigawa, 2000). The observed reduction in correlation dimension during seizures in these studies signifies fewer degrees of freedom, that is, less complexity in the dynamics (if this dimension is taken as one notion of complexity). The positive Lyapunov exponent reduction during seizures is also indicative of lower complexity (Iasemidis and Sackellares, 1996). Few degrees of freedom during absence seizures were reported in models (Friedrich and Uhl, 1996). Complexity determined by other type of analysis, such as linear complexity estimates, was also found to be lower during seizures as compared to control recordings (Bhattacharya, 2000). Even during interictal periods (without seizures), the linear complexity was lower in epileptic patients as compared with EEG recordings from nonepileptic subjects (Kim *et al.*, 2002). While there are quite a few studies that reported a decrease in signal complexity during seizures, it should not be a surprise that others reported the opposite. Thus, in a study examining the contributions to complexity of individual neuronal spiking activity, it was found that the algorithmic complexity of neuronal action potential firing increased during chemically induced seizures (Rapp *et al.*, 1994). Based on wavelet analysis, Rosso *et al.* (2006) found an increase in temporal complexity during tonic–clonic seizures. Still, a different type of complexity measure analyzing temporal complexity, suggested that complex partial seizures exhibit patterns of increased complexity (Jouny *et al.*, 2005), and that there is a continuous increase in complexity preceding seizure termination (Bergey and Franaszczuk, 2001). Entropy has also been used to assess complexity in brain activity, and van der Heyden and colleagues (1999) reported a decreased in the scaled entropy during ictal events (however, if not scaled, the entropy increased during the seizure, observations that, perhaps indicate the complexity in these "complexity" studies). Not surprisingly, changes in complexity measures have been proposed as possible seizure-forecasting algorithms (Jia *et al.*, 2005). These apparently opposite observations among the different studies aforementioned, of increased and decreased complexity associated with epileptiform activity, pose the question of whether the complexity estimations yield different results depending on the level of description: individual spike trains versus ensemble recordings. Or, whether the analytic techniques are so different that the results cannot be compared. One may wonder whether or not this state of affairs is an invitation to take some time off analysis to carefully think about what is meant by complexity in nervous system activity and function. One possible solution would be to clearly specify, in each study, what type and aspect of complexity is addressed and analyzed.

Most of these studies aforementioned have assessed the temporal complexity present in neurophysiological recordings, and indeed brain activity in general has been mostly studied through its temporal evolution rather than through its *spatial organization*. A recent study assessed the spatial variability of the phase synchronization patterns in neuronal activity associated with epileptiform events, exploring a relatively new description of brain synchronization through its spatial, rather than temporal, organization. Figure 3.10 shows the spatial complexity of the

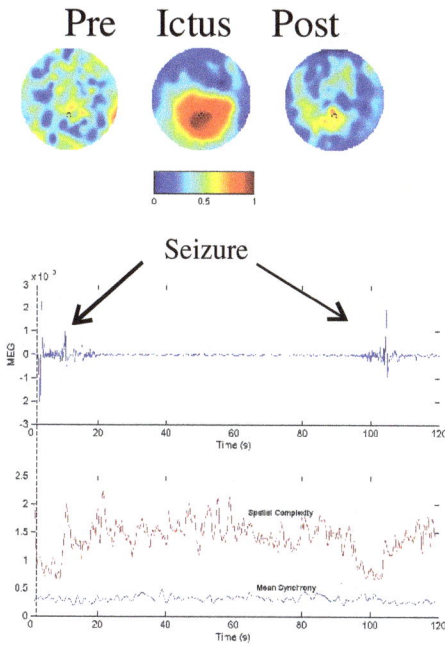

Figure 3.10. Spatial complexity of phase synchronization in epileptic recordings. Upper panel represents the color-coded phase synchrony on a schematic head a few seconds before (left), during (middle), and after the ictus (right). In the lower graph, the blue trace above shows the MEG recording (one sensor), with two seizures, and below, the spatial complexity (red) derived from the phase synchrony pattern is calculated, as well as the mean phase synchrony (blue), averaged over all MEG sensors. In this case, the seizures are better characterized by a change (decrease) in spatial complexity of the synchrony patterns than by any change in phase synchronization. The upper panel is reprinted with permission from García Dominguez *et al.* (2008).

synchronization pattern between and during seizures. Note the smoother pattern of synchrony during the paroxysm, which is quantified in the lower graph (those interested can find the details of the method to assess spatial complexity in García Dominguez *et al.* (2008)). It could also be the case that temporal and spatial complexities follow dissimilar trends in seizure activity. In reality, and to be precise, because the spatial pattern shown in the figure is based on a phase synchrony index (R) that depends on the temporal fluctuations in phase difference ($\Delta\theta$) over a certain time window, defined as $R = |\langle e^{(i\Delta\theta)} \rangle|$, it is clear that the spatial synchrony pattern contains information about the temporal variability of the relative phase, then it is not just the spatial but actually the spatio-temporal complexity that is evaluated in this study. More studies will be needed to assess these queries, but for the time being, the studies that have addressed the measurement of signal complexity in brain activity provide a general picture that indicates that trends in complexity exist as epileptiform events unfold in the spatiotemporal domain. However, it seems like the estimated complexity depends on the level of description used in the study and on the methodology to estimate it.

3.5. Anti-Epileptic Therapies

The examination of the efficacy in medications and other treatments to ameliorate epileptic phenomena provides examples of the sometimes very robust stability of these conditions, which is taken to be a consequence of the robustness of synchronous phenomena in neural activity. When addressing the theme of anti-epileptic treatments, it is worth considering that most of these will be efficacious at reducing the frequency of seizures, which does not mean the epileptic syndrome has been cured. Recall the words in the introduction to this chapter: epilepsy is a condition characterized by the presence of seizures. Reducing or halting seizures may not resolve the underlying neurophysiological abnormality. Having said this, however, stopping seizure occurrence will be undoubtedly beneficial because these represent one of the most disabling manifestations of the epileptic condition; it may not cure the syndrome but it will make the patient's life easier.

3.5.1. Anti-epileptic medication

While surgical resection of the epileptic focus is the most likely to effect a complete cure, any of the many available antiepileptic medication drugs will control seizures in temporal lobe epilepsy, with none of the drugs any more likely to be effective than any other in a given patient, despite their widely different mechanisms of action (cation channel blockers, GABA reuptake inhibitors, GABA agonists, NMDA antagonists, etc.). The reasons for the cases that remain intractable to medication are not known (Löscher, 1998), however, some ideas have been advanced. One hypothesis to explain the resistance to medication is the multidrug transporter hypothesis, proposing that the brain uptake of drugs from the circulation is diminished. Another attempt at explanation is the target hypothesis, claiming that acquired or intrinsic changes in the target of the drugs (ion channels, neurotransmitter receptors, etc.) prevent medications from acting (Löscher *et al.*, 2006). In spite of the efforts of combinatorial chemistry and high-throughput screening in the design of new antiepileptic drugs (Weaver, 2006), there still remain $\sim$30% of patients with pharmaco-resistant seizures (Regesta and Tanganelli, 1999). To be fair, there is still efficient control of seizures in the majority of patients ($\sim$70%), observation that prompts the question as to why there is such a success; this is the complementary question to the one most frequently asked: why some are intractable. The answer to the former question is framed in the dynamical system framework presented here and in Chapter 2, and starts to be revealed by the term that is used for almost all drugs used as anticonvulsants: "wide-spectrum" antiepileptic compounds. Before presenting an answer to this query at the end of this section, let us briefly look at the actions of the anticonvulsants. As a sample, a summary of the most commonly used antiepileptic drugs and their mechanisms is presented below (details in MacDonald and McLean (1986) and Löscher *et al.* (2006)). Aspects that deserve attention are some commonalities in the action of these drugs, and the numerous mechanisms of action of each drug.

Benzodiazepines, the modulators of inhibitory $GABA_A$ receptors, are very useful antiepileptics that act by promoting inhibitory actions mediated by the $GABA_A$ receptor (Raabe and Gumnit, 1977), decreasing sodium and potassium currents (Schwarz and Spielmann, 1983), and blocking presynaptic calcium uptake (Leslie *et al.*, 1980). Barbiturates like phenobarbital, also $GABA_A$ modulators and ineffective against absence seizures, enhance $GABA_A$-mediated inhibition and decrease presynaptic calcium entry (Seeman, 1972). Carbamazepine, ineffective against absence seizures (indeed it enhances these), reduces sodium and potassium currents, inhibits action potential discharges (Hershkowitz *et al.*, 1978) and adenosine uptake into cells (Skerritt *et al.*, 1982). Ethosuximide, very effective against absence epilepsy, blocks the T-type calcium current, sodium channels, and interacts with the sodium–potassium ATPase. Phenytoin has multiple actions as well, reducing excitatory responses, calcium entry into cells, sodium conductances, and enhances sodium extrusion and GABA-mediated inhibition (Mclean and MacDonald, 1983). Vigabatrin, ineffective against absence epilepsy, increases the levels of the inhibitory neurotransmitter GABA and inhibits GABA-transaminase, hence resulting in net potentiation of inhibitory transmission. Valproic acid, effective in the treatment of absence and generalized tonic–clonic seizures (Bruni and Wilder, 1979), increases GABA levels due to interactions with enzymes involved in GABA metabolism, diminishes repetitive spike firing and enhances GABA actions in general. Gabapentin increases GABA turnover and acts on calcium channels, increasing inhibition as the net result.

Two general principles are apparent from the previous list: each drug has a variety of actions, and enhancing inhibitory neurotransmission is beneficial in many epilepsies, except for the case of absence. The reason for this apparent paradoxical effect in absence epilepsy was addressed in the explanation of the thalamocortical network activity in Chapter 1 (end of Section 1.6). To recapitulate, inhibition becomes excitation in the thalamocortical network depending on the state of the thalamic neurons: if too hyperpolarized, a synaptic input will result in a rebound response with a relatively long burst of spikes and the possibility of developing a rhythmic behavior that entrains many cells, being maintained by the thalamocortical loop. In dynamical words, as discussed in Section 3.4.4.1, where the system's activity lies in state space will determine the outcome of the dynamics. As presented in Chapter 1, the thalamocortical system is a very clear illustration of the interplay between intrinsic biophysical properties of the constituent cells and their network connectivity in the generation of patterns of activity.

The efficacy of these medications with such a variety of actions suggests that being too specific may not be the best strategy to treat epilepsy. It is conceivable that the aforementioned drugs are efficacious due to their wide-spectrum actions, rather than operating via one specific action. A possible answer to the question posed above of why there is such a success in a majority of patients ($\sim70\%$) rests in the many actions of the drugs, which, as a result, alter significantly the brain state space dynamics: the more parameters are perturbed, the likely to change the dynamics. Other treatments that are considered in the next sections seem to operate, too, in a very nonspecific manner, like vagal nerve or deep brain stimulation (DBS). To close this section with even a stronger argument, let us comment on, perhaps, the most unspecific method in the treatment of epilepsies: the ketogenic diet. This is a high-fat, low-protein, and low-carbohydrate diet used in the treatment of refractory epilepsy since the 1920s, based on observations that starvation was effective for reducing the number of seizures (Geylin, 1921). It is very efficacious in children, with about 50% of patients (of that $\sim30\%$ population whose seizures are drug-resistant) experiencing reduction in seizure frequency, and 7–16% of children become seizure-free (Lefevre and Aronson, 2000). The mechanisms of action remain speculative, and most likely will remain like that in the foreseeable future, in spite of the attempts at finding "specific" effects of the diet based on actions of pH shifts or of ketone bodies such as acetone and related compounds. Because the metabolic consequences of the ketogenic diet are myriad, this variety of actions will most likely change considerably the state space which results, in the end, in less probability to generate paroxysms. In this sense, it can be conceived as another "wide spectrum" antiepileptic medication as those aforementioned, but vastly "wide" in this case.

The comments in this section revolve around a main theme of this monograph, that the dynamical features of nervous systems are not the sum of the individual properties of the constituents, but the result of the collective behavior. For this reason, there will be higher probabilities of a nonspecific approach for the efficacy to the treatment of these abnormalities. In the next sections and chapters, "nonspecificity" will be encountered numerous times.

3.5.2. Alternative strategies to control epileptiform activity: Electrical perturbations

The somewhat limited success in the pharmacological treatment of epileptic syndromes has aroused an increasing interest in the possibility of stopping seizures by brief direct electrical stimulation of specific brain areas. The idea substantiating the possible success of electrical perturbations at preventing seizures, or, if not stopping them at least shortening the ictal events, is based on the assumption that if the dynamics of the presumed abnormal synchrony that characterizes these paroxysms is perturbed by stimulations, then the ictus may not occur at all, or forced to stop if already initiated. If electrical engineers are routinely controlling the activity of electronic circuits, why not try to control the electrical brain activity. After all, most of the theoretical concepts used by engineers for this purpose have solid foundations. Considering the cellular brain networks as extensions of electronic circuits, gives hope that similar perturbations can be applied to brain activity to alter it at our own wish. Based on the notions of dynamic system theory related in previous sections, the main idea that has attracted the attention of quite a few investigators is easy to grasp: it is conceivable that once the dynamics of the transition to seizure are well characterized, then well-timed and properly-placed perturbations can alter the dynamics. For instance, if multistability or/and metastability is considered as representative of the brain dynamics, can the perturbations avoid the visit of the dynamics to some undesirable (epileptic) "attractors"? Or, if specific bifurcations in the route to seizures may be taking place, as discussed in previous sections, can the path toward these phase transitions be avoided?

The core of the concept is to use the internal brain dynamics, once characterized, to implement the control. But, how sure can one be of using the system's own dynamics? Let us imagine for a moment there were successful stimulation paradigms through which brain activity can be altered at will. It may still be difficult to ascertain whether the stimulation is using the system's own dynamics, for instance changing a possible attractor manifold, or just merely pacing the activity which is an altogether different story. Of course, this may be irrelevant for those interested only in the control of the activity, the clinical application, so who cares whether it is being forced (paced) or not. But here we are focusing on basic science enquiries, as these illustrate the possible subtle transition from one to another dynamical regime. However, there may be arguments related to this "subtlety," at least when talking formally from the dynamical system's perspective, for, as mentioned in Section 3.4.4.2, bifurcations may lead to a soft or a hard loss of stability, and in the latter case the transition is not subtle from the dynamical perspective because it involves a drastic change in state space. On the other hand, from the molecular, or microscopic perspective, the changes may be very subtle; and, from the patient's (and relatives) perspective, the alteration in behavior is quite dramatic (from having seizures to not having). This aspect of subtlety, then, seems to be relative to the viewpoint taken.

Considering the advances in the perturbation of cardiac dynamics that has led to the control of cardiac arrhythmias, the optimists may proclaim that it is quite probable to alter brain dynamics too. In this regard, neuroscience can benefit from these investigations on the control of cardiac activity. While cardiac dynamics may be simpler than that of brains, the successful control of some cardiac arrhythmias (Hall *et al.*, 1997; Christini *et al.*, 2001) as shown in Figure 3.9 (A and B), using theoretical frameworks of dynamical system theory is encouraging for neuroscience. To be fair, other attempts were not too successful (Gauthier *et al.* (2002); those interested should take a look at the focus issue of the journal Chaos on mapping and control of cardiac activity: Christini and Glass (2002)). Nevertheless, in systems endowed with complex dynamics, variability is the rule. Indeed, we will see that variability is a common circumstance that emerges from the studies of seizure prediction and control.

Related to the idea of seizure control is that of seizure prediction. Building on the concepts of subtle transitions between brain states and the implementation of small perturbations to control the brain's dynamics, the general idea, the "dynamical dream" (Perez Velazquez and Wennberg, 2004) is to stop the transition to the ictal event by a precisely timed brief stimulation of a certain brain area. Hence, rather than stimulating the brain for long times, what is sought is the anticipation of the seizure within a reasonable (short) time window such that it would allow a possible short, and hopefully subtle, perturbation to stop the transition to seizure. Scores of studies and papers have appeared in the past two decades addressing these two related matters of anticipation and control of seizures. This monograph is not the place to deeply enter into these matters, as they have been so much addressed, but just some few brief comments are in order, at least to introduce readers to an abundant and exuberant literature that relentlessly produces new results and controversies.

3.5.2.1. *On seizure prediction and control*

The field of seizure prediction has had a long history and recently it has been investigated with an unparalleled enthusiasm within the field of the dynamics of epilepsy, to the extent that a biannual meeting on seizure prediction is periodically organized (https://epilepsy.uni-freiburg.de/). Nevertheless, already in the 1970s, McDonnell Douglas Astronautics Co. funded projects on predictability of grand-mal seizures from EEG recordings (Viglione and Walsh, 1975). Projects were abandoned due to *variable outcomes across patients*; a first indication of what is (was) to come. As mentioned a few paragraphs above, variability will be the rule, rather than the exception, in all these experiments. Currently, a host of seizure prediction algorithms exist (probably more than 30, see Figure 1 in Le van Quyen *et al.* (2001), for the state of affairs around that time). Some of these have been reviewed in Iasemidis and Sackellares (1996), Litt and Lehnertz (2002), Winterhalder *et al.* (2003), and Ebersole (2005). These methods all seem to work to some extent, the prediction time windows ranging from a few minutes (Jerger *et al.*, 2001; Navarro *et al.*, 2002) to hours (Litt *et al.*, 2001). These results are based on a variety of variables: from variation in synchrony patterns (Mormann *et al.*, 2003; Le van Quyen *et al.*, 2005) to complexity measures and correlation integrals (Lerner, 1996; Elger and Lehnertz, 1998). Less optimistic studies include quite a few, like Maiwald *et al.* (2004), McSharry *et al.* (2003), and the exchange of correspondence between De Clercq *et al.* (2003) and Le van Quyen *et al.* (2003a). Negative results for specific methods have been reported in Aschenbrenner-Scheibe *et al.* (2003), Harrison *et al.* (2005), and Jouny *et al.* (2005) to cite a few. However, the failure to anticipate seizures is related to the generality of the validity of the different methods for all epileptic patients; otherwise, all these seizure-prediction (or "anticipation," as there is currently some discussion of the semantics regarding these terms, which we shall not discuss further, see Le van Quyen *et al.* (2003b)) algorithms succeed to some extent and only in some cases. It has even been proposed that one does not need to go "nonlinear," as a classical analysis of variance could account for the findings in previous correlation dimension studies in seizure anticipation (McSharry *et al.*, 2003). The question is, should one expect to find a general seizure-prediction rule that would anticipate seizure occurrence in the immense majority of epileptic cases? Our contention is that, most likely, not. As an illustration, let us look at one variable that was thought could be of value for seizure prediction: possible changes in synchronization before the ictal events. In fact, synchrony does seem to change as the ictus approaches, but it is patient-specific, as reported by Aarabi *et al.* (2008): the synchrony increases in 63% of seizures during a presumed "pre-ictal" state and decreases in 31% of the cases studied. Perez Velazquez *et al.* (2007c, 2011) described, based on MEG recordings, that phase synchrony decreases before (and after) seizures in restricted cortical areas, so depending on what MEG sensors were taken to be analyzed, a change would be appreciable or not. Schevon *et al.* (2007) found that the variation in synchrony was unique to each patient. Finally, Winterhalder *et al.* (2006) reported that 50% of the patients showed a trend, either increase or decrease in synchrony, and these authors judiciously recommend that "the prediction methods [...] have to be determined for each patient individually."

The astute reader will have noticed that, in all these discussions on seizure prediction something seems to be taken for granted: the predictability of ictal events.
But, are seizures really predictable? Or, in the words of Sonya Bahar "Are we
seeking to predict the unpredictable?" (Bahar, 2006). For, within the framework of
brain multistable/metastable phenomena discussed in previous sections, should a
fundamental role of noise be expected? And the result of noise-induced transitions,
would not it lead to unpredictability? Some investigators seem to think along these
lines, like Suffczynsky *et al.* (2004) who stated: "since random fluctuations in control parameters and/or dynamic variables can lead to [...] paroxysmal activity, this
implies that [...] this type of phenomena is *unpredictable* per definition." In addition, the presence of bifurcations and critical states in brain dynamics discussed
in the previous sections and Chapter 2, furnishes another reservation regarding
the possible predictability, since large-scale patterns of activity evolve from series
of interactions at smaller scales, a build-up of large-scale correlations leading to
crises (Sornette, 2004). As if this was not enough, the concept of noise-induced
transitions between "attractors" (recall Figure 3.6) does not make any easier the
problem of anticipating seizures. Investigations that may shed light onto this problem are those that have been undertaken to determine the temporal distribution of
ictal events. Hence, indications that seizure distribution obeys a Poisson process
have been obtained (Milton *et al.*, 1987), even though two other reports obtained
evidence for a nonrandom distribution (Balish *et al.*, 1991; Iasemidis *et al.*, 1994).
These different results could be attributed to the distinct patient populations used
in these three studies, or to different statistical methods. A Poisson process means,
basically, that the events occur independently, at random, distributed according to
the Poisson distribution (Koralov and Sinai, 2007). If this were the rule underlying seizure occurrence, there would be little predictability. Another study presented
evidence for seizures appearing in clusters in about half of patients monitored (Haut
et al., 1999), which may be interpreted along the lines of the transient stabilization of a metastable equilibrium point, discussed above. Like earthquakes, seizures
could have probability distributions with some statistical regularities, but the individual events may be unpredictable. To end this discussion about predictability of
paroxysms, just a brief mention that other studies have talked about the interplay
between randomness and determinism in brain activity (Lopes da Silva *et al.*, 2000,
2003; Andrzejak *et al.*, 2001), hence giving some hope (thanks to the "deterministic" side) for some success. Perchance, this underlies the aforementioned relative
success of the prediction algorithms in specific epileptic conditions.

If seizures are hard to predict, this poses a problem to the main approach to
seizure control by direct electrical stimulation of the nervous system. Recall that
the idea is to use a minimal, short and small perturbation so that the transition to

seizure is aborted or the ictus is halted. Indeed, some *in vitro* seizure-like models have been used to examine some stimulation strategies with more or less success, but in these systems the seizure-like activity is, normally, easily anticipated. If the situation *in vivo* is such that ictal events are not predictable, this represents an obstacle to try to replicate *in vitro* observations. The first solution that comes to mind, if paroxysms cannot be reliable anticipated, is to try relatively long periods of stimulation at regular, or irregular, intervals. This was the strategy used in the initial efforts to reduce the number of seizures in patients, and it is, currently, an effective strategy at least in some cases that will be discussed in the next section.

Experimental efforts to alter epileptiform activity by direct electrical stimulation date back to the mid-1950s, when it was shown that cerebellar stimulation shortened electroshock-induced seizures (Cooke and Snider, 1955). This led to the implantation of cerebellar stimulators in epilepsy patients, and some promising results were reported (Cooper, 1973). Continuing with the early series of successful perturbations performed on epileptic patients, we find a very illuminating case, today almost forgotten, reported by Penfield and Jasper (1954), where an spike-and-wave discharge was suppressed both by electrical stimulation of the neocortex and by a *cognitive task* (concentrating on a problem, see Figure 3.12 (D)). In the late 1960s, as primeval versions of the current sought-after approaches, headsets with earphones that monitored EEG and triggered a loud noise in the contralateral ear were tried on patients, reporting success in just a number of them (Forster, 1977). Thus, a variety of nonspecific perturbations seem to halt, transiently, paroxysmal activity. After these early periods, a wide variety of methods have been assayed *in vivo* and *in vitro*, including psychological methods (Schmid-Schönbein, 1998), proportional feedback (Osorio *et al.*, 2001; Gluckman *et al.*, 2001), chaos control techniques (Schiff *et al.*, 1994; Slutzky *et al.*, 2003), high-frequency stimulation (Lee *et al.*, 2005; Osorio *et al.*, 2007), low-frequency stimulation (Weiss *et al.*, 1995; Tergau *et al.*, 1999; Velišek *et al.*, 2002; Benini *et al.*, 2003; Khosravani *et al.*, 2003; Carrington *et al.*, 2007; Ozen and Teskey, 2009), and DBS in the centromedian thalamic nucleus (Velasco *et al.*, 1993). Equally diverse has been the area of the nervous system that is stimulated, from peripheral sites such as the trigeminal (Fanselow *et al.*, 2000; DeGiorgio *et al.*, 2006) or vagus nerve (Henry, 2002) to DBS of central sites such as caudate or thalamic nuclei (Velasco *et al.* (1993); reviewed in Parrent and Almeida (2006)). An account of the attempts at controlling epileptic activity using direct electrical perturbations can be seen in Durand and Bikson (2001).

Because the dynamics of the *in vivo* system are conceivably more complex than that found *in vitro*, it could be argued that *in vitro* seizure-like models are more amenable to perturbations. These systems rely on brain slices that contain a reduced network and thus the dynamical regimes in these "pieces of the mind" are thought to be simpler than the complex dynamics of the intact brain. Many of these *in vitro* models have spontaneous paroxysmal activity that resembles those found *in vivo*, and, therefore, they have been used as a playground to assess some perturbation methods to alter the spontaneous activity. Accordingly, reports have shown the control of interictal and ictal-like events in the high-potassium model (Schiff *et al.*, 1994; Jerger and Schiff, 1995; Slutzky *et al.*, 2003), anodic electric fields suppression of interictal-like bursting (Gluckman *et al.*, 1996), and low-frequency stimulation reducing seizure probability (Barbarosi and Avoli, 1997; Benini *et al.*, 2003; Khosravani *et al.*, 2003). That the dynamics of a brain slice *in vitro* can be relatively simple is observed in some of these seizure-like models. For instance, hippocampal slices submerged in a solution containing high potassium and low magnesium seem to exhibit, basically, two dynamical modes: either a rhythmic bursting pattern (which can be called interictal activity, and hopefully clinicians will not be greatly offended by our liberty in using this clinical term) or another mode in which the interictal-like activity leads to recurrent ictal-like events. The former, in the dynamical terms developed in Chapter 2, can be considered a limit cycle. The dynamical depiction of this scheme was shown in Figure 3.9 (C). This description is supported by the observation that, in one study, about half (54%) of slices showed spontaneous interictal activity but did not develop episodes of recurrent seizures, while the other half exhibited both interictal and recurrent ictal events (Khosravani *et al.*, 2003). Thus, it could be hypothesized that the seizure-like event in these slices is composed of a set of unstable limit cycles, which can temporally evolve, due to changes in control parameters, returning to a more stable limit cycle (or torus if synchrony is considered) representing the interictal activity. Moreover, brief periodic electrical stimuli applied during the preictal period immediately before the seizure-like event were capable of avoiding seizure manifestation in some slices, and resulted in the continuation of the rhythmic interictal bursting, as if the periodic stimuli was able to stabilize the interictal dynamics, supporting evidence that the interictal state may correspond to a stable "limit cycle" in this *in vitro* model.

The results of one of the pioneering studies that used chaos control techniques to alter paroxysmal activity is depicted in Figure 3.11, which is of interest as these methods rely on subtle perturbations of the activity, rather than forceful electroconvulsive-like therapy. The figure also illustrates the usefulness of the aforementioned first-return plots. The appeal in using these methods lies in the fact that, as opposed to model-based feedback control techniques originally developed by Norbert Wiener and others (Wiener, 1961), which require a detailed analytical model, chaos control methods are model-independent. That is, the dynamics of the system is used to develop means to direct the system's activity toward a desired

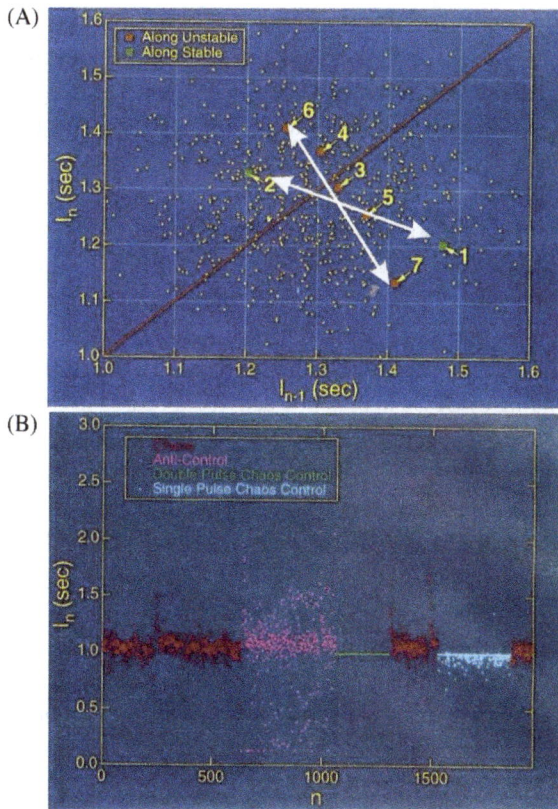

Figure 3.11. Representation of the method used to alter interictal-like bursting in a hippocampal slice using chaos control methods. (A), first-return plot of interburst intervals: the time interval I(n) is plotted against the previous I($n - 1$) taken from the spontaneous bursting of the hippocampal slice submerged in a high-potassium (8.5 mM) solution to make it hyperexcitable. Seven sequential points are shown, note how points 3–7 progressively diverge from the diagonal (the identity line where I(n) = I($n + 1$), that is, where fixed points are located which are periodic states because one interval is equal to the next), while 1 and 2 seem to be converging. This is taken as indication of the existence of a stable (1–2) and an unstable (3–7) manifold, shown in white arrows. These manifolds intersect somewhere in the diagonal, the intersection point representing the steady state which in these cases is an unstable saddle point. The chaos control method implemented consists of identifying the unstable fixed points and the stable and unstable manifolds and, when the system approaches the fixed point along the stable manifold, a perturbation is applied (a direct electrical stimulation to the slice), so that the next interval is placed back onto the stable manifold. In this way, a minimal intervention forces the system to approach the fixed points and thus become periodic, as shown in panel (B). In this graph, red points represent the interburst intervals without control, green and blue with chaos control, note how "tight" the distribution of intervals becomes indicating periodic activity. But this method can also be used to place the activity onto an area off the manifolds, in which case aperiodic activity follows as the system never reaches the (periodic) fixed points; this is represented by the pink dots (anti-control): the periodicity has been reduced compared with the unperturbed (chaos, red dots) activity. For more details, see the original article (adapted by permission from Macmillan Publishers Ltd.: *Nature*, Schiff *et al.*, 1994).

state by acting on a variable that is measured. Thus, the lack of the requirement for analytical models has obvious advantages for biological-inspired work, recall the comments in Chapter 2 about the difficulties in extracting equations from biological time series. The chaos control techniques used to alter brain activity are inspired in the classical work of Ott and colleagues that used small perturbations to control the system's activity (Ott *et al.*, 1990; Pyragas, 1992; Shinbrot *et al.*, 1993). These methods do require chaotic dynamics, because in nonchaotic activity a large effect requires a large cause (which means the requirement of strong control stimuli in these cases), while in chaos small causes lead to very large effects. The strategy is based on the stabilization of UPOs that are embedded in the chaotic attractor. Recall that a chaotic attractor contains an infinite number of UPOs. Comments on the finding of UPOs in epileptic activity appeared in Section 3.4.4.2, and Figure 3.11 depicts one method to find them.

Chaos control methods normally operate using a delayed feedback. Theoretical results have demonstrated how synchronization processes can be controlled by nonlinear delayed feedback methods, and that the major effect of these perturbations on a large population of coupled oscillators is a desynchronizing one. Consequently, this method is advocated for possible therapies to treat neurological diseases characterized by higher than normal synchrony (Popovych *et al.*, 2008). Not only that, but also the perturbations to achieve desynchronization, at least in these theoretical studies, were found to be minimal in some conditions, which, as explained in the introduction to this section, is a major interest in this field, lest electroconvulsive therapy be applied.

These chaos control studies demonstrate that subtle, minimal perturbations can alter the nervous system's activity, at least in some situations *in vitro*. To our knowledge, these methods have not been used in patients; however, other stimulation methods that are not based in chaos control have been reported to stop seizures once started, using *in vivo* rodent seizure models (Colpan *et al.*, 2007; Osorio and Frei, 2009). Another interesting possibility that has been profusely commented is the analogous to phase resetting, a concept used in many computer simulations and theoretical studies. It also entails the use of a minimal stimulation to alter the phase of the activity. Of course, the activity in this case should be relatively periodic so that the phase can be determined; hence, absence seizures (Figure 3.1) are a good candidate to try these methods. Consider the absence seizure waveforms depicted in Figure 3.12, taken from intracerebral recordings in rats. In principle and according to theory, one accurately-timed stimulation should suffice to alter the phase or even stop the seizure. In bi-stable systems with a limit cycle and a fixed point, transitions between the two can be triggered by one single perturbation, and the modeling study of Suffczynski and colleagues (2004) demonstrates how to stop a

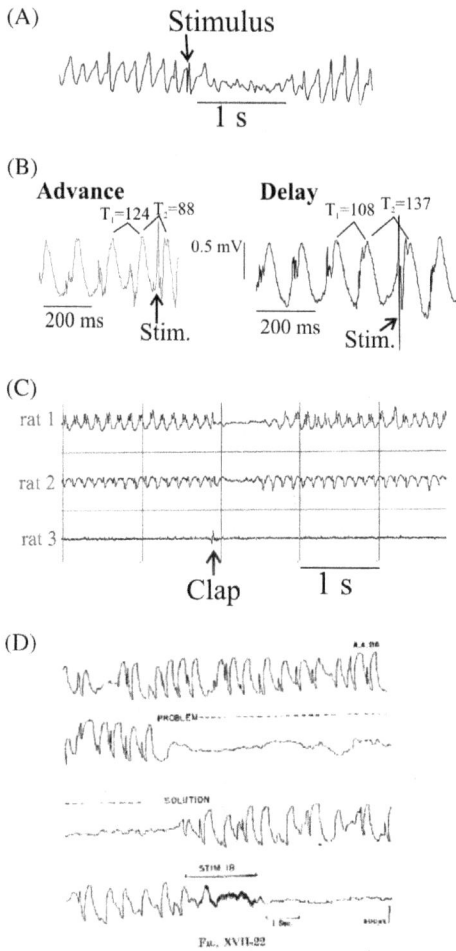

Figure 3.12. Stimuli and perturbations of *in vivo* seizures. (A), one single electrical stimulation in the cortex of a rat having absence seizures stops the spike-and-wave discharge very transiently (∼1 s). There was no noticeable phase preference of the pulse to observe this phenomenon. (B), stimuli delivered to the thalamus (or cortex) changes the phase of the spike-and-wave discharge in these rats. An advance and a delay of the phase by one single pulse (Stim.) are represented. T_1 and T_2 are the periods (in milliseconds) before and after the pulse, respectively. (C), simultaneous intracerebral recordings from three rats, two of which (rats 1 and 2) display absence-like seizures. Rat 3 did not, which was useful to detect the perturbation applied (arrow), in this case a hand clap by one of the investigators. Like occurred after the electrical stimulus in panel A, the seizures are halted transiently. (D), a seizure in a human patient is arrested briefly by a cognitive task (problem solving) or by electrical stimulation to a neocortical area (stim.). Panels A and B are reprinted with permission from Perez Velazquez *et al.* (2007b) and panel D from Penfield and Jasper (1954).

digital (simulated) paroxysm with a single pulse. However, at least in the rodent models of seizures investigated in Perez Velazquez *et al.* (2007b), halting a seizure could be accomplished in very few instances and only very transiently by one stimulation to the brain areas affected, regardless of the timing of the single shock, as shown in Figure 3.12 (A). On the other hand, the oscillating phase was consistently altered depending upon the phase of the perturbation (Figure 3.12 (B)), phenomenon that was used to generate phase resetting curves (Perez Velazquez *et al.*, 2007b). In general, this idea of phase resetting is very fine when playing with computer models, where the activity of the model network can be made as periodic as one wishes. The reason why this particular manipulation will be very hard to succeed in real life is that physiology is not a computer simulation. The method, as aforementioned, requires stable periodic activity, so that the phase may be extracted from the periods, a very uncommon and transient occurrence in normal brain activity. The almost periodic activity found in some seizures, such as absence, occurs too fast with frequencies changing continuously, so that any attempt to implement any sort of phase resetting will be unlikely to succeed, considering the time it would take to a computer to compute phases and stimulation times. Having said this, seizures can be halted briefly without the need for elaborate invasive intracerebral stimuli. Figure 3.12 (C) shows how one hand clap was enough to stop the spike-and-wave discharge in two rats for about 1 second, similar to the electrical stimuli in panel A. Compare these recordings with the classical result, already mentioned above, of Penfield and Jasper (1954) where a spike-and-wave discharge was suppressed both by electrical stimulation of the neocortex and by a cognitive task (Figure 3.12 (D)).

This transient halting of the seizure shown in the figure (panels A and C) for ∼1 second is certainly not very impressive, and probably indicates, along the lines of the discussion in Chapter 2, that the seizure "lives" in a relatively stable synchronization manifold, perhaps a torus that is the mathematical representation of synchrony as explained in the previous chapter. But then, how about trying the perturbation before the system gets trapped into the manifold? Could this be more successful in order to avoid the transition to seizure if the perturbation can alter substantially the dynamics in state space? Possibly the answer is already known. A phenomenon familiar to investigators using *in vivo* seizure models is that if the animal is handled too much, seizures will most probably vanish for a while (Figure 3.13). It is as simple as this: in our own experience, handling for just a couple of minutes rats that normally spend even up to ∼50% of the time having spike-and-wave discharges (absence seizure models described in Proulx *et al.* (2006)), will result in the rat having almost no seizures for about half an hour. The reason for this phenomenon probably includes the stress that rats undergo while being handled with the consequent activation of the brain stem reticular formation and the release of some "stress" hormones. The

Figure 3.13. A proven method to reduce number of seizures in animal models of epilepsy. Handle the animal, touch it, bother it for a few minutes, and seizure severity will diminish over the next few minutes to hours.

result of these cellular/molecular changes could be a considerable alteration of the attractor landscape in the state space of the brain's dynamics. This could also be the reason why some apparently very nonspecific stimuli can reduce the frequency of seizures, such as vagal nerve stimulation (VNS) and DBS, methods to which now we turn.

3.5.2.2. *Not-very-specific stimulation can alter epileptiform activity*

To devote a few words about these methods is fair, as these phenomena are related to the comments on "specificity" discussed in Chapter 2. It is also fair as there is a current enthusiasm in treatments of epileptic syndromes using direct electrical stimulation of the nervous system, exemplified in thalamic DBS or VNS, and recently trigeminal nerve stimulation (De Giorgio *et al.*, 2006). This enthusiasm is accompanied by a widespread acceptance that the mechanisms of action of these neurostimulation techniques are unknown. Nevertheless, these methods have been reported to be relatively successful at reducing seizures, both in animal *in vivo* models and in patients. This presumed success contrasts with the more limited achievement at controlling epileptiform activity *in vitro*, a rather intriguing fact considering that *in vitro* systems (brain slice preparations) are considered to represent a much simpler system than the whole intact brain, thus conceivable easier to control. Halting seizures or the transition to the seizure, using *in vitro* models where SLEs are observed (as opposed to interictal-like bursting like the example in Figure 3.11), has resulted in not much success, as some very strong direct electrical stimulations to hippocampal slices were ineffective, and only properly timed and placed electrical stimuli were able to arrest, sometimes ($\sim$65%), the transition to seizure (Khosravani *et al.*, 2003). If this observation about the proper timing has validity for *in vivo* seizures, then this raises the concern discussed above, in that we do need to know when the ictal event will appear within a certain time window, in order to impede its development. The recent observation in rats that single DC (direct current) pulses were effective at annihilating seizures that had already started, whereas AC (alternating current) stimuli were not as effective, points toward using this type of stimuli rather than the more usual alternating pulses (Osorio and Frei, 2009).

The typical VNS or DBS paradigm uses a regular, almost constant, electrical stimulation of the vagus nerve (in the former) or brain structures (in the latter). This manner of stimulation is far from the more desired subtle and short perturbation mentioned above, but the fact is that these long-lasting stimuli seem to be effective. As was mentioned in the previous section, if seizures are hard to anticipate, then stimulating at regular time intervals for long periods of time may work. It can be speculated that the reasons for the efficacy of those methods have something in common with the effects of handling of the rats on seizure frequency discussed at the end of the previous section, for the vagus nerve innervates, in addition to a number of muscles, a variety of cortical and subcortical brain structures.

Like VNS, DBS will alter the activity in many brain areas, perhaps in different manner depending on the location chosen for the stimulation. For example, in case of thalamic stimulation, the widely projecting thalamic nuclei can alter activity throughout the entire neocortex. For a discussion of the effects of DBS, the review by Kringelbach *et al.* (2007) is recommended. To be certain of the precise mechanisms of action of DBS is not a trivial matter because the electrodes will not only excite cells bodies in the neighborhood but also passing fiber tracts, axons from remote cells that travel along the electrode location will also be affected. Because some clinical effects of DBS are reproduced by lesioning brain tissue or by the application of inhibitory drugs, it is thought that DBS may primarily act by promoting inhibition. The important point for our discussion is that a general and relatively nonspecific alteration of brain activity results in the desired effect of reducing the number of seizures in patients. Furthermore, our own results with rats indicate that long periods of intracerebral stimulation either to the thalamus, amygdala, cortex, or hippocampus (at frequencies between 0.8 and 25 Hz) of rats experiencing kainate or tetanic-induced seizures, also decrease seizure frequency. It would not be surprising that the very strong perturbation in electroconvulsive therapy, a method with a long history (Kalinowsky and Kennedy, 1943), results in a substantial change in the dynamics and thus operates along the same mechanisms in alleviating psychiatric disorders that will be addressed in the next chapters.

But the observation that seems to be the most interesting in this field is that there is not even a need for stimulation, that is, no need to turn the stimulator on to reduce the number of seizures in patients! Seizure reduction was already observed in patients right after the surgery to implant the stimulator, before it was turned on (Hodaie *et al.*, 2002; Parrent and Almeida, 2006). Examination of a database of patients with implanted VNS devices revealed that the "benefits in seizure control with VNS in humans appear to have little specific to do with active stimulation of the vagus nerve" (Wennberg, 2004). Does this mean we are witnessing a placebo effect? The concept of neuroplasticity is globally accepted, but nobody said that the stimuli to cause this plasticity have to come from the outside. The celebrated placebo effect, while being a nuisance to pharmaceutical companies, is a very interesting phenomenon. In a following chapter, cognitive-based therapies will be discussed. The internal brain dynamics may be at work here self-altering its dynamics. The result in Figure 3.12 (D), and other psychological methods (Schmid-Schönbein, 1998), can be considered as cognitive-based approaches to stop seizures: the brain itself is perturbing its own dynamics, arresting the paroxysmal discharge while occupied in solving a problem. These findings and those of the well-documented biofeedback techniques raise the question of whether or not external perturbations, such as intracerebral electrical stimulation, are needed at all to control brain's activity. There is the substantial possibility that brain's activity can significantly change its own dynamics, so what is achieved by means of drugs could equally be accomplished, in some cases, from the inside, with self-instruction and motivation. Of course, these "cognitive" perturbations seem to be effective only in some cases, but after the discussions in this chapter it should be clear by now that there is nothing that will be effective in all possible cases, at least in contemporary science. Generalized 3-Hz spike-and-wave activity can be stopped by manipulations as simple as getting the patient to attend to something, but other seizures are not that controllable. Notwithstanding the financial interests of many companies involved in anti-epileptic medications and therapies, should not there be efforts to undertake alternative methods to complicated and invasive surgical procedures such as VNS or DBS, using simpler means such as "internal" perturbations? The aforementioned observations may supply a possible (cheap) solution to the problem of seizure anticipation and control, without the need for technology. Can patients be trained to stop the transition to seizure? Epileptic patients have auras, which are sensations that signal an imminent ictus. Thus, some patients can self-anticipate their seizures, which solves the problem of seizure prediction. At this time of the aura, the behavioral seizure has not started yet (some clinicians, though, consider the aura as part of the seizure), thus the patient is conscious and can, perhaps, try something to stop the incoming ictus, like concentrating on a problem as shown by Penfield and Jasper. Perhaps some patients may find unique actions that prevent their seizures from developing. We submit the idea that the control of brain activity, at least in terms of seizure control, can sometimes be achieved without the need of direct electrical, or any other type of intracerebral stimulation, because all needed may be already in the mind. To some extent, then, one moral of all these tales is that everything is in your mind... a hardly novel result!

What becomes apparent after reading the vast literature on seizure prediction and control is that all methods seem to work to some extent. However, the variability in the dynamics characteristic of a complex system precludes the finding of one technique of wide, general application. There is no reason why specific methods could not be tailored for each epileptic patient although it is acknowledged that the individual approach would require considerable time and effort from the clinical teams. A *patient-individual approach* to seizure prediction and control, we believe, has high probabilities to succeed. Indeed, classifiers derived from patient-specific machine-learning algorithms discriminated preictal from interictal EEG patterns and predicted seizures without false alarms in some patients (Mirowski *et al.*, 2009), hence indicating the feasibility of patient-specific approaches and the value of using machine-learning methods in this field. And, as a final recommendation: the more nonspecific the treatment, the more likely it is to relieve the patient from the burden of continuous seizures. At least according to what seems to be working in this day and age.

3.6.   Final Remarks on the Transition from What is Considered Normal Brain Dynamics to Epileptic Activity

This chapter is closed with final reflections derived from the evidence discussed here, accentuating the perspective from dynamic system theory. Most brains are not epileptic. However, it is not too difficult to make one so (Lopes da Silva *et al.*, 1994). The demonstrations that brain areas made "epileptic" with convulsive agents are able to make other (nontreated) regions "epileptic," as well as some subtle changes in coordinated activity before, during and after paroxysms in cellular networks that are seizing and in others that are not, draw attention to a main central theme: that it is not difficult to make a brain area seize provided that the interconnected networks can spread hyperexcitability by means of enhanced synchrony. Typically, in the nervous system, excitability brings out synchrony, and vice versa. Seizures are "cured" by reducing excitability or network synchrony. This is what most of the therapies here discussed, either of chemical, physical, or psychological nature, achieve.

It can hardly be overemphasized that the brain dynamics unfolds continually, and in particular the transition from interictal to ictal activity is continuous and ever changing, which explains the relative failure of seizure-anticipation algorithms that result in false-positives (to be expected as the algorithm tries to make sense of the variable and continuous alteration of the dynamics). On the other hand, this dynamical aspect can provide occasions for a variety of stimulation paradigms, external or internal, to stop or reduce seizure frequency. Because of the inherent variability, these methods should be tailored for each epileptic patient. As a corollary, our recommendation to the clinician is the implementation of patient-individual prediction and control approaches. The continuous spatiotemporal unfolding of the nervous system dynamics should preclude dichotomies common in this field, centered on attempts to define precise separations among interictal, preictal, ictal, and postictal periods. At different levels of description, one activity pattern becomes another in ever-evolving loops that can make the brain switch from normal to pathological function.

This chapter has presented some suggested dynamical regimes in epileptiform activity, but while the investigation of these specifics is fun (at least for the authors), a more comprehensive and intuitive understanding will benefit some readers not interested in details, lest they get stuck into technicalities. Thus, the global view that emerges about brain dynamics and epilepsy in particular, along the arguments presented in Chapter 2 and the comments in this chapter on intermittency, Šil'nikov-type dynamics and UPOs, is that the dynamics is best represented by transients evolving in the neighborhoods of stable steady states but never settling onto these. The dynamics of the epileptic brain display a greater tendency to fall into an attractor-like state and thus exhibit more stability in some steady states than what would be desired to avoid pathological outcomes. This dynamic feature may be shared by brains in other neuropsychiatric disorders, the variety of which can be possibly due to the distinct brain regions involved with different dynamics, whereas the collective behavior at the dynamic level is similar: a greater than normal tendency to fall into an attractor manifold; of peculiar interest, a synchronization manifold.

Within the framework of metastability, the initiating perturbations that result in epileptic seizures may be diverse, but the collective mechanisms underlying the manifestation of these events seem to be similar. Characteristic of dissipative systems is that large number of constituents give rise to coherent behavior. The microscopic routes to seizures may be varied, but the collective result is the same (Perez Velazquez and Wennberg, 2004), a reflection, perhaps, of the extraordinary tendency of brain cellular networks to synchronize their activities. Equally unsurprising is the fact that almost any perturbation to the brain, internal or external, alters epileptiform activity, consequence of the changes in the dynamical landscape. Indeed, the contrary would be a revelation, as it would indicate the presence of very stable and robust attractors in brain activity, something that nature does not seem to favor. It is tempting to speculate that the "success" of the aforementioned experiments using DBS, VNS, or any other treatment for that matter, results from the forcing of potentially many stable states present in brain dynamics, which do not result in the transition to the ictus. If all these considerations could be encapsulated in a sentence, this could be, perhaps, that the dynamical repertoire of cells and networks is constrained by the range of microscopic interactions that determine the final dynamical evolution of the nervous system. An effective control of the dynamics starts by knowing how to alter the interactions among the system's constituents.

CHAPTER 4

PARKINSON'S DISEASE: TRANSITION BETWEEN SENSORIMOTOR COORDINATION PATTERNS

Like epilepsy, Parkinson's disease (PD) has been considered a dynamical disease (Beuter and Vasilakos, 1995). The motor dysfunctions present in PD, and perhaps also other deviations of motor coordination in Huntington's disease and tardive dyskinesia, result from the alterations in the functional organization of brain and motor networks, thus providing another illustration of how changes in organized patterns of activity lead to deviations of normal, healthy, and adaptable function. The apparent regularity and periodic activities characteristics of tremors in these diseases should attract attention of those interested in the dynamical disease notion. Other, quite regular too, abnormal patters of organized brain activity (in some seizures such as absences) were considered in Chapter 3 but these were short-lived in general; on the other hand, the long-lasting and robust tremor oscillation in these motor dysfunctions makes it a fertile ground to apply and explore frameworks of dynamic system theory. For example, phase resetting methods, which need relatively stable periodic activity, can conceivably be evaluated for their usefulness in halting or at least perturbing the pathological tremor rhythmic activity; Section 4.3 presents past and current efforts at perturbing these abnormal rhythms with the purpose to stop them. Furthermore, inspecting the causes and mechanisms of PD enhances the understanding of the fine interaction of neuronal activity with muscle tissue that produces the variety of normal motor patterns and emphasizes the close relation among different systems in the body. As will be seen, while the initial insult to the brain that results in PD constitutes a strong perturbation of the normal dynamic pattern in sensorimotor activity, the consequent transitions between different oscillatory activities reflect more subtle perturbations of the dynamics, all these phenomena deriving from changes in the functional organization of sensorimotor networks.

**4.1. Clinical Features and Neurobiological Aspects of Altered Movement
Coordination in PD**

PD is a neurological disorder that has a wide spectrum of symptoms (Agid, 1991),
the most notorious are not only the dysfunctions in movement coordination, but also
including cognitive abnormalities such as dementia. Tremor, bradykinesia (slowed
movement), and rigidity are the three characteristics commonly associated with the
motor disorder in PD. Among many symptoms, PD patients are impaired at the
initiation of voluntary movements and present a slowing of reaction times. Tremor,
in particular, has been studied in great detail from experimental and theoretical
perspectives. It is an involuntary, almost periodic (as close as periodic as can be
seen in physiology) oscillation of a limb or any other part of the body that becomes
a disabling feature in PD and other syndromes, for physiologic tremor is present in
all individuals but to a degree that does not make it a disturbing factor in movement
coordination. This chapter focuses on motor control deviations specifically in PD
because it has been so much studied at the dynamic and network level compared
with other motor coordination disorders, of which chorea (sudden activity bursts in
muscles) and myoclonus (intermittent muscle jerking) are two examples. Much like
in the case of epilepsy, the molecular and the resulting network alterations in PD
provide an illustration of the dynamical transition from the so-considered normal
behavior to a divergence, or pathology as we normally term these deviations.

The molecular basis of PD is relatively well known: it has its origins in the
degeneration of the dopaminergic neurons of the substantia nigra, which causes
a dysfunction in the areas to where these cells project, and, importantly in PD,
is the dysfunction in the basal ganglia circuitries due to the lack of dopaminergic
inputs. Thus, we have a relatively simple trigger, the loss of the neurotransmitter
dopamine, leading to a variety of symptoms, of which the motor disorders in PD are
but one example. Along the lines of the dynamical concepts introduced in Chapter 2,
dopamine can be considered a (very important) control parameter in brain dynamics.
For this reason, Section 4.2.2 is devoted to exploring several abnormalities in the
coordinated patterns of sensorimotor activity at the level of both brain and muscles
caused by alteration in dopaminergic neurotransmission, and an attempt is made to
place those deviations in the context of dynamical system theory.

A summarized scheme of the motor control brain networks is presented in
Figure 4.1, including basal ganglia, subthalamic nucleus (STN), thalamus, and
motor cortex as crucial components. It may be surprising that the cerebellum is not
included even though it is fundamental for motor coordination, however, it does
not seem to be a key factor in PD; therefore, exercising simplification demands the
consideration of only the most fundamental parts of the sensorimotor circuitry in

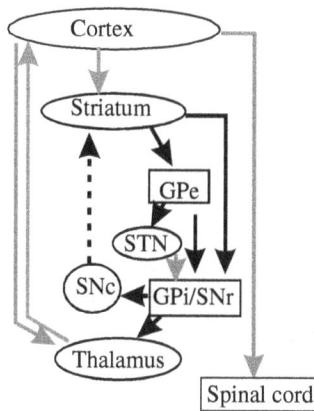

Figure 4.1. Simplified schematic diagram of the major cortico–striatal motor control loops, particularly of the basal ganglia. Inhibitory connections are shown as black and excitatory as gray, while the dopaminergic output from the substantia nigra is the dotted line. Loss of dopaminergic transmission in Parkinsonism due to degeneration of the substantia nigra leads to differential changes in two striato-pallidal pathways, which are associated with enhanced activity in the STN. This results in enhanced excitability in the internal globus pallidus (GPi) which in turn causes excessive inhibitory inputs to the thalamus. See text for more details. GPe, external segment of the globus pallidus; GPi, internal segment of the globus pallidus; STN, subthalamic nucleus; SNr, substantia nigra pars reticulata; and SNc, substantia nigra pars compacta.

PD ("The art of being wise is the art of knowing what to overlook," William James, in *The Principles of Psychology*).

The idea that brain oscillations dictate the tremor in patients has its origins, most probably, in the intracerebral recordings from PD patients performed by Guiot and colleagues in the early 1960s, where they reported that ventral thalamic cells discharge spikes rhythmically at the frequencies of the resting tremor in these patients. Currently, a host of more detailed observations has been gathered and the general scenario can be summarized as follows. Specifically in PD, after the loss of dopaminergic transmission in the basal ganglia, the net result is an increase in inhibitory (GABAergic) transmission to the thalamus from the GPi. The enhanced inhibition to the thalamic neurons will have distinct results depending upon the conditions in which the thalamic cells are: recalling the properties of thalamic cells described in Chapter 1, it is conceivable that if the thalamic neurons are relatively depolarized, then the inhibitory potentials will decrease their firing and thus there will be a net diminution of excitatory drive to the cortex; however, if the thalamic cells are relatively hyperpolarized, they may respond with a low-threshold rebound burst of spikes mediated by the T-type calcium conductance (described in Section 1.6 of Chapter 1), with the bursting pattern occurring at specific frequencies depending upon several factors (Figure 1.7 of Chapter 1 presented two examples), and now the net result will be an enhanced excitation of the cortex because of the strong input due to the many spikes in the synchronous thalamic bursts. This rhythmic excitation will force a particular oscillation in the cortex that may disturb normal motor control, as proposed by Volkmann *et al.* (1996) to explain the 3–6 Hz resting tremor in PD patients. According to this description, the T-type conductance could be important in this pathology. Indeed, the role of the thalamic low-threshold calcium conductance in the genesis of the resting tremor has been explored by Paré *et al.* (1990), particularly as a result of the firing patterns displayed by anterior ventrolateral (VLa) thalamic cells. As stressed in Chapter 1, the thalamocortical networks represent an excellent example for a variety of network phenomena arising from the combination of intrinsic cellular and synaptic properties. Perhaps here, as described in the aforementioned study, another example of emergent phenomena is found: Figure 4.2 depicts a frequency transformation by a VLa thalamic neuron. Note that whereas the stimulus, in this case hyperpolarizations introduced into the cell by an electrode, occurs at a frequency of 12 Hz, the cell bursts at ~4 Hz, due to the summation of the inhibition that is needed to deinactivate the T-type calcium current and generate a low-threshold rebound response. Here, then,

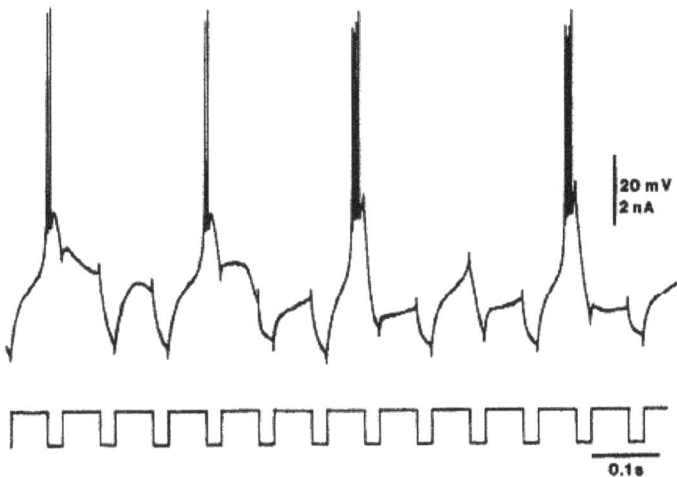

Figure 4.2. Responses of one thalamic neuron to rhythmic (12 Hz) hyperpolarizing pulses. The rebound burst firing occurs at ~4 Hz (upper trace), and the brief hyperpolarizing pulses injected into the cell can be seen in the bottom trace. Thus, the intrinsic biophysical properties of the cell result in an emergence of a different rhythm from the one that is imposed on it. This frequency transformation has been hypothesized to be of importance in the generation of tremor oscillations, as described in the text. Reproduced with permission from Paré *et al.* (1990).

is found the emergence of a different rhythm from the one that is imposed on it, this frequency transformation being due to the intrinsic biophysical membrane characteristics of the cell, a phenomenon which has the features of emergence according to a definition of emergent phenomena: the resultant activity is not directly imposed by the forcing stimuli (to be precise and punctilious, it can be argued that the stimuli in this case is not as nonspecific as, for instance, the temperature in the Bénard convection discussed in Section 2.3, Figure 2.8 of Chapter 2, because the hyperpolarization pulses are imposing, or forcing, an activation of the T-type channels). This frequency transformation is of significance when the following observation made in a primate model of PD is considered (this model achieved by causing neurodegeneration of dopaminergic cells using 1-methyl-4-phenyl-1,2,3,6-tetrahydropyridine (MPTP), a compound that is toxic to cells containing dopamine): neurons in the internal segment of the globus pallidus, that sends direct inhibitory input to the VLa, not only increase their excitability by firing spikes at higher rates but they also tend to fire in a bursting pattern at 12–15 Hz in the lesioned monkeys, rather than the tonic firing most commonly observed in non-Parkinsonian states. This observed enhanced excitability at the cellular level parallels what was found in epilepsy, particularly the study of Babb *et al.* (1987), discussed in Chapter 3, Section 3.4.1, a parallel that leads to envisaging a common principle in these diseases, that of an initial enhancement of excitability leading to synchronous activity. Thus, Paré *et al.* infer, the inhibitory input that VLa cells receive from the internal segment of the globus pallidus at 12–15 Hz may be transformed, due to similar biophysical mechanisms as observed in their experiment shown in Figure 4.2, into the slower frequency characteristic of the tremor at 3–6 Hz. This resultant thalamic activity is then passed onto the premotor and motor cortices that ultimately send the rhythmic activity to the muscles via the spinal cord. Underlying these phenomena here mentioned we find, again, the tendency toward maximal synchronization due to mass action: an increased excitability of neurons in the globus pallidus results in enhanced synchronous bursting in the thalamus. Further, in analogy to the cases of seizure therapies discussed in Chapter 3, it is thus conceivable that reducing the probability of synchronous activity will ameliorate the PD symptoms. It is not too surprising then that in experiments where Parkinsonian primates were treated with the T-type calcium channel blocker ethosuximide (mentioned in Section 3.5.1 of Chapter 3), that reduces the thalamic rebound bursting response, the tremor was reduced substantially (Gomez Mancilla *et al.*, 1992).

Can these brain oscillations related to pathological tremor be a natural consequence of what already is the normal, perhaps background mode of activity in motor cortices? The presence of different types of oscillatory activity in motor cortex is known, and of significance is a study that reported the association between the intermittent motor control of slow finger movements that occurs at 6–10 Hz with synchronous oscillatory activity in the motor system encompassing the cerebellar–thalamocortical loop (Gross *et al.*, 2002). In this regard, perhaps the notion proposed by W.W. Alberts (1972) several decades ago is of significance, that the resting tremor is a manifestation of underlying motor programs used in the production of voluntary movements. There is evidence that cortical areas develop a preferred oscillatory mode. For instance, some regions seem to respond with a preferential oscillatory activity upon direct stimulation: transcranial magnetic stimulation (TMS) tends to evoke alpha oscillations when applied to occipital cortex, beta rhythms when applied to parietal, and gamma/beta in the frontal cortex (Rosanova *et al.*, 2009); these experiments were done using TMS, so one can ponder whether the same occurs during the processing of normal sensory stimuli, and here we again take refuge in the notion that sensory stimulation is but a perturbation, or modulation, of the ongoing background brain activity. Nevertheless, the TMS experiment suggests that there could be a sort of fixed, or preferred, activity in neural nets ready to be exposed when properly stimulated, which supports the previous discussion on deviant cortical motor programs in PD, deviant in the sense that they manifest themselves at the wrong time or with inadequate intensity.

More specific details of the intricate basal ganglia circuitry involved in the disease are not of importance now, and interested readers are invited to consult other works, like the *Handbook of Tremor Disorders* (L.J. Findley and W.C. Koller, editors, Marcel Dekker, New York, 1995). Details appear in the following discussions, but now the aspect to be emphasized is the dynamic switching between the normal network dynamics for movement control with functional dopaminergic neurotransmission and the abnormal dynamics resulting from the alteration (lack) in this control parameter. It is dopamine depletion that results in the alteration of the interactions between cortical, striatal, and substantia nigral networks giving rise to abnormal functional coordination patterns, as discussed in Section 4.2.2 where dopamine is considered a major control parameter of brain function. Particularly, as dopaminergic transmission decreases, neurons in the substantia nigra become more excitable to cortical inputs and there is an enhancement of synchronous activities in the nigral and basal ganglia–cortical networks. This scenario should already sound familiar: an increase in excitability in a specific brain region, for whatever the specific reasons, results in enhanced tendency toward synchronization in connected networks. It is observed in epilepsy, as described in Chapter 3, and continue to appear in subsequent chapters dealing with phenomena as varied as normal sleep or schizophrenia. No doubt that other factors besides dopamine depletion contribute to the PD symptoms, but the major cause is the degeneration of dopaminergic cells that causes a significant change in brain dynamics and consequently in sensorimotor coordination dynamics. Thus, this initial insult to the brain cannot be considered too subtle. However, once the new dynamics develop, further transitions between coordinated activities may be the result of much subtler variations in a variety of factors. To better inspect these alterations in the sensorimotor functional organization, and now that some physiological details have been considered, it is time to simplify the anatomic and physiologic heterogeneity and focus on the high-level network viewpoint using dynamic system concepts.

4.2. The Dynamical Perspective of Sensorimotor Function and Dysfunction

To achieve a global comprehension of the motor dysfunctions in disease, it is advisable to look first at the coordination in the brain networks and associated muscles that lead to normal functioning. The first question that could be open to debate would be how to best describe the degree of coordination among the many and varied components of the sensorimotor system, from the central nervous system networks to the muscles. Taking the perspective of synergetics (Haken, 2002) may be favorable, as there is no need to introduce mechanistic interpretations of the transitions between movements such as in terms of motor programs fixed in brain networks; rather, at this level of description, the switching between one and another activity pattern arises from the intrinsic dynamics of the system characterized by metastable (or unstable) states, as discussed in previous chapters. Therefore, an adequate order parameter should be found that allow one to study the dynamics of these phenomena. This order parameter, introduced in the next section, is, once again, the relative phase.

4.2.1. Sensorimotor coordination dynamics

It has been proposed that the cooperative phenomena of sensorimotor coordination can be captured by the relative phase, that is, the phase differences between the almost-rhythmic limb movements that sometimes appear in particular tasks (Schöner *et al.*, 1986; Kelso, 1995; Bressler and Kelso, 2001). Thus, continuing with the dynamical systems vocabulary developed in Chapter 2, "phase difference" is used as a collective variable, or order parameter, to investigate the changing of the relationship between components of the system, in this case the motor components but, as already discussed in previous chapters, it can be useful too to assess relationships among neuronal populations, provided the phase can be extracted from neurophysiological signals. Thus, once again, we are using the approach emphasized throughout this book, by taking advantage of reducing a multidimensional system composed of myriads of neurons, muscles, and joints needed for sensorimotor coordination, to a one single order parameter that captures the dynamics. Consequently, dynamic system approaches have been employed to assess the coordination dynamics in sensorimotor activities, and specifically in relatively very simple motor actions: bimanual rhythmic coordination. These investigations date back to a few decades, perhaps the original studies (at least the published ones) were those of two groups, Yamanishi *et al.* (1980) studying the stability of finger tapping patterns, and those of Kelso and associates (Kelso, 1981, 1984, 1995; Haken *et al.*, 1985). It was originally noted that when two fingers (one in each hand) move rhythmically at the same frequency, there are only two modes that seem to be stable: the in-phase or antiphase modes, the former indicating when the two limbs are moving against each other and the latter when they move in the same direction. When this simple finger motion is paced by a metronome, as the frequency increases, a switch from the antiphase to the in-phase takes place at a certain frequency (depends on the individual but tends to occur between 2 and 4 Hz). Readers can easily experience this phenomenon by themselves without any metronome, just start moving rhythmically the index fingers in the antiphase (same direction, when one moves toward the right, the other does the same), first at low frequency (once a second) and then start increasing the frequency of the movement: the in-phase motion will appear after reaching ~3 Hz (again, it depends on the individual, check your own bifurcation value!) and remain stable for higher frequencies. Thus, the in-phase mode seems to be more stable than the antiphase. Using the dynamical system terminology, the metronome's frequency acts as a bifurcation parameter: it makes the system switch between two fixed points at a particular value (Kelso, 1984). Detailed accounts of this saga can be found in Kelso's monograph *Dynamic Patterns* (1995). These empirical observations led to a dynamical model known as the Haken–Kelso–Bunz (HKB) model, which accounts for the dynamics of the dependence of the order parameter, the relative phase between the two oscillating limbs, on the control, or bifurcation, parameter, the frequency of the oscillation (Haken *et al.*, 1985).

An indication that the relative phase in these motor coordination phenomena can serve as an order parameter is that its fluctuations increase on approaching the bifurcation, as shown by Schöner *et al.* (1986), Kelso *et al.* (1986), and by Schmidt *et al.* (1990), the latter in the case of interpersonal coordination of rhythmic movements which suggests that the consideration of phase relations is of value in a wide variety of natural phenomena, from interacting cells to interacting individuals. Recall that, as already discussed in Section 2.7 of Chapter 2 with regards to relative phase in oscillatory brain signals (particularly, see Figure 2.15 of Chapter 2), "critical fluctuations" in the order parameter may increase near the transition points, depending upon the nature of the bifurcation, whether it be supercritical or subcritical (these concepts are extensively treated in Sornette (2004)). Hence, to observe an enhancement of fluctuations on approaching a transition is always a good sign that a collective variable functions as an order parameter. Other behavioral motor studies have used the phasing of the alternating steps as the order parameter to investigate motor development in humans, and, as a control parameter, muscle tone was used (Thelen and Ulrich, 1991).

The specifics of these experiments can be found in the cited literature and for our purposes a brief description of the experimental set-up and the main results of the analysis are described in the following paragraphs, if only to illustrate the main concepts and the strategy used in the investigation. Basically, the experiments measured the elongation of the fingertips while performing the (almost) rhythmic movements, and based on the phase of each finger's oscillation, the relative phase, or phase difference, was obtained and its temporal evolution described using the following differential equation

$$\frac{d\theta}{dt} = -\frac{\partial V}{\partial \theta} \tag{1}$$

where θ is the relative phase and V a potential function. This represents a general form to describe the temporal evolution of the order parameter in synergetics (Haken, 2002). To find the form for the potential, after a few reasonable assumptions (periodicity and symmetry), the authors assumed a simple form as the superposition of two cosine functions: $V = -a \times \cos\theta - b \times \cos 2\theta$ and, following the stability analysis methodology described in Chapter 2, the fixed points of the system can be found by solving $(d\theta)/(dt) = 0$, or equivalently, finding the extrema of the potential function by calculating the derivative of V with respect to the phase: $(dV)/(d\theta) = 0 = a \sin\theta + 2b \sin 2\theta$.

Proceeding with the usual computations, the roots of the equation, that determine the fixed points, are found at $\theta = 0$ and $\pm\pi$ radians, the former is the in-phase state, and the latter the antiphase, while the other set of roots is given by $\theta = \arccos\left(-\frac{a}{4b}\right)$ and as can be seen it depends on the parameters a and b, thus these parameters determine the transition between the maxima. Figure 4.3 depicts the landscape of the potential function V for varying values of a and b. This landscape, constructed from theoretical grounds, portrays the main behavioral finding: the transition from antiphase (π) to in-phase (0). The transition occurs by varying the two model parameters that can be related to the movement frequency in the experiments. There is a critical value of the parameters when the transition starts and the system falls into the minimum at $\theta = 0$ (third row in Figure 4.3). This transition, in the model, is due to a specific b/a ratio, whereas in real life it occurs

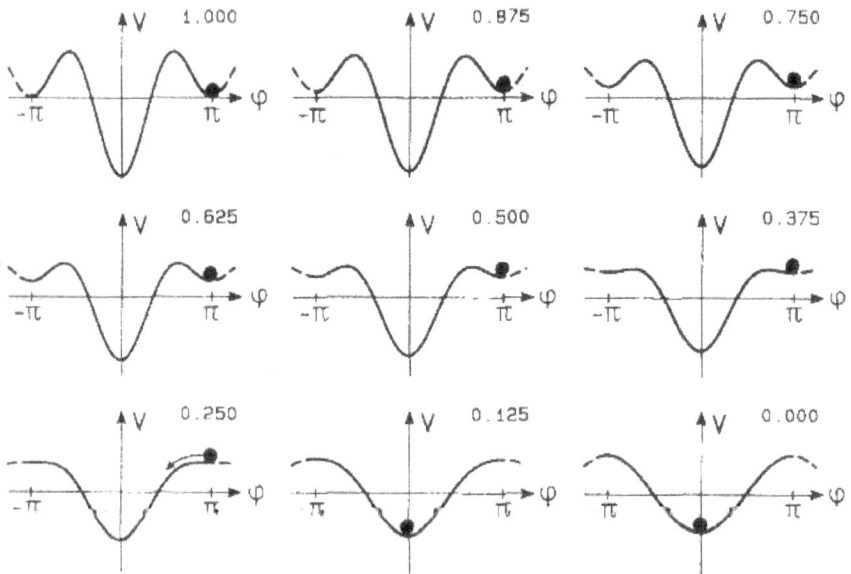

Figure 4.3. The potential function V varies according to the values of the parameters a and b, represented as the ratio b/a at top right in each panel, starting from $b/a = 1$ and ending (bottom right) when $b = 0$ and there is only one stable fixed point. The relative phase is represented in the horizontal axes (φ). The black ball represents the state of the relative phase, in this case, the initial conditions make it start at π radians (antiphase according to the nomenclature in the text) and moves toward 0 (in-phase) when $b/a < 0.25$. Reprinted with permission from Haken *et al.* (1985).

after a particular movement frequency, hence the equivalence between the model parameters and the frequency parameter.

These experiments illustrate a general and accepted approach to study these phenomena: the construction of a potential function derived from reasonable assumptions that need not be extremely quantifiable; the posterior analytical study of the dynamical equations; and the comparison of the analytical results with the empirical observations. Readers with some interest in theoretical neuroscience should notice that this approach is different from the computational neuroscience approach, which would entail the making of a detailed model describing with more or less precision ionic currents involved and other microscopic mechanisms. As emphasized throughout this monograph, both approaches are complementary, the latter more able toward a mechanistic interpretation and the former adept at capturing the collective, macroscopic dynamics.

Further studies using this motor coordination paradigm have sought to relate brain activity to the relative stability of the aforementioned fixed points (in-phase and antiphase) and specifically to the transition in the behavior, perhaps to glimpse at another signature of bifurcations in brain activity (Wallenstein *et al.*, 1995; Kelso and Fuchs, 1995). The main result of these studies is one that is familiar to us already: the *intermittent stabilization* of behavioral or brain states, once again alluding to the notion of *metastability*. To sum up, these empirical and theoretical studies on sensorimotor coordination reported many typical characteristics of the physics of complex systems in terms of phase transitions driven by control parameters, instabilities, and the enhancement of critical fluctuations of the order parameter before transitions. A great emphasis has been placed on phase relations as an adequate collective variable, or order parameter, and just to be fair let us stress that other observables can serve as order parameters, and in the specific case of motion coordination some have used the amplitude of grasping movements to study transitions in grasping behavior, from one-handed to two-handed grasping as the objects to be grasped become larger (Frank *et al.*, 2009).

Why are these studies on sensorimotor coordination mentioned here? First, because this type of combination of empirical observations and theoretical analyses exemplify that brain–behavior relationships can be investigated in detail even without complex behavioral tasks (those who read contemporary cognitive science papers will appreciate and take delight in the simplicity of these aforementioned tasks, as many behavioral manipulations used today are so hard to understand that it is remarkable that the experimental subjects can perform them!), and thus it is of pedagogic value; and second, because the results support the notion that there are relatively stable states in motor coordination that can be switched depending upon the values of certain control parameters. In principle, similar approaches that rely on relative phase and analysis methods can be applied to signals from nervous systems, some examples were presented in previous chapters and continue to appear in the following discussions. However, it is fair to note that measuring the relative phase between limbs is perhaps easier than measuring phase differences form neurophysiological signals, and the specifics of this sensorimotor task with regards to transitions between states makes life much easier when constructing models: the convenience of having two relatively stable states in this bimanual coordination task (one more stable than the other) and a precise control (bifurcation) parameter as the metronome's frequency, cannot be underestimated; then, differential equations can be constructed and thus the model can be studied in detail, with the added advantage that it affords a rigorous analysis of the brain–behavior–environment relationship, or how information from the environment (the metronome if you will) serves to drive behavioral states (Jirsa *et al.*, 2000), and facilitates the very global and more ambitious approach to characterizing brain activity under external influences as described by Frank *et al.* (2000). On the other hand, the matter of how to describe the stable states in brain signals and how to capture transitions between them is a harder problem to tackle, even though efforts abound, as they are illustrated in this monograph. For those focused on brain signals, it is almost a source of envy that such neat sensorimotor experiments can be carried out, but of course someone had to think about them.

Yet, some readers may ask, what is the relation between these experiments of sensorimotor coordination and PD or other deviations of coordination of movements? Is there any specific insight to be gained by considering those empirical–theoretical results in healthy participants? In reality, the same underlying principles at a certain (high) level of description may be at play in those experiments aforementioned and in PD. The basic difference is that, in those experiments, it is the will of the individual that causes the limb motions, that increases the frequency of movement (forget the external metronome, the "internal" one does it well too) and experiences the bifurcation between motion patterns. Thus, the commands are coming from some brain networks, a sort of intentional dynamics in the words of Scott Kelso (1995); in PD and other pathologies, there is also a command from brain networks (basal ganglia-thalamocortical) that results in tremor and motor discoordination phenomena, obviously this command is not "willed" but yet, in the final analysis, traveling from brain networks to the muscles just like in the experiments above in healthy conditions. The question of consciously versus unconsciously generated signals is another story, about which a whole encyclopedia could be written (there will be just a few words about this in Chapter 7). The possibility thus exists that there are other stable or metastable states at the level of the brain networks or the muscles that can be switched accordingly in attempts to ameliorate tremor and motor disorders, possibly by direct electromagnetic stimuli, as explored in Section 4.3. To illustrate this point, let us inquire whether the stable states derived from this theoretical HKB model presented above can be forced to switch, in real life, by direct perturbation applied not to the muscles but to the brain. The observations by Meyer-Lindenberg *et al.* (2002) seem to provide an affirmative answer, and to support the model. In this study, TMS was used to transiently and locally perturb the cortical activity in the premotor and supplementary motor cortices while subjects performed the metronome-paced finger movements aforementioned, and this perturbation caused the transition from the less stable state (fingers moving in antiphase) to the more stable (in-phase), thus corresponding to the theoretical findings mentioned above in that the antiphase pattern is not as stable because at higher metronome frequencies it switches to the in-phase motion. These authors also report that the TMS perturbation did not induce the opposite transition (from the more stable to the less stable), and hence they take these results as an indication that sensorimotor dynamical bifurcations can be forced by perturbing the brain with direct electrical (magnetic) stimulation of certain areas. These results could be of significance when devising strategies to improve the movement coordination in patients; Section 4.3 explores this theme.

4.2.2.   Global scenario of motor disturbances in PD: Altered dynamics due to dopamine deficiency

Reminiscent of the epilepsy cases discussed in Chapter 3, it is conceivable that, in PD, the intermittent manifestation of tremor and other motor symptoms is a reflection of a very much altered state space landscape, in this situation due to the loss of dopamine in the neural circuitry involved in motor coordination aforementioned in Section 4.1 (basal ganglia–STN–thalamus–motor cortex). The altered "attractor" landscape in the state space may result then in the emergence of enhanced synchronous activities in this circuitry, as described below, and the resulting tremor and other motor dysfunctions. Hence, the initial alteration, the loss of dopamine, seems to be a strong enough perturbation to substantially alter the dynamical structure of the spate space and, once this alteration has taken place, more subtle variations in other control parameters cause the emergence of motor dysfunctions perhaps by favoring the dynamical route toward an attractor-like synchronous state and the tendency of neural tissue to exhibit limit cycles. For example, the 3–6 Hz resting tremor in PD, which is so regular that can be considered a limit cycle in dynamical terms, appears and disappears spontaneously at rest (it completely stops during purposeful movement, this is why it is termed "resting" tremor), and ceases during sleep. The emergence and passing away of the limit cycle is a sign of metastability: the system's trajectories, which represent the measured neurophysiological activity at this dynamical level of description, move around metastable states and may be transiently locked onto one or another. The coexistence of aperiodic and periodic behaviors in dissipative systems and their alternation due to some features of the state space have been treated theoretically in numerous studies (Franceschini, 1983). The global scenario is then similar to the one of epilepsy discussed previously: the epileptic brain has some particular anatomic–physiologic alterations that cause a substantially different from "normal" state space dynamics, where the tendency toward synchronous activity is enhanced by, perhaps, the more frequent appearance of attractor-like relatively stable synchronization manifolds. In summary, the initial perturbation to alter the dynamics (neurodegeneration in a brain area, mutations in ion channels with profound alterations in cell properties, etc.) may be relatively strong in the sense that the dynamics is considerably reconfigured, whereas the consequent perturbations to generate synchronous activity that manifest themselves as seizures in epilepsy or motor dysfunctions in PD (affecting the striatum–motor cortex networks) may be much subtler: it may suffice a small increase in potassium concentration or any other change that increases excitability therefore enhancing cell depolarization and spike firing, to trigger the phenomenon. It was mentioned in Chapter 3 that enhanced cellular excitability associated with seizures has been observed. Is there evidence for an increased neural excitability in PD? The answer is in affirmative, obtained in conditions where dopaminergic transmission is reduced.

Dopamine, a widespread neurotransmitter in the nervous system, is considered to be involved in the gating and regulation of brain networks in general (O'Donnell, 2003). Evidence from rodent models of PD, where the dopaminergic fibers have been damaged and thus the models simulate the loss of dopaminergic synapses in PD (this is normally achieved by injection via cannulae into targeted specific brain areas of 6-OHDA which is toxic to dopaminergic cells), revealed a tendency toward hyperexcitability in the substantia nigra pars reticulata and the consequent enhancement of rhythmic synchronization between the basal ganglia and cortical networks after dopamine depletion, and, more specifically, the response of basal ganglia cells to cortical inputs was found to be enhanced after the dopaminergic lesion (Belluscio *et al.*, 2007; Dejean *et al.*, 2008). Likewise, increases in firing rates of neurons in the GPi were observed in a monkey model of PD. Other studies using mice genetically engineered with an inducible dopamine transporter knockout method have revealed that, while the neural firing rates remained similar after dopamine depletion, the synchronization in cortico–striatal networks increased (Costa *et al.*, 2006). Further, to better complete this story with a somewhat different observation, let us mention that a reduction in cell firing in the external globus pallidus was reported in one study using the aforementioned rat model of PD; however, once again the synchronized activity was potentiated upon dopamine depletion (Cruz *et al.*, 2009). Thus, in spite of some variability in the published results with regards to neuronal firing rates, these observations taken together indicate that dopamine depletion alters the interactions between cortical, striatal, and substantia nigra networks resulting in higher synchronization, and that there is a tendency toward hyperexcitability at least in some parts of this network when dopaminergic inputs diminish. The crucial point is, once again, that the coordination among cells, the organized activity patterns, are altered as a result of changes in the control parameter, dopamine. For these reasons, it is conceivable to consider dopamine as a control parameter of nervous system's function. The molecular mechanisms for its actions at the collective level are varied and beyond the scope of this work. It may be of interest to note that if dopamine is present in excess, other disturbances in behavior are known to occur, perhaps the best characterized is drug addiction, as dopamine plays an important role in the reward networks and reinforces pleasurable behaviors. In general, dopamine is considered a modulator of a variety of behaviors perhaps due to the wide dopaminergic innervation throughout the brain, especially the fronto-striatal and limbic areas. A recent review is presented by Cools (2008).

Section 4.2.1 introduced results on nonpathological sensorimotor coordination with the presence of fixed points in the dynamics. Can these results be extended to tremor rhythms? Are these oscillations manifestation of other fixed or metastable points as was suggested above in the first paragraph of this section? Unlike normal physiologic tremor, whose function has been proposed to provide an almost constant oscillating state to gate timing of motor events, pathological tremor is characterized by higher intensity and perturbed coordination of the antagonistic muscles involved, the extensors and the flexors. An illustration of the altered muscle coordination is found from the results of Boose *et al.* (1995), who reported the greater-than-normal regularity in the values of the relative phase between these muscles of patients with essential tremor; in other words, higher synchronization between muscles was found. The relation between tremor and higher tendency to synchronous muscle activity has been reported in other studies. Indeed, a side effect of amitriptyline (used to treat depression) illustrates the transition from the normal, physiologic tremor to a pathologic deviation, in that it causes a disabling postural tremor. Not surprisingly (from the perspective of "too much coordinated activity–more pathology"), an increased synchronization in the electromyographic (EMG) signals between flexor and extensor muscles was enhanced under amitriptyline (Raethjen *et al.*, 2001); there are other drugs that induce tremors and in this sense provide additional control parameters, some of them like amitriptyline seem to act exclusively at the central nervous system level. Of course, there is coherence too between flexor and extensor muscles in healthy individuals during physiologic tremor, but this is a rare finding. Raethjen *et al.* (2000a,b) found significant coherence between these muscles in the physiologic tremor in only $\sim$10% of healthy participants but in 70% of PD patients (to be precise, coherence between muscles within the same limb). These observations refer mainly to muscle coordination, but how about synchronous activity between brain and muscle? An occurrence of brain–muscle coordination in normal physiology is found in the case of discrete slow limb motions. Contrary to the subjective experience of continuous motion, normal slow finger movements are discontinuous, characterized by 8–12 Hz steps, or discontinuities, in finger acceleration (Vallbo and Wessberg, 1993), which our perception cannot grasp and thus the brain creates the illusion of continuity, as the brain is the master of illusions. In reality, psychophysical and biophysical data seem to indicate that most perception is discontinuous (VanRullen and Koch, 2003; Palva and Palva, 2007). Perhaps as expected, the discontinuous limb movement is thought to depend on the activity in a distributed network that includes sensory feedback and motor components. Hence, this oscillation can be considered as yet another example of an emergent phenomenon (a recollection of the discussion on central pattern generators (CPGs) in Chapter 1, Section 1.2 is useful at this point because of the similarities in the general notions of CPG resultant activity and the one presented now). The empirical evidence for the coordination between central nervous system and muscle activity during these slow movements and tremor represents sufficient illustration of a collective pattern of organized activity spanning several organ systems. For example, inspecting intracortical recordings in monkeys, it was found that individual neuronal activity in motor cortices was synchronized to the peripheral discontinuity in these slow finger movements (Williams *et al.*, 2009). In a study performed in humans, coherence at 6–9 Hz between MEG signals over the sensorimotor cortex and EMG signals of the extensor muscles was reported during slow finger movements (Gross *et al.*, 2002). With regards to pathological tremors, evidence exists for enhanced tendency to synchronized activities in a diversity of cortical areas (including motor cortex and basal ganglia measured as EEG, MEG, or intracerebral recordings in patients and monkeys) and the resulting pathological tremor normally recorded by EMG methods (Volkmann *et al.*, 1996). Based on phase synchronization analysis of cortical (MEG) and muscle (EMG) signals in a Parkinsonian patient, Tass *et al.* (1998) concluded that the temporal evolution of the peripheral tremor reflects the abnormal synchronization between motor cortical regions. This study found multifrequency locking between cortical signals and muscle activity, specifically 1:2 phase locking, whereas cortical regions were 1:1 phase locked. Recall that multifrequency locking can be conveniently described on a torus, the geometric representation of synchrony (Figure 2.5 of Chapter 2).

But there is more to this story. The brain–muscle connection is not the whole system operating in the emergence of tremor patterns. There are indications that normal resting tremor results from the interaction between muscles and other physiological variables like respiration rate and blood pressure, with no special contribution of the brain. Hence, this indicates that it is a manifestation of the coupling between different physiological rhythms. The tremor model of Beuter and Vasilakos (1995), consisting of a coupled pair of delay differential equations, was used to explore the parameter space in which tremor arises, specifically using blood pressure and respiration rate directly recorded from individuals, as parameters that modulate coupling weights in their model equations. Changes in these parameters resulted in tremor that mimicked the pathological condition, therefore the proposal was put forward that pathological tremor occurs from a dynamic shift in the functional organization of several interacting rhythms; in other words, it is a "dynamical disease." The interesting aspect of this study is that the parameters came directly from the patients, through recordings of various types. As well, changes in the state space configuration dynamics of tremors in other diseases, like tardive dyskinesia, has been inferred from the lower dimensionality of the finger tremor time series compared to normal tremor (Newell *et al.*, 1995), thus suggesting a "simpler" state space dynamics perhaps a manifestation of the tendency toward a limit-cycle dynamics or toroidal synchronization manifold as depicted in Figure 7.3 of Chapter 7. These studies are of peculiar interest because they emphasize the fact that oscillations in biology are linked within levels and among organ systems, as already noted a few decades ago by Cardon and Iberall (1970), who also observed that rhythms are marginally stable (they could have very well used the term metastability). The moral of these tales on motor (dis)coordination is conceptually similar to that presented in Chapter 3 on epilepsy: the parameter space where the dynamics occurs is wide, and changes in a variety of parameters will result in comparable dynamical regimes, more particularly enhancing tendencies to synchronous and limit cycle activity.

4.3. Deep Brain Stimulation and Dynamical Systems Approaches to Control Abnormal Motor Activity

Chapter 3 presented evidence that epileptic activity can be altered or even stopped by several perturbations, including direct electrical stimulation to the nervous system. Considering that PD is characterized by more robust and very periodic oscillatory activities compared with the complex and not-so-rhythmic appearance of some seizures, it is conceivable to expect more success in halting these PD pathological rhythms by direct electrical stimulation to the brain, if only because of the phenomenon of phase resetting for which a periodic oscillation is needed (if it is not too rhythmic, the phase is hard to establish). The work of P.A. Tass and others on phase resetting in a biological scenario is worth considering for those interested in these themes. There is a wide array of theoretical studies demonstrating, using computer simulations, the effectiveness of feedback signals in the suppression of collective synchronization (Popovych *et al.*, 2008; Tukhlina and Rosenblum, 2008). The advantage of these methods is that no precise knowledge of the system's constituents is required, only the macroscopic properties are of relevance in these approaches. The aforementioned experimental observations of Meyer-Lindenberg *et al.* discussed in Section 4.2.1 provide support for the feasibility to perturb dynamical states in motor activity by direct brain stimulation (TMS was used in that study).

Compared with the debates surrounding the mechanisms of action or even the efficacy of direct nervous system stimulation in epilepsy, discussed in Chapter 3, the scenario is much more unambiguous in the context of PD. To our knowledge, the successful application of these methods to arrest PD symptoms is well accepted in the community of investigators and clinicians, at least in current times. The reason for this approval is derived from the relative clarity of the results, which can be inspected in the recordings of Figure 4.4. Albeit this obvious success does not really help in understanding the mechanism of action because the deep brain electrodes

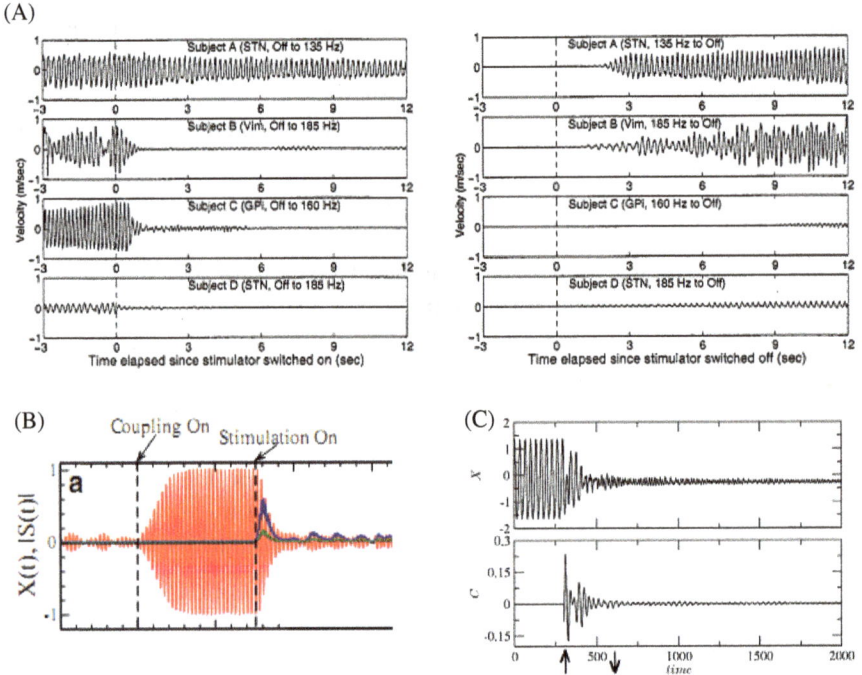

Figure 4.4. (A) Switching on deep-brain stimulation arrests the periodic limb movement (left), while there is a fast transition to the tremor after turning off the stimuli (right). The recordings represent the Parkinsonian rest tremor finger velocity. The left-hand side panels show the halting of the tremor in four patients upon turning the stimulation on at the dashed vertical line. The traces on the right represent the transition to the pathological tremor oscillation after switching off the deep-brain stimulation at the dashed vertical line. The location of the deep-brain stimulation was, for each subject, in the STN, ventrointermediate thalamic nucleus (Vim), and GPi. The frequency of the stimulation is in brackets. Subject A needed a longer time to halt the tremor. (B) and (C) show the qualitative similarity between the experimental traces of A and those from theoretical studies. Both represent the time courses of the mean field (X) of coupled model oscillators, upon coupling them (in B, "coupling on," note how they synchronize, resulting in a high amplitude rhythmic mean field) and after a perturbation (in (B) the amplitude of the perturbation is the blue line, and in (C) the time course of the feedback stimulation (C) is in the bottom panel). (A) is reprinted with permission from Titcombe *et al.* (2001), (B) from Popovych *et al.* (2008), and C from Tukhlina *et al.* (2007).

excite not only cell bodies in the neighborhood but also passing axons from remote areas (for those interested in these details, please consult Kuncel and Grill (2004) to explore the selection of stimulus parameters for deep brain stimulation discussing anatomical targets as well).

Deep-brain stimulation in Parkinsonian patients (Benabid, 2003) stops the periodic activity for a while, with a return to the pathological rhythm a few seconds after turning off the stimulator. Figure 4.4 depicts some examples of the relatively fast transition from periodic tremor activity, measured at the fingers, to aperiodic (basically cessation of the tremor) upon switching on the stimulator, which was inserted in different brain areas (detailed in the figure legend). Also represented in the figure is the quick return of the tremor after turning off the stimulation. According to the concepts presented in Chapter 2 regarding attractors, the scenario depicted in these experiments is reminiscent of manifestations of limit cycle attractors in limb movement, and the transition (the qualitative change in the finger dynamics so clearly observed in the traces in the figure) can be conceptualized as a bifurcation. More specifically, the transition between periodic and quiescent activity is reminiscent of a Hopf bifurcation, described in Section 2.3. Hence, is this a demonstration of attractor dynamics in physiology? Can the pathological oscillation be considered an attractor (limit cycle-like or a toroidal synchronization manifold) in PD patients? Yet, even if left unperturbed the resting tremor comes and goes spontaneously, which would not be expected if the activity was to be dictated by a limit cycle attractor in the sense that there would not be an easy escape from its area of attraction, so conceivably this is a manifestation of metastable dynamics, already discussed in several sections of this work and presented in more technical detail in Section 2.2.3 of Chapter 2. The result of the perturbations induced by deep-brain stimulation can be thought of as the destabilization of fixed points, which in PD and related pathologies are high-amplitude periodic tremor rhythms, whereas in the epilepsy studies mentioned in Chapter 3 the metastable fixed point may represent the seizure onset (recall Figures 3.9, 2.13, and 2.14 of Chapters 3 and 2, respectively). Thus, at the collective level of description, same phenomena seem to reappear in these deviations from the considered healthy conditions. We leave it to the reader to ponder more in depth on these matters. The suppression and initiation of limit cycles has been scrutinized in a number of theoretical studies. For example, using a global, field model of scalp electrical activity, it was found that a limit cycle could be temporarily suppressed by perturbations if the system had only a limit cycle attractor, whereas in systems with a limit cycle and a fixed point the perturbation was able to permanently stop the oscillation in case it was strong enough to push it toward the fixed point (Kim and Robinson, 2008). It is worth considering the similarities between the physiological traces (Figure 4.4(A)) and those from computer simulations, two of which are shown in the figure (Figures 4.4(B) and 4.4(C)), but more graphs, equally similar, can be found in the literature.

In addition to the experimental demonstration of tremor abolition, the study by Titcombe *et al.* (2001) used a network model to assess the possible mechanism, at this network level of description, by which high-frequency stimulation halts tremors. Their model consists of an oscillating system that is perturbed by periodic stimulation, and simulates the arresting of the oscillation by a perturbation that leads to the destabilization of the abnormal, or pathological, limit cycle through a Hopf bifurcation. As stressed in Chapter 2 when discussing bifurcations, a bifurcation parameter is needed to explicitly demonstrate the existence of a dynamical bifurcation, although a qualitative change in activity already gives a good indication. In their model, the bifurcation parameter was related to a time-dependent release of neurotransmitter during the stimulation. The amplitude and frequency of the stimulation can serve too as bifurcation parameters because very low amplitudes or low frequencies do not have any effect, sometimes even worsening the motor symptoms. Modolo *et al.* (2008) used stimulation frequency as the bifurcation parameter to study bifurcations in a population-based model of the subthalamo-pallidal networks. Their exploration of the stability of the (pathological) Parkinsonian limit cycle on the frequency of deep-brain stimulation suggests that the limit cycle becomes of vanishing amplitude as the stimulation frequency increases, in what the authors describe as a quasi-stationary state that perturbs the limit cycle regime.

The standard electrical stimuli currently used in deep-brain stimulation protocols for PD are based on high-frequency pulses, note that in Figure 4.4(A) all frequencies are over 100 Hz, in a similar fashion as the previously discussed vagal nerve stimulation in epilepsy. Yet, perhaps other stimulation paradigms can be equally or more effective and be more gentle than the almost continuous brain perturbations imposed by the current devices, since high frequencies may have nondesirable secondary effects. There is theoretical evidence that methods based on nonlinear delay feedback and phase resetting are efficient at desynchronizing populations of cells. As opposed to the brute-force approach of the on–off, almost constant high-frequency stimulation, these "nonlinear" stimulation protocols selectively counteract the synchronization mechanisms. Once again, this is another version of the comparable story in the case of brain stimulation to stop seizures: brute-force approaches (vagal nerve stimulation) versus more subtle paradigms already discussed in Chapter 3. Accordingly, a recent study presented a deep-brain stimulation-recording apparatus that sends appropriately timed stimuli to selected brain areas in order to cancel out pathological synchronous activity leading to PD tremor. The device relies on the recordings of EEG and EMG activity (among other variables) in the individual and sends adequate pulses to deep-brain sites. These devices, provided they work on patients, could represent major advances in the field of deep-brain stimulation in PD and related disorders; full details can be found in the original article by Hauptmann *et al.* (2009). Figure 4.4(C) depicts the amplitude of the feedback stimuli in computer simulations ('C', in the lower panel), illustrating how the feedback signal becomes smaller and then almost vanishes when the synchronous mean-field activity of the coupled oscillators (X, upper panel) is progressively reduced.

The important point from this monograph's perspective is that, regardless of whether the protocol consists of high-frequency or delayed-feedback stimuli, the net result of the perturbations is a tendency toward reducing the synchronous activities in the nervous system. A perturbation of the synchronization that underlies the tremor is always feasible since this synchrony is a collective phenomenon of multiple coupled oscillators, as concluded by Raethjen *et al.* (2000b) while studying Parkinsonian and essential tremor. Recall the comments in Chapters 1 and 2 regarding the possibility to perturb collective synchronization of coupled neural oscillators. But, of course, there is no need to reduce brain synchronizing phenomena in general because of the importance of the transient coordinated activity for efficient information processing, thus the site of stimulation is crucial and should be chosen as specifically as possible. Most of the currently targeted sites are the STN and the GPi, in accordance with the PD network model presented in Section 4.1. Other movement disorders, such as essential tremor, target different thalamic areas. There are a few reviews on the principles of deep-brain stimulation that may be pertinent for those interested in the details of these matters (Kringelbach *et al.*, 2007). The STN stimulation, developed in the early 1990s, desynchronizes basal ganglia networks that results in the motor units firing more independently and thus reducing the higher synchrony characteristic of the tremor, a phenomenon also reproduced by treatment of the patients with dopamine (Sturman *et al.*, 2004). Thus, two very different therapeutic methods — pharmacological and biophysical — result in a similar amelioration of the symptoms, phenomenon that finds explanation under the dynamic framework exposed in previous chapters: changing different control parameters results in same dynamical alteration in the state space, perhaps by reducing the area of attraction of the synchronization manifold or by pushing the trajectories away from it (Figure 7.3 of Chapter 7). Indeed, the therapeutic effects of deep-brain stimulation do not exceed those of pharmacological medications that target dopaminergic transmission. The inspection of the brain regions that change activity, measured by positron emission tomography (PET), as a result of either subthalamic stimulation or levodopa therapy (that restores dopamine to some extent) revealed great similarity between the two protocols, which indicates a common mechanism of action (Asanuma *et al.*, 2006). Both treatments appear to decrease the activity of the STN resulting in less excitability in the GPi (Dostrovsky *et al.*, 2000). Invasive recordings, especially local field potentials recorded in the globus pallidus and neuronal spike recordings in the STN of PD patients, revealed more clearly than any neuroimaging recording the decrease in the activity with stimulation frequencies greater than $\sim$75 Hz applied to the STN. Coming back to neuroimaging, and for those interested in details, the common PET-measured metabolic reductions observed after these two treatments included areas such as the cerebellum, putamen/globus pallidus, and sensorimotor cortex. In general, any approach that reduces excitability in the network (and thus makes synchronous activity less probable) will have the desired results. Therefore, in addition to stimulation or pharmacological therapy, lesions of the STN (subthalamotomy) in patients reduce disease symptoms too. This may seem paradoxical: an irreversible damage to the STN causing same effect as its high-frequency stimulation. However, the net effect in both cases is the decreased excitability in connected regions, particularly the globus pallidus, due to the intrinsic biophysical characteristics and the connectivity pattern between the STN and other areas, the internal and external globus pallidus more importantly. The side effects of subthalamic stimulation are not of great concern currently since it does not seem to lead to cognitive dysfunctions; however, some deficiencies in specific cognitive tasks have been reported in patients with deep-brain implants. Once more in this context, it is worth recalling (Section 2.7 of Chapter 2) the evidence presented by Diesmann *et al.* (1999) using computer simulations of integrate-and-fire units, where the necessity of a minimum number of cells to maintain synchronous activity was demonstrated. In this study, cell number was taken as the bifurcation parameter. Hence, reducing the cell number by any method will hinder the spread of synchronous activity. Deep-brain stimulation, according to the observations mentioned above in patients, achieves the effective reduction of the number of cells that are active or synchronized.

The approaches aforementioned relied on the destabilization of synchrony, the destruction of limit cycle activity. Conversely, other motor abnormalities in PD may still necessitate a relatively stable periodic dynamics, such as gait dynamics. There are considerable gait disturbances in PD patients, such as freezing (termed episodic gait impairment) and greater than normal variability in stride length and interval (continuous gait impairment). In general, walking is not just a strictly periodic phenomenon, but shows small stride-to-stride fluctuations that result in fractal and multifractal properties of gait dynamics. This fractal structure implies the existence of long-range correlations in stride time that tends to disappear in PD where the fluctuations between strides increase and become random (Hausdorff,

2009). One interpretation of this breakdown of long-range correlations is that each stride becomes unrelated to the previous one, as if each stride were to start as a new process, so the loss of fractality can be understood as a loss of memory in the locomotor system. Hence, in these cases there is a need to maintain a less variable and almost-periodic gait activity, and thus some periodic forcing could in principle be applied. This in fact has been done using the so-called rhythmic auditory stimulation (RAS), which improves gait in PD. This auditory input serves as an external clock to set the pace and thus substitutes the altered internal rhythmicity in patients (McIntosh *et al.*, 1997). The limit cycle, in this case, has been restored or stabilized by the auditory pacing (periodic forcing). Treadmill walking similarly is useful in PD patients as it provides external cues to restore the impaired rhythm, being, in the final analysis, another manner to implement periodic forcing. But, while the stride variability is reduced in the patients and their gait improves considerably, a detailed study revealed that the gait fractal scaling did not change upon the auditory stimulation or treadmill walking; it still remained lower than in normal gaits. It did, however, change (became like that of healthy individuals) when patients were administered methylphenidate (also known as ritalin), a drug that blocks catecholamine uptake and enhances central nervous system function in general; in addition to recovering normal fractal scaling, this chemical treatment accompanied an improvement in the gait and postural control. These observations have been interpreted as indicating that the fractal structure, or long-range correlations, in gait dynamics may reside within the central nervous system because ritalin acts directly within the brain while the RAS is a more general stimuli acting from the periphery. Nevertheless, the periodic forcing imposed by RAS ultimately acts through the auditory system and the motor cortex in the brain itself. The commonly used RAS is strictly periodic. On the other hand, considering that normal gait is not strictly periodic but displays fluctuations, perhaps adding some variability about the mean to the "rhythmic" auditory stimulation may be even more beneficial to the patients. As a matter of fact, there is preliminary evidence for this notion, in that "nonrhythmic" auditory stimulation (this is meant as periodic stimuli with added noise in order to more closely simulate the real gait dynamics) prior to walking improves gait dynamics in patients with moderate PD (Filer *et al.*, 2008).

This section is closed with a further reflection on the comments above comparing the state of affairs in the deep-brain stimulation protocols in PD with those used in epilepsy. It is perhaps unfair to compare the success at controlling tremor in PD by brain stimulation with the less apparent and much debated control of seizures in epilepsy discussed at length in Chapter 3. First, the tremor is normally present almost continuously in patients, whereas seizures are not, nor are they easily anticipated. Second, the circuitry in PD is relatively well known, offering clear targets to desynchronize the circuitry, while each epilepsy has its own anatomy. Third, the pathological motor discoordination, especially tremors, is a quite periodic rhythm so that attempts at phase resetting can be assayed, but seizures display a wide array of waveform morphology and nonperiodic activity. For these and other reasons, treatment of motor dysfunctions by direct brain perturbations is more probable to succeed than the control of seizures in clinical settings.

4.4. Concluding Remarks

The observations and comments presented in this chapter have provided illustrations of the alternation between periodic and aperiodic behaviors in sensorimotor coordination as manifestations of synchronous activities in the networks of the central nervous system as well as between muscles in the limbs affected with tremor in PD and related pathologies. This chapter has also introduced further support for the usefulness of phase relations in the study of the dynamics of interacting networks of muscles, brain regions, and even in interpersonal transactions (recall the study of Schmidt *et al.* mentioned in Section 4.2.1). The transient nature of tremor oscillations constitutes a phenomenon where dynamical regimes like intermittency can be invoked, much like in cases of epilepsy mentioned in Chapter 3.

The intermittent manifestation of tremor and other motor symptoms in PD is a reflection of an altered "attractor" state space landscape due to the loss of dopamine in the neural circuitry involved in motor coordination aforementioned in Section 4.1. The resulting deviant dynamics due to the abnormal state space leads to enhancing the tendency toward emergence of synchronous activities in these networks and thus determines the pathological tremor and other motor dysfunctions. Hence, the initial alteration that is the loss of dopamine seems to be a strong enough perturbation to substantially alter the dynamical structure of the spate space and, as a result, variations in other control parameters, that need not be strong, cause the emergence of motor dysfunctions perhaps by favoring the dynamical route toward an attractor-like synchronous state. This scenario assumes a synchronization manifold (torus) relatively stable, a notion that was proposed also in the case of epileptic activities. The passing away of the periodic rhythm can be envisaged through period doubling bifurcations, which, as expounded in Chapter 2, is a common route to chaos and was invoked to explain the end of the seizure shown in Figure 3.7 of Chapter 3. This is just a possible dynamical scenario. The coexistence of aperiodic and periodic behavior in dissipative systems has been treated theoretically in numerous studies.

Nevertheless, it is hard not to note basic similarities in the collective network activity in PD and in epileptiform phenomena discussed in Chapter 3. Two basic principles include: (1) states of more extensive excitability that promotes higher rates of neuronal spike firing, in the case of PD this is restricted to the specific circuitry discussed in this chapter while in the cases of epilepsies the circuitries vary depending upon the seizure types; and (2) the tendency toward synchronous activity consequent to the enhanced excitability that will spread the activity patterns throughout connected networks. Thus, as explained in Section 4.3, the success of DBS in halting tremor is derived from its net effect, which is an effective reduction of activity in key structures thus reducing excitability and thereby opposing synchronizing tendencies. Whether this occurs with high-stimulation frequencies as opposed to low frequencies is due to the details of the interaction between these stimulation frequencies and the intrinsic cellular biophysical characteristics and the network connectivity. After all, it is the basic physiology that determines the network, linear or nonlinear, dynamics.

In the final analysis, while in some instances the dynamical landscape that results in brain activity and behavior may seem restricted to neural tissue, it should never be forgotten that all these "attractors" or bifurcations exhibited in sensorimotor coordination dynamics exist in the individual–environment system, rather than being an exclusive characteristic of one individual's body (Warren, 2006). Thus, the state space where the dynamics takes place is multidimensional and of considerable complexity: even the apparently simple motion that is the tremor at the fingertips requires a coordinated activity of several degrees of freedom of the body involved in kinematic relations so as to constraint the tremor (perhaps, this is the reason for the high dimensionality in finger tremor found in the study of Newell *et al.* (1995)). In general, these oscillations are derived from the interaction between several physiological rhythms (Beuter and Vasilakos, 1995), which should be kept always in mind when thinking about control parameters and the possible degrees of freedom required to model, or to capture, the whole system's functionality. Undoubtedly, many control parameters can be found but their net effect on the dynamics will be identical, stopping or enhancing tremors, hence posing a question mark on the significance of searching for "specific" control parameters, question mark that will reappear in subsequent comments regarding other deviations. The fact that DBS and other manipulations may not be too "specific" should not deter us from using them and from unraveling their dynamical consequences.

CHAPTER 5

SUDDEN AND SUBTLE TRANSITIONS IN ALTERED STATES OF CONSCIOUSNESS

The investigation of patients who have lost, completely or partially, consciousness in syndromes such as the minimally conscious state (MCS), is providing fundamental information about the transitions in brain activity and their behavioral correlates with regards to awareness. This is fundamental not only for clinicians working in settings like critical care units, who have to handle ethically challenging cases of coma or vegetative states (VS), but also from the basic science perspective in terms of the nature of consciousness and awareness. Along the lines of the rest of the chapters, the main concern in this chapter is the analysis of the experimental data and theoretical approaches that endeavor to characterize brain dynamics at the collective level in states of altered states of consciousness to comprehend the nature of the transitions in neural dynamics and how these derive from the local and global aspects of brain function. The term "altered states of consciousness" is used here to describe situations that depart from the usual awake states with awareness of self and environment, lest we become trapped in semantic arguments because we are aware that some do not approve of that term to describe unconscious states in general, and indeed in some of these altered states there is partial awareness as described below. Not only pathological conditions such as coma, vegetative, and MCSs are reviewed here, but also more natural transitions in awareness, such as sleep and anesthesia. To our advantage, theoretical efforts have been and are currently being done trying to frame these transitions within the context of dynamical bifurcations, and offer avenues to explore at a more abstract level the collective dynamics of brain networks that result in either subtle or abrupt transitions in consciousness. Previous chapters have included comments on how a widespread and long-lasting synchronization may not be profitable for the brain to process sensory inputs and it may be associated with diseases, and examples included the higher than normal synchrony during seizures and Parkinsonian tremor. This chapter provides further evidence in support of this view.

Before reflecting on these conditions of changing awareness, two comments are in order as these experiences advise us to consider consciousness not as an all-or-none phenomenon, but as a continuum. Thus, Carl Jung thought that "we must accustom ourselves to the thought that the conscious and the unconscious have no clear demarcation, the one beginning where the other leaves off," and similarly, Victor Lamme wrote more recently: "Any sharp division between conscious and unconscious cognitive functions will again be arguable, often resorting to taste rather than scientific argument" (Lamme, 2006). We will not try to define consciousness, perhaps because declaring any particular definition is bound to offend some investigators and individuals in general. In this way, we avoid embroiling ourselves in the very ancient debate on what consciousness is and what is not. Prominent scholars throughout the centuries have dedicated enough deliberation on this topic. We just mention that defining consciousness is like trying to define life, which perhaps is best left undefined and what could be more constructive, or practical, is to describe it in terms of characteristics. Why should there be a problem with enumerating characteristics to define phenomena? Nonlinearity, for instance, a major concept in modern science, has basically no strict definition, other than that a system that is not linear (equations whose effects are not proportional to their causes), and linearity is defined by characteristics (proportionality and superposition). Simply put, as Colin McGinn (1989) does, consciousness is a phenomenon naturally arising from certain organizations of matter, with certain features that can serve to describe it depending on the context. For our present purposes in this chapter, loss of consciousness means a lost awareness of self and surroundings.

5.1.   From Wakefulness to Sleep and Back

The transition from wakefulness to sleep is the most common alteration of states of awareness experienced by all of us. The transition is characterized by specific changes in the electrophysiological patterns of activity, yet another illustration of how distinct patterns of brain coordinated activity result in various behavioral states. Typical waveforms observed in the electroencephalogram (EEG) during sleep are slow waves (<1 Hz), δ waves (1–4 Hz), and spindles (7–14 Hz) and are used to define the different sleep stages. The appearance of distinct oscillations motivated a classification of several stages of vigilance and particularly of the sleep stages as they correlated with behaviors (Dement and Kleitman, 1957). A typical transition seen in the EEG is the following: drowsiness is accompanied by slowing of the frequencies concomitant with an increase in the amplitude of the waves (stage 1), high-voltage bursts then appear (K-complexes) as well as spindles (stage 2), which progresses to a mostly δ wave pattern and slow wave at frequencies lower than 1 Hz (stages 3 and 4). These are the so-called nonrapid eye movement (NREM) sleep patterns. At some intervals, the slow wave activity disappears and higher frequencies in the γ range take over, basically indistinguishable from the EEG during awake states. These periods are associated with rapid eye movements (REMs) and muscular atonia, and normally individuals experience dreams during this time. Figure 5.1 shows some examples of these waveforms during sleep stages. It should be remarked that, while these are the main oscillations at every sleep stage, rhythms are intermingled many times, so that, for instance, during slow wave sleep (SWS), brief periods of fast oscillations can be seen occasionally, riding on the depolarizing component of the slow oscillation. By taking a look at how cells fire during SWS, the presence of the fast oscillations in EEG signals can be understood. In Figure 5.1(B), a simultaneous recording is shown to depict how a cortical neuron in an anesthetized cat fires in bursts that are synchronous to the downward deflection of the local field potential (LFP). The high frequency spike firing during the burst makes it possible to detect, at the EEG level, some fast activities, in this case in the figure at a frequency of about 8–10 Hz. Thus, the fact that natural events are almost never all-or-none phenomena is emphasized once again.

The progression of the EEG pattern mentioned above indicates that falling asleep is associated with a more pronounced synchronization of brain activity. The increase in the signal amplitudes, depicted in the figure for stages 2 and 3, during drowsiness and some sleep stages already suggests that there are more synchronous inputs arriving at the brain areas near the recording electrodes. While the spatiotemporal extension of synchrony may be related to pathological states as proposed in previous chapters, this depends on the context. Consequently during sleep, a long-lasting and widespread synchrony is not pathological. Evidence for this enhanced and extensive synchrony during sleep stages includes the studies evaluating coherence between EEG signals. Thus, coherence between intra- and interhemispheric EEG signals was larger during SWS than in REM episodes or wakefulness (Achermann and Borbély, 1998; Duckrow and Zaveri, 2005). Specifically, the coherence found in these studies was more broadly distributed in the low-frequency range, which can be expected as these are the typical sleep waveforms. In animals, intracerebral recordings have also revealed higher and widespread synchronization of activities in slow (<1 Hz) and 1–4 Hz frequencies during SWS, while synchronization at higher frequencies was localized in small neighborhoods and restricted to short time windows (Destexhe *et al.*, 1999). In a more detailed analysis of the sleep synchronization patterns, Ferri *et al.* (2005, 2006) reported that each slow wave is correlated with a maximum in the magnitude of a synchronization likelihood index and a large-scale extension spread over the scalp.

(A)

(B)

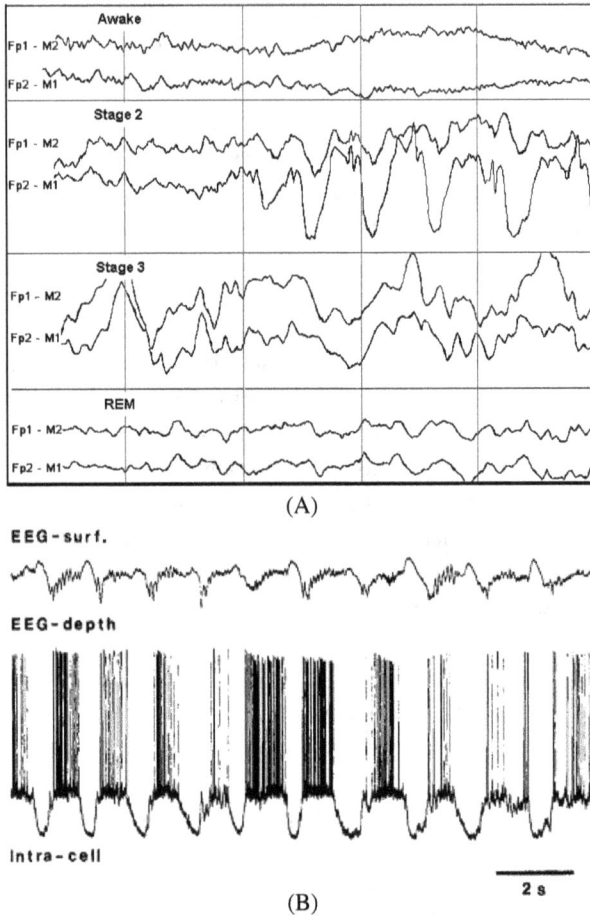

Figure 5.1. Sleep waveforms. (A) scalp EEG signals in a subject during wakefulness, and sleep, stages 2 and 3 (NREM), and REM. Notice the lower frequencies and the increase in amplitude of the signal, reflecting a higher synchronous activity during sleep, except for REM events. (B) simultaneous intracellular (bottom) and extracellular depth recording (EEG depth) in an anesthetized cat, during slow wave rhythmic activity, where some higher frequencies (spindles) are grouped by the slow oscillation at ~0.5 Hz. B is reproduced with permission from Contreras and Steriade (1995).

The decrease in signal complexity during SWS, measured as estimates of correlation dimension and Lyapunov numbers, further indicates a more extensive coordination among brain networks during sleep (Fell and Röschke, 1994). However, not all consists of more extensive synchronization during sleep. Gamma band activity was found to be less coherent in sleep as compared to wakefulness (Cantero *et al.*, 2004), perhaps a natural consequence of the cellular activities in these higher frequencies to process sensory stimuli during awake states, and therefore it is not surprising that coherent 40 Hz oscillations appear in REM episodes (Llinás and Ribary, 1993).

The cellular mechanisms that account for this higher synchrony in the low-frequency range during SWS that underlies the transition from wakefulness to sleep were sketched in Chapter 1 (Section 1.6). To recapitulate in very few words, the main mechanism is based on the change of the thalamocortical activity pattern due to the hyperpolarization of the cells, principally the thalamic neurons, because of the decreased cholinergic brain stem input to the thalamus. This hyperpolarization of thalamic neurons promote the rebound low-threshold calcium spike at lower frequencies than γ and the consequent excitation of the cortical cells that in turn project back to the thalamus and thus start a particular activity pattern in the thalamocortical loop characterized by slow oscillations. Without going into details that are not needed for the present purposes, the transition from the thalamocortical fast rhythms during wakefulness to the slow oscillations characteristic of sleep is gradual, and several waveforms appear in the EEG recordings that herald this transition, like the aforementioned spindles, brief oscillatory bursts ($\sim$1–2 s) at frequencies between 7 and 14 Hz that can be seen in lighter stages of sleep. With the thalamus and cortex involved in a slow oscillatory synchronized activity, processing of external sensory stimuli becomes impossible and thus the individual is unconscious, until the possibility to process stimuli (but now internal ones!) emerges during the REM episodes because now the cortical cell ensembles can exchange information almost as it is achieved during wakefulness due to the break up of the highly synchronous thalamocortical loop that results in the appearance of higher frequencies in the EEG. When cholinergic transmission starts to become prevalent, signaling the end of sleep, and depolarizes thalamic and cortical neurons, the tendency to express rebound bursting is diminished and therefore the slow wave rhythms in the thalamocortical loop tend to fade away, sensory inputs can be relayed by the thalamus, and the individual will wake up, ready to process sensory inputs thanks to the now open thalamic "gate." There are many works where readers can inspect the minutiae of the transition between these states, and the similarity of the neurophysiology of sleep with the mechanisms that result in absence seizures, for, as detailed in Chapters 3 (Sections 3.4.3) and 1 (Section 1.6), analogous cellular events occur during spike and wave discharges and during the sleep rhythms (the monograph by Destexhe and Sejnowski (2001), presents a comprehensive account of cellular and network events).

From the above discussion, it is clear that REM events that include dreams represent a very comparable brain state as during wakefulness, the main difference being that, since the thalamus is still not transferring the external sensory inputs, the cortex processes, or perhaps better to say mixes in aberrant ways, the stored information in the memory banks, and this constitutes dreams. At this moment, the coordinated cortical activity results in anomalous thoughts, dreams or nightmares, because, while some cortical regions coordinate their activity to process information, there is another coordination of activity that is lacking and fundamental for nonaberrant cogitation: the coordination with the thalamic nuclei and with other deep brain networks. Thus, we have a situation that may teach us something about pathological states like schizophrenia or psychoses in general: the cortex is doing what it can when some type of coordinated activity is present. In the sleep condition the result is a dream, perhaps a nightmare, in the nonsleep situation it may result in a hallucination. On the subject of the bizarre imagery of dreams, the transmitter/neuromodulator dopamine is again invoked. It turns out that dopamine neurotransmission is at the waking level during all sleep stages, but the brain is in a different state because, among other factors, it is depleted from other monoamines (serotonin and norepinephrine) but it is cholinergically stimulated form the brainstem during REM, and this results in the hallucinations we call dreams. This molecular/cellular scenario has been proposed to share some similarities with psychoses where there is, too, an abnormal sensitivity to dopamine (Hobson, 2004). Whatever the molecular mechanism, the consequence is an altered brain coordination dynamics that result in a variety of experiences that, depending on the context (sleeping or awake), are considered normal or deviant. More on this in Chapter 6.

5.1.1. Dynamical bifurcations in the transition to sleep and in the sleep cycle

Steyn-Ross *et al.* (2005) reported a growth of the fluctuation in the voltage recordings (EEG) in the transition between SWS and REM episodes. This observation suggests that the transition could be represented by a dynamical bifurcation. As commented several times in previous chapters, the standard deviation of the order parameter diverges as a critical point is approached, thus a growth of the amplitude of the "noise" (fluctuations of the order parameter) is indicative of a bifurcation. To be more precise, it is suggestive of a supercritical bifurcation because in the subcritical case, there may or may not be a precursory growth of the noise, these technical details are expounded in Sornette (2004). To assess the specific dynamics, the authors then created a mean-field model (which in this case means that the fine detailed anatomical structure is ignored, thus it is a "global" model) of the cortex to describe changes in the EEG patterns during sleep stages, basically representing a manifold of cortical steady states. The model included two main modulatory factors: acetylcholine and adenosine (the former promoted cell depolarization and the latter hyperpolarization), and they modeled the effects of these substances as slow changes in synaptic efficacy and cell resting potential. The results from the analysis of this model indicate the presence of a bifurcation in the transition from SWS to REM sleep. Specifically, they claim that a first-order phase transition occurs, which is (the thermodynamic) equivalent to a (mathematical) subcritical bifurcation. Their computational model is of interest as they used a pair of coupled Langevin equations, and, as mentioned in Section 2.4 of Chapter 2, these describe the time evolution of a parameter (voltages in this case) with a noisy term added, and it was argued in that section that this formalism may be more adequate to model nervous system dynamics because of the omnipresence of fluctuations at all levels. The stochastic simulation of their coupled Langevin equations showed an increase in the fluctuations of the cell resting potential on approaching a transition in the membrane potential, which the authors use to draw a parallel with the observed fluctuations of power in the experimental EEG recordings.

Other computational models of the sleep cycle exist, where more detailed cellular mechanisms are investigated (for instance, Diniz Behn *et al.* (2007)), but these are not reviewed here. Rather, it is the transitions at the collective level that are of interest in the present discussions. In this regard, and using a similar mean-field model of the cerebral cortex, other transitions between stable and unstable states of cell firing in the cortex have been inferred to explain the K-complexes, bursts detected in sleep EEG mentioned in Section 5.1 that are considered to be propagating cortical waves. This study also revealed that the cortex can be considered to be a self-organizing system close to the edge of stability (of a low firing rate, as detailed in Wilson *et al.* (2006)). Once again, the concept of itinerant transitions between (meta)stable states is implied. Signs of a self-organized critical system in sleep has been obtained in another study that analyzed the duration of sleep and wake episodes in rats (Comte *et al.*, 2006).

The subcritical bifurcation derived from the aforementioned analyses dictates that sometimes the transition is not gradual but abrupt. That result was obtained in the SWS–REM transition but, as discussed below, a similar bifurcation has been proposed on theoretical and experimental grounds in the transition from consciousness to unconsciousness during anesthesia. Perhaps, then, such a subcritical bifurcation may generally occur in the transition from wakefulness to unconsciousness. Recall that when a subcritical bifurcation takes place, the consequence is a completely reorganized dynamics (Section 2.3 and Figure 2.9 of Chapter 2). Another pictorial way to represent a subcritical bifurcation, or first-order phase transition, is shown in Figure 5.2. In this representation, the stable state of the order parameter X becomes metastable after the bifurcation parameter μ reaches a certain value μ_{met} and from that moment to the bifurcation value μ_c the system can suddenly jump (marked by arrows) to one of the stable states above or below represented by thick

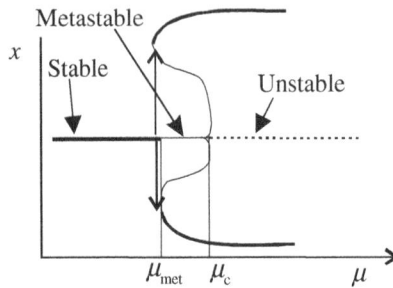

Figure 5.2. Bifurcation diagram of a subcritical, or first-order bifurcation. Axes are the same as those in Figure 2.9 of Chapter 2: X is the order parameter and μ is the bifurcation parameter. Stable states are in bold. The open arrows point to the possibility of sudden transitions as explained in the text.

lines, or, in the absence of fluctuations, the order parameter will abruptly switch to one of the branches when it reaches μ_c. In this situation, we find abrupt transitions. It is then conceivable to conjecture that the transition between wakefulness and sleep, in those occasions when it is so abrupt as we all have experienced from time to time, may be described by such a subcritical bifurcation. But, what could be the order parameter X at this mental level? For the time being, we leave it to the readers to consider this question.

While previous chapters have presented the idea of dopamine as a control parameter in brain function, acetylcholine can as well be acting as a control parameter at least in the sleep–wake dynamics because it influences the resting membrane potential of thalamic cells, is absent in SWS, and reappears in REM and wakefulness, and was used in the mean-field model described above as one of the two main modulatory effects.

5.2. Brain Coordination Dynamics of Altered States of Consciousness: Anesthesia, Coma, and Other Global Disorders of Consciousness

The phenomena that are reviewed here provide us with other scenarios of loss of consciousness and serve to examine the sometimes gradual transitions between consciousness and unconsciousness. As in the studies on the sleep patterns described above, anesthesia has been characterized in terms of dynamical bifurcations too. Anesthetics have a multiplicity of actions at the cellular level, and no single action at one specific molecular entity can account for all effects of these drugs (Pittson *et al.*, 2004), which reminds us of the comments on specificity in Chapter 2, Section 2.8. In general, the main, net effect of anesthetics is to reduce excitability by hyperpolarizing cells (Nicoll and Madison, 1982). The changes at the EEG level are similar to those of sleep onset: waning of fast frequencies and appearance of slow and higher amplitude waves. Kuizenga *et al.* (2001) reported changes in some aspects of the scalp EEG during induction of general anesthesia with a variety of agents, and found an increase in the low-frequency fluctuations (at delta band) during the transition to unconsciousness. Empirical studies found a decrease in spectral entropy caused by general anesthetics, a phenomenon that was predicted by the mean-field cortical model mentioned in Section 5.1.1, results that suggest that the collective patterns of brain activity become "simpler," or more organized (coordinated) during loss of consciousness (Sleigh *et al.*, 2004). Is then brain synchronization more widespread during anesthesia, as was reported during sleep? The literature on brain synchrony patterns during anesthesia may seem controversial upon a cursory reading, as some studies indicate higher than normal synchrony, whereas others report a decrease. But let us consider the main findings.

Coherence, evaluated in 10 patients undergoing propofol anesthesia, was found to be of lower amplitude during unconsciousness (Lee *et al.*, 2009), but this study was limited to the evaluation of synchronization between frontal and parietal areas, that is, the finding was that long-range coherence was decreased during anesthesia. This study also reported the directionality of frontoparietal coupling and compared it before, during, and after the anesthetic administration, and found a patient-specific hierarchy in the functional organization of the frontoparietal connectivity, which supports the notion of the fundamental importance of these networks in the determination of states of consciousness. While the frontoparietal coherence was found to diminish during unconsciousness in the aforementioned study, experiments in animals have revealed an enhanced synchrony within the thalamocortical networks. In a study focusing on the fast rhythmic activity, γ oscillations were found to be synchronized during ketamine–xylazine induced anesthesia in cats (Steriade *et al.*, 1996) (recall that fast rhythms occur during sleep too, and also during anesthesia, riding on the depolarizing component of the slow oscillation, Figure 5.1(B)), but the synchrony of fast rhythms tended to be local, whereas there was a large-scale synchronization of slow oscillations. Other studies reported a decrease in γ band synchronization between cortical areas during anesthesia derived from EEG signals in humans (John and Prichep, 2005) and in rats (Imas *et al.*, 2005, 2006), the decrease being more noticeable when evaluated between distant cortical areas (at least, those that are not neighbors). Using a different technique, functional magnetic resonance imaging (fMRI), the anesthetic midazolam was found to increase the temporal correlations in the BOLD signals at very low frequencies (Kiviniemi *et al.*, 2005), and, similarly using fMRI data, the temporal correlations were found to decrease upon induction of anesthesia in the areas of the default brain network (Deshpande *et al.*, 2010), which implies a decreased long-range synchrony.

A few observations are then in order, trying to sum up all these contributions with some apparently contradictory results. First, in these last two fMRI studies, the "synchrony" evaluated is of a different nature from the one assessed by fast electrophysiological recordings. Second, distinct anesthetics were used in all these studies, but we think this may not be a problem because, after all, the net result caused by all of them is a loss of consciousness and the likelihood that this will be reflected in similar synchrony patterns, we believe, is high. Third, the general trend that emerges from these studies is a tendency to increase long-range coordinated activity at low frequencies in thalamocortical and cortico-cortical networks, and a tendency toward decrease in long-range synchrony of high-frequency oscillations mostly in the cortex, while maintaining local high-frequency synchronization in the thalamic and thalamocortical networks. In summary, the picture that becomes apparent is that it is plausible that local neighborhoods may maintain fast rhythms synchronized even during loss of consciousness but the long-range γ synchrony characteristic of wakefulness and sensory processing is lost, and it is taken over by a more widespread coordination at lower frequencies during unconscious states, just as reported in Section 5.1 for sleep. Perhaps at this point a figure is worth many words. Figure 5.3 shows the cortical local synchronous firing in an anesthetized rat during the slow rhythm at a frequency within the δ range ($\sim$2 Hz). Let us imagine for a moment that this is a typical situation during anesthesia, and that phase synchrony, for instance, is evaluated in those recordings. The analysis would probably reveal phase locking

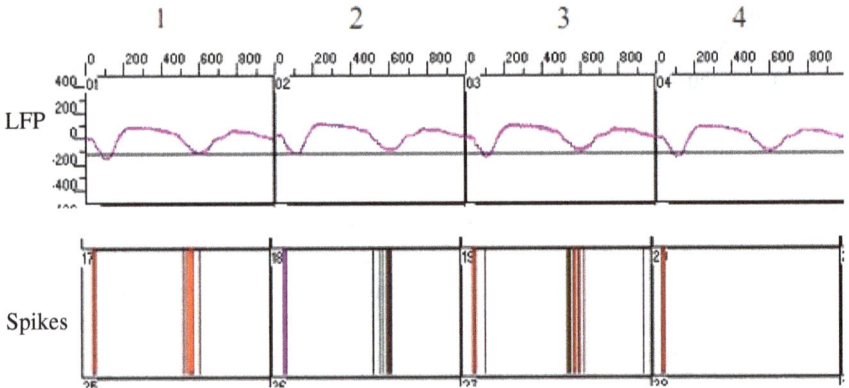

Figure 5.3. Synchronous firing of bursts during anesthesia. Recordings (each box is 1 s) of LFPs (upper panel) and identified spikes from several cells (lower panel) in four sensors of an 8-trode (ALA Scientific Instruments) inserted into the cortex of a rat anesthetized with gamma-hydroxybutyric acid. Note the slow collective activity recorded as the $\sim$2 Hz LFP oscillation. The spike sorting algorithm detected a few spikes from, possibly, three different cells, that were more or less synchronous, at least the bursts of spikes were synchronous, and coincided with the downward deflection of the LFP.

(phase synchrony) between the LFP recordings, but when evaluating the synchrony between spikes (at high frequencies) the phase locking may be lost due to the fluctuations in the timing of each spike, even though the bursts of spikes may be phase locked but this bursting would correspond to the low-frequency pattern at about 2 Hz. Same considerations can be made of the recordings in Figure 5.1(B). It is then entirely possible to find a higher synchrony in the slow oscillations and lower in the case of fast rhythms, a result that has been obtained in recordings of *in vitro* seizure-like events and that was commented in Section 3.4.2 of Chapter 3, particularly the study where periods of synchronization at different time scales was reported during experimental seizures precisely because the analysis of the synchrony of individual spike firing did not lead to high values of synchrony due to the jitter in the action potentials within or between bursts (Netoff and Schiff, 2002). It all depends on what type of synchronization is evaluated, recall the comments in Chapter 1, Section 1.3.

How are these observations of brain coordinated activity related to possible dynamical bifurcations as was reported in the case of the sleep patterns mentioned in Section 5.1.1? Using, again, a mean-field continuum model of the cortex, Steyn-Ross and colleagues provided evidence for an abrupt transition to unconsciousness caused by anesthetics described by a first-order phase transition, or subcritical bifurcation, very much as these authors found for the transition between sleep stages discussed above. The bifurcation occurs, in their model, at some critical concentration of the anesthetic agent, and further computer simulations revealed that the voltage fluctuations of cell populations in the unconscious state should be more correlated than in wakefulness, results that complement the empirical observations (Steyn-Ross *et al.*, 2004). In the model, the bifurcation was evaluated by monitoring the population average voltages, but this can be translated to the collective dynamic patterns that have been described above. Essentially, a transition toward a synchronization manifold can be postulated, once again the dynamics expounded in Section 2.2.3 may have relevance here, where a wandering of the dynamics from attractor to attractor results in the transient stabilization of possible metastable states; the main state that is stabilized in these situations would be the synchronous state that can be represented by a toroidal manifold. The proposal of this specific type of bifurcation in anesthesia came from a model, not surprisingly because, as perhaps overemphasized in this book, it is only from analytical models that precise dynamical analysis can be performed. However, signatures for the occurrence of a bifurcation, both in anesthesia and in sleep, have been obtained by the same group in experimental recordings from anesthetized rats (Wilson *et al.*, 2008), and from scalp EEG recordings in humans during sleep (Steyn-Ross *et al.*, 2005). These signs included precursors of an incoming phase transition that were observed in the increase of fluctuations in the recordings, because, as was noted in previous sections and chapters, the growth of the fluctuations is a sign of an impending phase transition. This is, to our knowledge, as far as the empirical evidence has gone so far in detecting specific bifurcations in the transition from consciousness to unconsciousness in sleep and anesthesia. Chapter 3 introduced other empirical evidence with regards to seizures, which also cause loss of consciousness after all.

More detailed work on the specifics of individual neurons provided further evidence for a dynamical bifurcation. It was observed that a common neuronal firing pattern in the slow sleep rhythms and during anesthesia occurs, consisting of up (depolarized) and down (hyperpolarized) states (these were already alluded to in Section 1.2 when discussing central pattern generators). These two states are considered to represent stable solutions for the cell membrane potential equation describing its dynamics. Many conditions can result in these two states, and there is already an extensive literature on this topic (Steriade, 2001; Destexhe, 2009). During sleep and anesthesia, these states occur with a frequency of about 1 Hz, this is the slow rhythm, and one example can be seen in Figure 5.1(B). A study of the temporal fluctuations of these states recorded in neurons during the cortical slow oscillation in rats revealed some indications of a subcritical bifurcation, now at the cell level (the previous studies were done using a mean-field model), but a note of caution, added by the authors, was that it is not easy, in experimental settings, to identify clearly the position of the transition (Wilson *et al.*, 2008, 2010). This cautionary thought, perhaps a limitation of these empirical studies, was already considered in Chapters 2 and 3 when discussing the transition to seizures: how can the bifurcation be best identified, by collective or by individual cell recordings? Further, where to mark the exact moment of the transition? This limitation applies to all comments that appear in the following chapters when discussing dynamical bifurcations in other conditions. The study by Wilson *et al.* (2010), in particular, found no great evidence for the phenomenon of bistability in the emergence of the up- and down states. Recall that cellular bistability was discussed in Chapter 1 in the context of rhythmic activity in central pattern generators. Rather, the authors suggest that the transition between these states during slow cortical rhythms is due to synaptic inputs coming from the thalamus. If this were to be the case, then, the bifurcation, or control, parameter would be a synaptic one. In conclusion, there are signs of a dynamical bifurcation in brain activity at the cellular- and network level, and because it is proposed to be of subcritical nature, the possibility of sudden transitions exist (Figure 5.2) which may account for the commonly experienced abrupt loss of consciousness upon sedation.

In general, then, anesthetics disrupt the coordinated activity that is needed for information processing, and the typical scenario that was seen when discussing seizures and Parkinson's disease materializes once again: more widespread synchronization, particularly in the slow oscillations, is associated with loss of consciousness, in anesthesia and sleep as well. As it has been stressed many times throughout this book, the fact that there are more synchronous activities does not mean that the networks are interacting in a meaningful manner, meaningful in the sense of sensorimotor processing. For this reason, one could say that during anesthesia and sleep (and seizures) the brain is less functionally connected, but more synchronized. Or, using the functional coordination notion described in Chapter 1, the *brain is less functionally coordinated in unconscious states*. It seems that conscious awareness needs a particular functional coordination, specific coordinated activity patterns that do not involve widespread and long-lasting synchronous activities. It is not a matter of more or less activation of brain areas (and thus neuroimaging results based on metabolic activations have to be carefully interpreted), but of the coordination of those activities. Higher than normal activity in certain brain regions occurs during seizures and yet patients lose consciousness (Blumenfeld and Taylor, 2003). According to this view, this situation can be expected in other altered states of consciousness, and to them we turn now.

5.2.1. Brain dynamics in coma, vegetative, and MCS

Global disorders of consciousness that merit consideration from the perspective of the continuum in states of awareness include the conditions of coma, VS, and MCSs. The locked-in syndrome is not really a disorder of consciousness and is not discussed here. The reviews by Adrian Owen (2008) and Steven Laureys *et al.* (2004) are recommended for an introduction to these pathologies. Here, these states are briefly considered in order to illustrate the importance of the brain coordinated activity in the emergence of awareness, and also to reveal the graded nature of consciousness, although the insult (for instance, an impact to the head) may have resulted in a robust perturbation to the system.

Whether a result of traumatic brain injury (TBI), ischemic episodes, or any other insults to the nervous system, the individual may lose consciousness. This poses no small problem in settings like critical care units, where clinicians have to treat patients with whom they cannot communicate. Not surprisingly, misdiagnosis is common between states like coma, VS, and akinetic mutism. It was reported that the first person to realize that a patient is conscious is a family member in

~55% cases, and only in 23% cases it is the physician who makes the observation! (León-Carrión *et al.*, 2002).

The pathophysiology of TBI, including not only impact to the skull but also ischemic episodes, is relatively well known. The first result of the injury is a depolarization of cells which brings about a period of hyperexcitability, followed by a prolonged period of hypoexcitability where phenomena like spreading depression may take place (Shaw, 2002). The altered neurophysiology is reflected in the EEG as a generalized slowing of brain frequencies in the δ and θ ranges (Bricolo and Turella, 1990). These bandwidths predominate while patients remain in coma, which is a common pattern similar to the likes of those described during SWS or anesthesia in the previous sections. Detailed electrophysiological analysis of patients with cerebral trauma and concussion was first reported in the 1970s, and of special attention was the appearance of slow waves, which had already been noted in the 1940s (Omaya and Gennarelli, 1974; Gloor *et al.*, 1977). The slow oscillation appearance of the EEG already suggests a pathological scenario, specifically that a long-range and long-lasting synchronization may be occurring. The synchrony of EEG signals after these injuries has been evaluated in a few studies. One of the first studies that evaluated EEG coherence in a very large patient population, that of Thatcher *et al.* (1989), proposed that the analysis of the coherence of post-traumatic EEG can detect and quantify diffuse axonal injury and is a good predictor of outcome post-injury. This study indicated an increased coherence in the EEG signals in frontal and frontotemporal regions. To our knowledge, there have been very few studies assessing the brain synchronization in the VS or MCS. One study has examined coherence in EEG signal in a VS patient, but these results were not compared to control noninjured subjects. The investigators detected very different coherence patterns depending on the cerebral hemisphere due to, the authors propose, the unilateral damage to the brain in that patient (Davey *et al.*, 2000). Another study found an increased power in the δ band in several MCS patients, which, as noted in previous chapters, indicate a synchronization of synaptic activities (Leon-Carrión *et al.*, 2008). Other measures, somehow related to synchrony of neural activity, have demonstrated impaired functional connectivity between the parietal association cortex (the precuneus, in the medial area of the parietal lobe) and the reticular activating system of the brainstem. This indicates that without meaningful reticular input to the cortex, the brain reverts its operation to its origins: slow and synchronous rhythmic activities (Silva *et al.*, 2010). In another very recent study, lower entropy of EEG signals in VS patients compared with controls also suggest a decreased "complexity" in this state, which can somehow parallel a more synchronous cortical state (Sarà and Pistoia, 2010).

The evaluation of changes in cortical phase synchrony in the brains of adults with TBI revealed a variety of brain dynamics as patients emerged from coma, with both increasing synchrony in cortical areas adjacent to the injury, and decreasing synchrony within brain hemispheres as the level of consciousness increased, more particularly the synchrony in the δ band tended to be higher in the more unconscious patients, the level of consciousness being measured with the Glasgow Coma Scale (Shields *et al.*, 2007). The decrease in global synchrony with recovery of consciousness is related to other findings of increased temporal variability in phase synchronization, derived from scalp EEG in children, as patients recovered and emerged from coma after an episode of TBI (Nenadovic *et al.*, 2008). It was described in Chapter 3 how the spatiotemporal complexity of the brain synchrony pattern is decreased during epileptic seizures, and now we find a similar scenario after TBI. Of interest to the clinician, the evaluation of the variability of EEG signals may be of prognostic value, and, more specifically, the latter study indicated that a persistent decrease in the temporal variability at δ frequency could be associated with poor outcome post-TBI. We have recently obtained similar observations in other cases of coma not associated with TBI, but with cardiac arrest (Nenadovic *et al.*, 2009). This may represent a novel diagnostic and prognostic approach in the evaluation of TBI and related conditions, but still more data are needed to validate this procedure. Incidentally, the idea that coherence analysis of EEG signals could be one of the best predictors of outcome may have been originally advanced by Ducker *et al.* (1982).

One point that could be raised is that most of these patients are sedated after severe neurotrauma, and thus the changes in synchronization detected could be due to the sedation rather than the comatose state. The study by Shields *et al.*, mentioned in the previous paragraph, evaluated the phase synchrony related to the amount of sedation in some patients, and found no significant correlations. In any event, the consequence of medications in this type of analysis is not easy to control. One aspect that is apparent is that the EEG waveforms in nonsedated comatose individuals and in anesthetized subjects are very similar, regardless of the, obviously very different, specific mechanisms. Equally true is the fact that the individuals are unconscious in all these conditions. It is not surprising then that the analysis provides similar results. Another point that may seem peculiar is the existence of higher cortical synchronization after TBI because in these injuries, cortical axons tend to be damaged, thus interrupting communication between cortical networks, so why are they more synchronized? The answer could be that the remaining axons still are enough to cause an extensive synchronization, but another, more probable factor, is that the cortical networks are receiving synchronous inputs from deep structures, like the thalamus, so with more drive from other brain regions, the analysis of the cortical signals will reveal synchronization. This would be the typical case of a common drive from one source to several areas. It was mentioned two paragraphs above that less coupling between the brainstem reticular formation and the cortex was found, but there are other deep structures that can drive the cortex into slow synchronous rhythms. As a matter of fact, anesthetics tend to suppress excitatory synaptic transmission but there is still higher cortical synchrony. It was suggested in previous chapters that the cortical circuitries allow for the avoidance of the brain becoming entrapped in long-lasting limit cycle dynamics that is more characteristic of other deep-brain structures. Once the cortex is knocked out by either anesthetics or injuries, the perennial and ancestral neural slow oscillations materialize.

After many comments in previous chapters and in this one, the idea that reduced variability in brain synchrony is associated with disease sounds almost trivial. It is, in fact, no more than a reinterpretation of the concept of metastability and relative coordination: a transient formation of cell assembles by coordinating their activity is needed for efficient sensorimotor processing. This transience should be reflected to some degree in fluctuations in synchrony, thereby it can be hypothesized that the lower the variability, the less the information processing. As to how to measure the degree of variability and fluctuations and how to correlate these measures with brain information processing, is a complicated problem. To start with, is there any upper- or lower bound to the variability? When is too much or too little neural synchronization unhealthy? McIntosh *et al.* (2008) concluded that "In a certain way it could be stated that a noisy brain is a healthy brain," but, we wonder, noisy to what extent? We provide some comments in this regard in Section 5.3.

5.2.2.   Global brain deactivations in disorders of consciousness

The distinction between these states of altered consciousness is based on the behavior of the patients, and it can immediately be appreciated how delicate this matter is in terms of making an accurate diagnosis, for patients in these states cannot reliably communicate or exert almost any behavior. Can it be the case that the mental life of some of these patients is comparable to the normal individual, even though they are somehow restricted in behaviors? Recent imaging and neurophysiological evidence indicates that this may be true in some instances, as described below. Figure 5.4 represents a depiction of the cognitive and motor impairments that are used to classify these disorders of consciousness. The main characteristics of these states are explained in the figure legend.

To summarize a wide spectrum of experimental observations, it can be stated that the brain in coma has less metabolic activations throughout, but, on the other hand, there are specific deactivations in different brain areas in the vegetative and the MCS. This is shown in Figure 5.5, the brain metabolic activity in coma, VS,

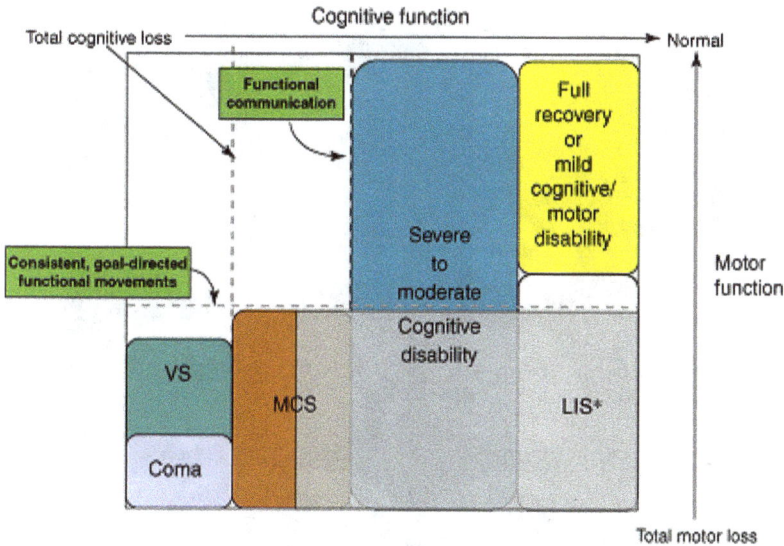

Figure 5.4. Disorders of consciousness following severe brain injuries. The *x* axis represents the degree of cognitive impairment while the *y* axis is the degree of motor (dys)function. Coma and VS are the more impaired states (lower left-hand corner) and are considered unconscious brain states, and therefore are placed to the left of the line "total cognitive loss." The MCS is characterized by demonstrations of inconsistent evidence of awareness of self and environment. In the locked-in state (LIS), patients are fully conscious but cannot move. Reproduced with permission from N.D. Schiff (2010).

sleep, and anesthesia as compared with resting wakefulness. The common regions that are "deactivated" include the frontal and parietal cortices, two main association areas that are thought to be fundamental for self-awareness and attention in general. The lower metabolic activity is normally taken as a surrogate of less synaptic and cellular function in general. If this is true, then what is seen here is basically a representation of the concept of the global workspace theory. This framework proposes that multimodal perceptions, emotions, memories, and anticipations become subjectively integrated in a continuously changing flow of consciousness, and the bringing together of activity in different brain modules into a conscious workspace favors that information become available for inspection by other processes/modules distributed throughout the brain (Baars, 1988). Interruptions in the broadcast of information from the primary sensory cortical areas to the association areas (parietal and frontal mostly) results in the subject not being aware of the sensory input that has reached those primary sensory areas. In other words, it is expected that conscious events (normally understood as those reportable by the individual) are associated with widespread cortical activity, a concept for which there is empirical evidence (Lamme, 2006). But there must be some coordination in

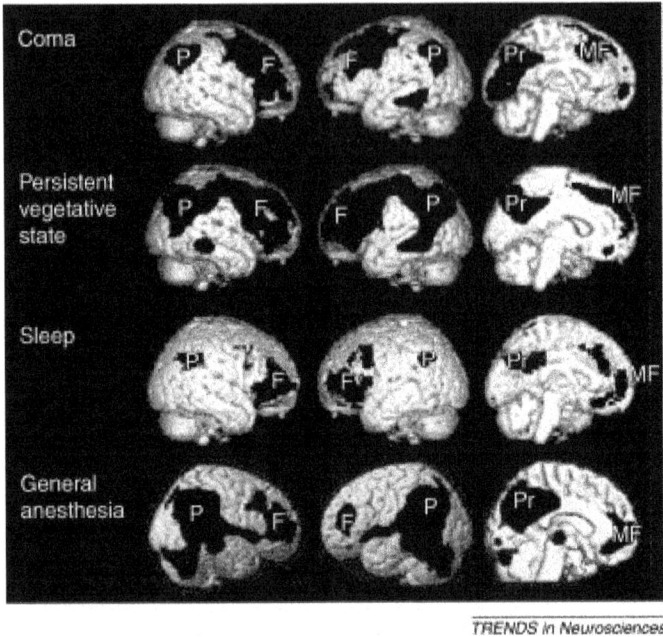

Figure 5.5. Brain metabolism is decreased in altered states of consciousness. Positron emission tomography (PET) scans show the regional decreases (in black) in metabolism when unconscious states are compared with resting consciousness. All show decreased metabolism in frontal (F) and parietal (P) areas. Pr is posterior cingulated/precuneus and MF is mesial frontal cortex. Reproduced with permission from Baars *et al.* (2003).

this long-range spread of activity, which is the most important factor as continuously highlighted in this book. The wide broadcasting of information to reach awareness is not only characteristic of sensory perception, but also of motor actions. As an illustration, a report on one individual with alien hand syndrome showed more localized brain activations (mostly in contralateral primary motor cortex) measured by fMRI, during the performance of movements of the alien limb, that is, when the patient is moving the limb but he declares that is not working according to his will and many times is not even aware of these movements. On the other hand, the activations were more distributed when he performed voluntary movements, the activations including those of the motor and premotor cortices and inferior frontal gyrus (Assal *et al.*, 2007). Thus, awareness of movement seems to be associated with widespread brain activations in various regions. The observations in this patient are consistent with the reports on the generation of consciously produced movements by direct brain stimulation. Stimulating motor cortices makes the patient move a limb but without awareness, while stimulations of parietal cortex, which probably causes a more distributed broadcast of the signal, makes the movement conscious. In addition, the readiness potential (an EEG waveform that precedes self-initiated movements) recorded by scalp EEG along motor cortices is reduced in patients with parietal damage who also have distorted conscious experience of willing to move, which indicates the importance of interactions between numerous brain areas for not only sensory but also motor awareness, at the behavioral level, and for normal neurophysiology (the readiness potential) at the cellular level (these accounts were reviewed by Desmurget and Sirigu (2009)).

Clinicians must be very careful and should not take for granted that someone in VS does not really perceive the sensorium and is not aware of anything at all, because there is a lack of activation of cortical association areas. Brain imaging revealed residual activity in brain networks in some VS patients (Plum *et al.*, 1998) and, even large-scale networks, in the MCS (Schiff *et al.*, 2005). This residual brain activation measured using neuroimaging methods suggests that some brain areas could be processing information. Electrophysiological recordings are more adequate to assess this aspect. Processing of sensory stimuli, at least a partial processing, in VS and MCS patients has been inferred from magnetoencephalography, in that the MEG recorded waveforms in response to auditory stimuli were similar to the expected waveform in a noninjured brain (reviewed by Laureys *et al.* (2004)). Going back to neuroimaging evidence, the brain response to sensory stimuli in the VS normally remains localized to the primary sensory areas, but there was a wider activation of areas beyond primary cortices in patients in MCS. These observations have led to the notion that processing in the VS does not lead to a complete integration of the information received, and thus it is not consciously perceived, but in MCS things are different because of the wider activation of networks and even a hope for an improvement arises here. After all, if some areas tend to become activated by sensory input in this state, why not try to improve this activation by some means? It was already known since long ago that, after bilateral destruction of the midbrain reticular formation that left patients in coma, repetitive stimulation for prolonged time periods of 20 days with direct current (DC) in the lamella pallidi and in the intralaminar thalamic nuclei resulted in arousal, observed both behaviorally and in the EEG recordings (Hassler, 1969). Accordingly, Schiff and (many) colleagues used deep-brain stimulation (DBS) in a patient who had been for six years in MCS following TBI. Some aspects of DBS were commented in Chapter 3 in the context of the control of epileptiform activity and in Chapter 4 on the improvement of motor dysfunctions. Neuroimaging done on this patient revealed the preservation of a large-scale activation of brain areas, in comparison to control subjects and other VS or MCS cases, and therefore an attempt was made to enlarge and improve the existing communication among those brain areas. The DBS electrodes were inserted into the intrathalamic nuclei and adjacent regions, which is a good target because these nuclei project widely to the cortex and thus provide a source of excitatory inputs and the possibility to enhance communication between brain regions. The DBS treatment restored some behavioral responsiveness, in that the patient recovered limb control and other traits (Schiff *et al.*, 2007).

While this was a forced recovery, spontaneous recoveries of patients have also been known. Take, for example, that of the recovery of motor function in a patient who had remained in MCS for 19 years (Voss *et al.*, 2006). Neuroimaging performed in the patient's brain revealed a recovery of activation in brain regions of the midline cortex, areas that have been proposed to be important for self-awareness. Brain imaging in the few patients who regain consciousness after VS showed an increased metabolic activation of specific brain regions such as the precuneus and a wide frontoparietal association areas (Laureys *et al.*, 1999).

The previously mentioned observations of cases of wider brain activation upon sensory input in these patients make it very probable that there are brain networks partially active and thus processing information. Naturally, the question of whether they are aware of those sensory stimuli can never be answered, at least currently, because that necessitates the first-person report that can hardly be obtained from these individuals. Perchance if frontoparietal cortices become active, there is the possibility that the patients do experience the stimuli, but this is only an educated guess.

5.3. Main Conclusions and a few Speculations

The phenomena reviewed in this chapter have provided more indications of the common theme that seems to arise when thinking about efficient information processing by the brain: the tendency toward widespread and long-lasting synchronization will manifest when given the opportunity and this pattern of activity is not suitable for an information processing that confers adaptive value to the individual. These phenomena provide the opportunity to start pondering about how to characterize, perhaps even quantify, adaptive information processing. These considerations remain speculative, but all research is an exploration, is it not? Thus, we shall speculate a bit. Those grant reviewers who disapprove of exploratory projects are invited to leave the chapter at this point.

The neurophysiological data and the behavioral manifestations of patients with the disorders of consciousness here examined indicate that unconsciousness is characterized by less widespread activation of brain regions, activation understood in the metabolic sense (glucose or oxygen consumption, since these are the foundations of the neuroimaging methods used in these studies). At the same time, widespread brain synchronization has been estimated from most cases when conscious awareness is absent, either in sleep, anesthesia, or coma. This long-range synchronous activity indicates that, in fact, most of the brain is working in a correlated (but not coordinated or organized) fashion, which means the brain networks are exchanging information, at least understood in terms of correlated synaptic activity. The problem resides in the fact that to process a wide variety of sensorimotor processes, the synchronous activities must be very transient and varied (and organized), this is just a variant of the principle of integration and segregation of information discussed in previous chapters. Hence, one possible solution to the problem of how to go about determining efficient and adaptive information processing lies in the quantification (if at all possible) of the spatiotemporal variability in synchrony, rather than its magnitude: it is fine to have a very high synchrony for a very short time and in the networks required when performing a task; it is not fine if the synchrony remains for a long time and spreads over other areas that are not supposed to be involved in that task. Hence, it seems that the quantification of the fluctuations in neural coordination is of significance when addressing this problem. Studies have been carried out in which the variability in synchrony has been evaluated in some pathological conditions, and is mentioned in Chapter 6 when discussing about autism and schizophrenia principally, and in this chapter was examined in Section 5.2.1 in the context of TBI. According with this line of reasoning, it is conceivable that less fluctuations in synchrony will be associated with neuropathologies. This is perhaps a particular case for the more general phenomenon of less variability in physiological signals that has been many times, and more clearly in heart dynamics, found correlated with different diseases. Loss of complexity in disease and aging was proposed and has long been known (Lipsitz and Goldberger, 1992; Pincus, 2001). The altered interaction among physiological systems in pathologies due to either external inputs or changes in coupling between organs or networks implies that the complexity will change, and Lipsitz and Goldberger submitted the notion that the changes in complexity in physiological systems reflect underlying alterations in structural and functional organization. Biological entities are characterized by adaptability, which requires flexibility in responses to a changing environment. This is one reason as to why the nervous system has evolved. Section 5.2.1 was closed posing the question of whether there are boundaries of the variability in brain function to be adaptive, healthy. Those interested in the general theory of fluctuations in dynamical systems can consult many works in the field of physics, perhaps the studies of Prigogine and colleagues in nonequilibrium thermodynamics are clear illustrations where the crucial role of noise and fluctuations on the emergence of organized patterns is demonstrated (reviewed in Prigogine and Stengers (1980)).

These investigators taught us that fluctuations lead to instabilities, which in turn will determine new dynamical regimes (recall the description of the Rayleigh-Bénard instability in Section 2.3 of Chapter 2). In these situations far from equilibrium, the fluctuations drive the system to different states, so it is in the nervous system activity, where fluctuations in the activity allow for a more complete search of the dynamical repertoire within the state space where its dynamics reside. The signs of a dynamical bifurcation in brain activity in sleep and anesthesia (and epilepsy as described in Chapter 3) accentuate the importance of fluctuations. Recall that a subcritical bifurcation was proposed (Sections 5.1.1 and 5.2), and a brief look at Figure 5.2 reveals how fluctuations can make the dynamics "jump" toward different states, bringing about the possibility of sudden transitions which may account for quick loss of consciousness on sedation or in the transition to sleep patterns.

The neuroimaging data reviewed above that has been correlated with the behavior in patients with disorders of consciousness are just that, a correlation between behaviors and data. Does it imply a causal relation? Some argue that this is not the case (Nachev and Hacker, 2010). These matters are surely debatable, mainly when thinking about the possible control data with which to compare the patients' data. Nevertheless, the data from patients must be compared to something, and the best "something" are data from healthy subjects, and the differences are apparent. Once again one faces the question of whether the measurements obtained from nervous tissue underlie, or cause, the observed behaviors. Let us just say that, to some degree, there must be a relation, and, as someone said, relations are all we can hope to uncover. The view taken here, already mentioned in Section 1.8, is that the dynamics patterns in the nervous system are inseparable from the behaviors, the behaviors are the patterns themselves, these are two aspects of the same phenomenon.

Taken together, the observations reviewed here strongly suggest that consciousness is not all-or-nothing, rather it seems to be a gradual process. The clinical experience in critical care medicine with the assessment of recovery after TBI also reinforces this point. Consciousness is a graded, continuum process and depends on the dynamic coordinated activity among many modules. There can be moments of sudden transitions too, abrupt loss or recovery of consciousness, these mainly related to sleep or anesthesia, but, it may be fair to declare that the gradual transitions are more common. It is not only that the demarcation between the conscious and the unconscious is unclear, as Jung remarked and was quoted in the second paragraph of this chapter, but also the distinction of conscious and unconscious cognitive processes is not easy to delineate, as pointed out by Lamme and also cited at the start. The two comments by these influential investigators of the mind should always be taken into consideration.

5.4. Epilogue: On the Concept of Brain Death, or How Many Parts of the Body Have to Die for One to be Really Dead

Addressing the phenomena of unconsciousness after injuries provokes a few comments on death, brain death principally, and more because the question has risen in modern times, especially after the advent of life-support technologies that can maintain "life," such as mechanical ventilators that assist respiration, in spite of a flat EEG signal. In this case, is the patient alive or dead? The brain seems to be dead, but some muscles in the thorax are mechanically animated. Much as consciousness has been proposed in this chapter to be a gradual process, death is also gradual (Laureys, 2005). The concept of brain death is far from clear, something that may sound surprising since, after all, an isoelectric EEG trace surely indicates absence of neural activity (or, more precisely, absence of synchronous neural activity, particularly if scalp EEG is measured). There are a few definitions of brain death, from the "whole-brain death" concept (the irreversible cessation of the critical functions of the entire brain) to the "brain stem death" (the loss of the capacity for consciousness combined with loss of capacity to breath). But the reviewed data in this chapter have shown that some aspect of consciousness may be preserved but can be hard to be clinically detected, even after destruction of large parts of the brainstem. There are even other brain death concepts that would "kill" some patients, for example, the "higher-brain formulation," the loss of higher function served by neocortex, a designation which implies that patients in VS are dead by this definition. Currently, the "whole brain death" notion seems to be more widely accepted in the world, but there are new papers constantly appearing on this theme, so opinions may change or even new definitions may arise. Nevertheless, the crucial aspect is that the artificial dissociation of vital functions that has become apparent with the technological advances makes life no easier for the critical care physician. But in general, death, like life, is a biological phenomenon hard to define, and one feels that its designation is often constructed around artificial moral aspects and legal policies. So, when J.L. Bernat (1992) asked: "How much of the brain must die in brain death?" in the title of one of his papers, he may have had a point.

CHAPTER 6

PSYCHIATRIC DEVIATIONS: TRANSITIONS WITHIN THE BEHAVIORAL CONTINUUM

Behaviors that escape the mean region in the Gaussian distribution are normally assigned the label of "deviations." This chapter presents some either considerable or subtle deviations in behaviors and the corresponding brain dynamic patterns at least as it is presently known. The analysis of the brain collective activity and the coordination dynamics in these syndromes have not reached the depth of those studies presented in previous chapters on epilepsy and motor disorders, nonetheless, some general considerations can be already envisaged. Slight deviations in behavior are those found, for example, in high-functioning autism that is treated in Section 6.1. Larger behavioral deviations are present in schizophrenia, low-functioning autism, and a variety of disorders that will be the topics in subsequent sections. Throughout history, various views on behavioral normality have been advanced: normality as health, as utopia, as average, and as process represents four perspectives, and most likely, each individual may have his/her own views on this matter. For instance, Sigmund Freud defined it as the ability "to love and to work." To a large extent, the finding of a precise definition is, in the end, an exciting semantic exercise for those who care.

Regardless of the precise demarcations, a view of psychosis as a continuum has been advanced, from unipolar through bipolar disorders to schizophrenia, even though other, earlier perspectives, do not share this notion of a smooth continuum (these aspects have been discussed, from the genetic standpoint, by Crow (1986)). Today, it is acknowledged that different molecular alterations converge on common phenotypes associated with behavioral anomalies. As an example, a failure in neuronal homeostasis has been proposed to underlie autism spectrum disorders (ASDs) and mental retardation syndromes (Ramocki and Zoghbi, 2008). On the other hand, sometimes, same biochemical factors result in different psychiatric phenotypes, for instance, common gene defects are found in schizophrenia and bipolar disorders. As many times repeated in this volume, there are different levels of the continuum, molecular, behavioral, and even semantic.

Indeed, many of the syndromes occur simultaneously in the same patient such as depression showing comorbidity with anxiety, or Parkinson's disease with other cognitive disorders, cutting across the disciplines of neurology, psychiatry, and psychology. In this regard, the high-level perspective may be useful to comprehend the global nature of these mental phenomena, and perhaps as a consequence, the application of dynamic concepts to neurology and psychiatry is increasing in current times (for instance, see the special issue of the *Journal of Biological Physics* edited by Braun *et al.* (2008)).

6.1. Autism Spectrum Disorders

Autism and related syndromes (ASDs), not uncommon "disorders" that occur in $\sim 0.6\%$ of individuals, are accompanied by a different style of brain information processing, often reflected in the behavioral features of individuals with ASD. The ASD is mainly characterized by difficulties in social interaction and narrowed interests. The Austrian psychiatrist L. Kanner originally described autism as "the inability to experience wholes without full attention to the constituent parts" (Kanner, 1943). It would be perhaps unfair to attempt to confine the wide spectrum of this condition to one specific autism phenotype, for some individuals with this syndrome lead almost normal lives, for instance those with the so-called high-functioning autism including the Asperger syndrome; others, on the other hand, are severely cognitively impaired. Therefore, the autistic phenotypes are very heterogeneous that, to complicate matters, have associated comorbid disorders. It is also tempting to try to expose a commonality in the brain mechanisms that result in the broad repertoire of autistic behaviors, from the high- to the low-functioning individuals. Can one reasonably expect to find any common neurodynamics, or even neuropathology, in such a varied set of individuals? The following paragraphs summarize empirical evidence for altered brain dynamics, but the prominent lingering question is, is there common dynamical patterns in the autistic brain that differ from the nonautistic?

Some have suggested that the behavioral characteristics of autism may be reflected in a bias toward local, detail-focused processing rather than global processing in what has been termed the "weak central coherence" theory (Frith, 1989; Happe and Frith, 2006). The distinct behaviors of these individuals (which of course do not escape from being considered anomalous in spite of the advantage that the "local" approach may offer, perhaps causing the superiority of these subjects in some tasks (Shah and Frith, 1983)) result from possible events at molecular, cellular, anatomical, and biophysical levels (this last term comprising cellular and network characteristics, hence it overlaps the others). However, these alterations may be very subtle and hard to identify, as exemplified by the numerous post-mortem studies in autistic brains, where the clearest evidence, to our knowledge, for a microscopic pathology was found only in the cerebellum. Recent years have seen more studies being done trying to understand the collective, global dynamics of brain function in autistic disorders, because the information-processing characteristics of the autistic brains, conceivably, will be reflected in different patterns of brain activity.

Brock *et al.* (2002) have suggested that there may be a temporal binding deficit in neural activity in autism, as a sort of neurophysiological correlate of the weak central coherence theory. For example, if this hypothesis is true, it could conceivably be reflected in differences in brain synchrony patterns. Nonetheless, while some psychological studies have found support for the weak central coherence concept (Grinter *et al.*, 2010), others have not supported it (Ozonoff *et al.*, 1994; Brian and Bryson, 1996), and the universality of a local processing bias in autism continues to be debated (Behrmann *et al.*, 2006). Proposals of disruption of coordinated timing in neuronal activity in autism have been recently advanced (Herbert, 2005), along with the possibility of reduced brain synchronization (Uhlhaas and Singer, 2007; Rippon *et al.*, 2007). Recent neuroimaging evidence supports the concept of reduced functional connectivity in autism, for instance, in visuomotor performance (Villalobos *et al.*, 2005) and sentence comprehension (Just *et al.*, 2004), while more extensive task-related activations between frontal areas were detected in control participants but not in those with autism, in a functional magnetic resonance imaging (fMRI) study using the embedded figures task (Ring *et al.*, 1999); underconnectivity in inhibitory networks of the frontal–parietal cortices has been documented (Kana *et al.*, 2007); as well as less connectivity in the baseline resting state of cortical networks (Cherkassky *et al.*, 2006; Murias *et al.*, 2007). An interesting piece of information from the study of Ring *et al.* was that subjects with autism showed a more restricted activation pattern than control subjects during the performance of a perceptual task, the embedded figures task. In this task, a relatively simple target shape is presented, followed by a complex figure containing the simple target, and the participant is then asked to find the target shape embedded in the complex figure. In this task, individuals with ASD have been reported to perform better than control participants perhaps because, it has been hypothesized, the autistic bias towards processing local details that allows these individuals to find the specific figure within the complex picture (Shah and Frith, 1983). The more restricted brain activations observed using these neuroimaging methods support the general idea aforementioned of less, or at least different, global brain coordination, but of course does not imply that brains with this characteristic may not perform better in some behaviors, as seems to be the case in this task. Almost all of these studies have relied on metabolic measurements, fMRI or positron emission tomography (PET) characteristically. Section 6.1.1 presents a more direct approach to the inspection of brain dynamics in ASD.

6.1.1. Different patterns of brain coordination dynamics in autistic brains

A more direct approach to assessing functional coordination is facilitated by electrophysiological recordings that have greater time resolution. One of the studies mentioned above used scalp electroencephalogram (EEG) signals and reported a decreased long-range coherence in the α frequency range within frontal regions in the resting state in participants with autism, and at the same time an enhanced local coherence at θ frequency was also found (Murias *et al.*, 2007). This observation is of interest within the context of the results that are mentioned below, because the participants were seating quietly in this study, a situation normally described as "resting" even though the brain, as repeatedly said, never rests. The studies that are described in the following paragraph depict similar observations of reduced global synchronization when subjects with ASD perform certain tasks. These parallel results indicate that it is the background, ongoing brain dynamics that is somewhat different in the autistic brain, conditions that may become more apparent during behavioral actions.

So, to address the hypothesis that the patterns of cortical coordinated activity of autistic brains could be different from those found in control participants while performing particular behaviors, one study assessed phase synchronization derived from magnetoencephalography (MEG) signals during the performance of executive function tasks. Executive functions are those that enable the individual to select actions on the basis of external and internal goals. These tasks were chosen because individuals with autism tend to be impaired in, at least some, if not all domains of executive functions (Ozonoff and Jensen, 1999; Hill, 2004). Examples of these tasks are card sorting tests and the Stroop task. The former is supposed to measure cognitive flexibility, normally related to frontal lobe functions, in which participants sort different symbols printed on the cards by a specific feature (either color or shape of the symbol), and the rules for sorting change continuously. This has been used as a test for frontal lobe (dys)function (Milner, 1963). The latter is a commonly used test of inhibition, invented by J.R. Stroop in the 1930s (Stroop, 1935). In the color Stroop interference paradigm, the participant has to name the colors of the ink in which words are written. It consists of a list of color words written in congruent color (e.g. the word "green" written in green color), and follows with a list in incongruent color (e.g. the word "green" written in red color). It is well established that processing the content of the word is more automatic than processing the color of the word. Therefore, in the incongruent condition, the individual needs to inhibit the response of word naming (just reading it) that competes with the response of color naming. Overall, these tasks demand operational executive functions that are related to mental flexibility and cognitive inhibition. Individuals with ASD tend to persevere in behaviors and perhaps that is why they tend to perform worse at card sorting tasks since they do not show the mental flexibility needed to adapt to the new rule and they persist using the old rule (Ozonoff, 1995; Ozonoff *et al.*, 1994). It is also known that patients with prefrontal cortex lesions persevere on sorting the cards by the initial rule (Barceló and Knight, 2002). With regards to the Stroop color word task, some studies have found no impairment in autism (Bryson, 1983; Ozonoff and Jensen, 1999).

The evaluation and comparison of the phase synchronization patterns from MEG signals in children with high-functioning autism and nonautistic children while performing those two aforementioned tasks, revealed a disruption of long-range phase synchronization among frontal, parietal, and occipital areas in the ASD group (Perez Velazquez *et al.*, 2009b). More specifically, the long-range brain synchronization tended to increase in control children with a more significant prefrontal synchronization, mostly at the frequency range between 16 and 34 Hz (incidentally, the increase in prefrontal synchrony was observed in all but one control male and in only one of the control girls). The ASD group did not present any significant enhancement of synchronization associated with task execution, and most of the significant changes in synchrony in this group were desynchronization (a decrease in synchrony compared with the baseline condition for each task). This is a pattern too, for there is no reason to consider only the increases in the magnitude of synchrony. In terms of task performance, there was a (slight) tendency for the ASD children to commit more mistakes in these tasks than the control participants even though the differences were small. However, it is not the case that there is lack of synchrony in the autistic brain, in fact a very robust enhancement in phase synchrony was observed in the parietal cortex (evaluated between left and right parietal) in the ASD children relative to controls, but this high synchrony was always present, regardless of task operation, and did not change significantly during task execution. Hence, to conclude that there is lack of synchrony in the autistic brain would not be accurate, it is *the pattern of synchronization that is different*. We have also obtained evidence for very distinct patterns of cortical synchrony during the performance of an auditory attention task (Adam Teitelbaum, MSc thesis, University of Toronto, 2010). The difference in the mean values of synchronization indices between the ASD and control group was enough to correctly classify patients and controls within their own class, using linear discriminant analysis (details are available in the original publication). The fact that these individuals can be separated based on the magnitude of their brain synchrony, by the way, may be of clinical, prognostic value.

It is not too surprising that individuals with ASD and controls show distinct patterns of neural coordination in the frontoparietal networks during executive behaviors, for it is thought that these areas compute sensorimotor transformations. Which is to say that they are involved in almost all cognition, for cognition consists, in the final analysis, in sensorimotor transformations. The more significant enhanced prefrontal synchrony in controls during these tasks can be expected by considering the established role of the frontal cortices in cognitive demands in general and in executive functions in particular (Fuster, 1995; Miller and Cohen, 2001), as well as by the neurological observations in patients with lesions of dorsolateral frontal cortex in that they show impaired performance in the Wisconsin Card Sorting task (Milner, 1963; Barceló and Knight, 2002). This has been attributed, in part, to less attentional control exerted by prefrontal lobes in these pathologies. Neuroimaging studies using PET and fMRI in the Stroop task demonstrate higher activation in frontoparietal networks (Adleman *et al.*, 2002), and particularly increased activity in the right prefrontal and bilateral parietal and occipital regions is common to many studies. Disruption of prefrontal neural networks associated with errors in card sorting tasks has been detected using event-related potentials (Barceló, 1999). Anatomically, reciprocal frontoparietal connections, not only intra- but also interhemispheric, have been described (Cavada and Goldman-Rakic, 1989). Thus, it is not too surprising that there is an enhancement of synchrony among most of these areas in the performance of our tasks by the control participants.

Other observations supporting alterations in brain coordination dynamics in autism were provided by studies with autistic patients that showed abnormalities in cortical activation sequences during imitation of facial expressions in autism (Nishitani *et al.*, 2004). Different patterns of the amplitudes of the power in the γ frequency associated with perception of illusory Kanizsa figures (Brown *et al.*, 2005) or faces (Grice *et al.*, 2001) were noted in autistic participants as compared with controls, but the controls used in the former study were mentally retarded children (or, in the words of the authors, a group with "moderate learning difficulties"). A different cortical pattern of power in the θ frequency range was also reported in children with autism while visualizing videos showing body motions (Martineau *et al.*, 2008). The lower power during the perception of human actions in the control children was interpreted as "desynchronization," which, as many times already commented in this book, is open to interpretation, for a change in power in two connected brain areas does not necessarily mean that their phase synchrony is altered. There are other manners best suited to assess synchrony or correlated activity. Nevertheless, the aberrant patterns at this level of power spectra analysis still suggest distinct dynamic patterns in autistic brains during certain perceptual tasks. Another study that indirectly assessed synchronization, based on entropy information rates, found less synchronization in ASD too, even though this work studied exclusively sleep EEG patterns (Kulisek *et al.*, 2008).

6.1.2. The case of high long-lasting synchrony in parietal cortex of individuals with ASD

The preceding section mentioned in passing that high intraparietal phase synchrony was found in the ASD group, which occurred at all frequencies studied, from 10 to 32 Hz, and regardless of what the individual was doing: either task performance or control recording. This phenomenon was observed in the children with high-functioning autism recruited for the aforementioned study, while doing the two tasks described above and, in addition, an auditory attention task. How high was this parietal synchrony? Let us just say, lest the reader become weary of raw data that can always be found in the original article (Perez Velazquez *et al.*, 2009b), that the magnitude of the synchrony evaluated among parietal sensors was extraordinarily high when compared to others, even though, of course, not as high as that found in seizures, for instance. The magnitude of the parietal synchrony accounted for a great part of the total synchrony measured in frontal, occipital, and parietal sensors, both within and between these three regions, and in any combination possible (temporal cortex signals were not studied). Previous chapters have accentuated the perils of a widespread long-lasting synchrony for adequate information processing. But the diseases and injuries examined in those chapters were substantial: epilepsy, brain trauma, and Parkinson' disease. Do we have here, in autism, a subtler example of ongoing high synchrony in a particular brain region being detrimental to cognition? This cognitive detriment, in that study, was rather subtle though: the individuals with autism performed the tasks just a bit worse than the control participants, but in general there are other more challenging cognitive disturbances in the autism spectrum.

Why did these children with autism have such an elevated basal synchrony in the parietal areas? May this be related to parietal lobe abnormalities? There is the possibility that the higher intraparietal synchronization found in that study may be age dependent, because it was not observed in half of the children under eight years, but in all of the rest between 9 and 16 years of age. Anatomically, parietal lobe abnormalities in autism have been noted: either gray matter reduction (McAlonan *et al.*, 2005) or enlargement (Ashtari *et al.*, 2007), and cortical volume loss (Courchesne *et al.*, 1993) have been reported. Incidentally, there have been numerous reports on neuropathological findings in autistic brains, but in many instances the results were not reproducible from study to study, which perhaps is not surprising due to the already mentioned broad autistic spectrum (for a recent review, consult Amaral *et al.* (2008)). It is also known that greater activation of parietal cortex occur during development, from childhood to adolescence, as measured by fMRI (Adleman *et al.*, 2002). How these anatomical abnormalities may explain the reason for the high intraparietal synchronization is not clear. A more detailed inspection of the phase synchrony results may shed some light into this matter. This is something that can be scrutinized with this type of electrophysiological recordings, and not with neuroimaging based on metabolic activation. Remember, all these recording methodologies are complimentary, some better than others at certain aspects. Linking the data from neuroimaging techniques that allows a full brain analysis of activity, with results from electrophysiological recordings, is the best approach to understand important features of brain cognitive processes despite the different time scales of the techniques.

Specifically, when estimating the differences in oscillating phase, it is always advisable to take a look at the values of this difference. The reason is that angles around 0 could be due to summation of signals in the MEG sensors (and hence nothing to do with real synchrony between two regions). To give an idea of what is meant by the phase difference angle, basically a measure of the phase lag between two signals, Figure 6.1 shows the polar diagram that can be used to estimate it.

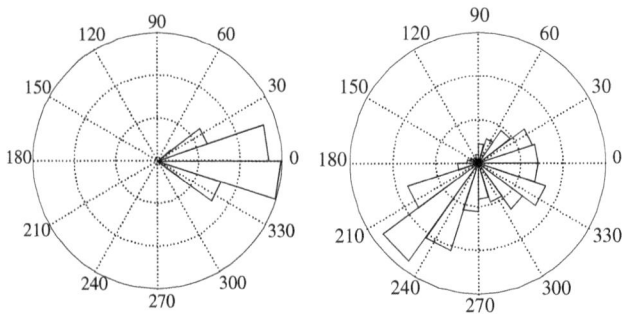

Figure 6.1.   Polar depictions of the phase difference angles associated with a relatively high-synchrony index between two cortical MEG signals. The phase synchrony index derived from the phase differences may be similarly high for the left- and right-hand side graphs, note the clustering of the angles around 0° and 220°, respectively that would result in a high value of the index in both instances. However, clustering about 0° rises suspicion. It could indicate either summation of signals at the sensors due to strong neural sources, or that the two areas are receiving a common input from a third connected to both, in addition to what could be considered the "normal," or "real," synchronization scenario, where the two neural areas under the sensors are interacting and synchronizing their activity with 0 phase lag. It is not feasible to discern what of these three events are taking place from these data alone. If the sources within the neural tissue could be reconstructed from the sensor space and the synchrony between them evaluated, that could resolve the problem of signal superposition, but still the uncertainty about a common drive by a third, hidden network would remain.

Signal superposition can more likely occur when a few high amplitude signals are present that can be picked up by many sensors (García Dominguez *et al.*, 2007), which may be a concern in seizures, for instance, because of the high-amplitude waves indicating strong sources, but perhaps not so much trouble in recordings during normal brain cognitive activity where high-amplitude signals are unlikely to occur (here the assumption is made that high-amplitude waveforms denote strong neural sources). Regardless, the mean phase difference among signals in the parietal cortex was inspected and, in those autistic brains that had high synchronization in this area, were concentrated around 0°. At the same time, there were not noticeable differences among the amplitudes of the MEG signals in frontal, temporal, occipital, and parietal sensors; in fact, the signals from the parietal sensors had slightly lower amplitudes than those of the others. This does not support the fact that there could be a summation of a strong source in several sensors that could account for the 0° difference, which would be revealed by large-amplitude recordings in some parietal, or neighboring, sensors. Another fact that probably rules out this possibility of signal superposition is that the high synchrony was observed between not only neighboring sensors but also distal right and left parietal signals, because signal superposition is more likely in neighboring sensors.

One possibility then is that it is the connections between the parietal networks that are synchronizing these areas. However, if the elevated synchrony was due to direct intraparietal communication, we would expect to see angles different from 0, since there must be transmission delays between the distant right- and left parietal regions. A quick and rough estimation: adding the synaptic delay ($\sim$2 ms) and the delay due to the action potential velocity traveling along axons of the corpus callosum (an average velocity of 2.6 m/s was measured from neurons in cortical layers 3 and 5 that send callosal axons) would surely result in a delay of about 40 ms traveling a distance of 10 cm between right- and left parietal cortices, which for a frequency of, say, 20 Hz (thus with a period of 50 ms), would give a phase angle difference of $40 \times 360/50 = 288°$. This rough estimation assumed that the action potential on one side and the resulting post-synaptic potential on the contralateral are measured, which may not be the situation for EEG and MEG, but in any case serves to illustrate the point that some delays in transmission would result in angles not centered about 0. Of course, for this estimated time delay, the angle becomes smaller for lower frequencies (for example, it is only 28° for 2 Hz, already relatively close to 0), but the high intraparietal synchronization was detected at frequencies between 10 and 32 Hz, which was the frequency range considered in that report.

So, what is left as an explanation for this suspicious observation? It is tempting to conclude, based on these data, that the angles around 0 strongly suggest that there is a common input that is arriving at the same time to all areas of the parietal cortex, perhaps from deep structures or from other cortical areas that are providing a synchronous input to the parietal cortex. A similar cortical drive from deep-brain structures (thalamus especially) was considered in Chapter 5 when discussing the high cortical synchronization detected in unconscious states. Recall that in MEG or EEG signals, it is the synaptic potentials that constitute the major contributors to the signal, therefore, assessing synchronization in these types of recordings is studying the correlation between synaptic activities. It would be then reasonable to declare that the synchronization measured in these experiments reflected the arrival of synchronous volleys of synaptic activity throughout the parietal regions. Where these inputs are coming from, is a matter of no easy assessment unless one has multiple electrodes located in a variety of cerebral regions. In this regard, there is evidence for an enhanced thalamocortical connectivity in high-functioning autism, based on fMRI data (Mizuno *et al.*, 2006), and for stronger connnectivity in specific cortical areas (cingulate and temporal) at rest (Monk *et al.*, 2009). These findings emphasize the fact that it is not a matter of less connectivity, or coordination, in autism but of a different style of coordination, and Mizuno *et al.* concluded that subcortical input to the cortex may be hyperfunctional in autism. Whereas these results may explain in part the high intraparietal phase synchrony, they do not explain why high synchrony is not detected throughout the whole cortex if the subcortical inputs are arriving in large numbers to all cortical regions. Could it be that they are not synchronous in other cortical regions? Or perhaps, that the anatomical connections of the parietal areas with subcortical structures are more numerous or physiologically stronger? All these are, for the time being, mere speculations, conjectures that can potentially be resolved (research funding agencies willing). Provided that these findings of elevated synchronization in specific cortical regions in ASD hold (to our knowledge, up to now, there has been only the study here discussed), it may be of interest to understand the reasons for it, because it may reveal aspects of brain information processing. If, according to current thought already expounded in Chapters 1 and 2, transient dynamical states in specific neuronal networks contribute to information processing, with the prominence of coordinated activity between separate cortical regions and specifically between the frontal and parietal association areas, then it can be envisaged that any disruption of this coordination may be harmful. Long-lasting high parietal synchrony due to intraparietal communication would disturb the coordination of this area with other cortices, frontal cortex especially; this is analogous to the so-called "cocktail party effect," where one listener pays attention to only one speaker and ignores the rest around, these two individuals having the verbal intercourse being the parallel, in this example, to two parietal areas. But it was mentioned above that the intraparietal transactions are unlikely to be the scenario that explains the high parietal synchrony, which would have been very convenient. Nevertheless, even if the case is that synchronous synaptic volleys arrive from subcortical or other cortical regions to the parietal, still a difficulty of this area to functionally communicate with others can be expected. Imagine a set of football players passing themselves a ball and, suddenly, many balls fall upon them at the same time, synchronously; if they pay attention to these, we can be sure that their quality of ball passing will be affected. If, conversely, they ignore the incoming set of footballs, then they can go on with their passing game. But neurons do not have the luxury of choosing between ignoring or accepting inputs. In any case, it is plausible that the high and long-lasting parietal synchrony does not support flexibility in neural coordination. Predictably, there is a decrease in the rate of fluctuations of phase differences in this cortical region of the children with ASD as compared with that of controls (Teitelbaum *et al.*, 2009), this fluctuation rate being determined as was shown in Figure 2.15 of Chapter 2, as the time derivative of the instantaneous phase difference in order to quantify the fluctuations in phase. The higher magnitude of parietal synchrony already suggested less variability, and this is not supportive of flexible pattern formation. It is relevant to note that, recently, Greicius *et al.* (2007) speculated along the same lines when they found an abnormal increased functional connectivity between thalamus and areas of the default network in the resting state in patients with major depression, arguing that the enhanced interactions between these areas occurs in depression at the expense of executive function performances.

6.1.3. Concluding remarks

To make complete sense of the results reviewed here is complicated by the intrinsic variability in the subjects, starting with the distinct ages and behavioral dispositions, which, adding to the limitations of the recording and analytical methods, make the correlation of biophysical signatures of brain activity with behavioral responses a complicated matter. It is somewhat encouraging that the main findings using distinct experimental and analytical methods seem to be providing comparable information. These fast emerging novel and interesting empirical results in the increasingly abundant neuroscientific literature should not make one fall into adapting an insufficiently critical attitude.

Whereas most autism studies based either on neuroimaging or electrophysiological data have indicated reduced functional connectivity/coordination within the cortex, the possibility exists that the connections of the cortex with subcortical networks are enhanced in autism, recall the study of Mizuno *et al.* discussed in the previous section. The important aspect is that the coordination patterns are different, rather than the magnitudes of synchrony indices (although it is that magnitude that determines the pattern). To put these findings of different patterns of brain coordination dynamics in autism within the context of normal or deviant brain function, a couple of comments are in order. The first aspect to consider is to what extent the synchrony patterns can be associated with the autistic brain in general or, more specifically, with the performance of the task and the making of errors. This is a necessary consideration because non-autistic individuals commit mistakes too, in, for instance, card sorting tasks, and one study reported differences in event-related potentials recorded in the frontal cortex of normal subjects erring (Barceló, 1999), thus indicating that these errors may reflect similar transitory discoordination of activity in frontal areas in ASD and non-ASD individuals, or in those with other psychiatric disorders. Along these lines, similar synchrony patterns in the autistic brain (more specifically, the lack of cortical coherence) during the resting state have been reported, so it is tempting to conclude that the less brain coordinated activity in autism could be a general phenomenon, and not associated to one particular task, for it seems to be present already at rest. If we denote as background neural activity the recordings performed at rest, then it can be said that task-specific activity is measured superimposed on the background activity. This is in line with current views of task-specific event-related brain responses as transient perturbations of the "default state" of background activity, which historically was regarded as noise, an opinion that the numerous studies on the default brain network and background resting activity that are appearing very often these days seem to be about to dispel.

A second aspect to ponder is whether changes in coordinated activity that deviate from the "control" population have to be interpreted as adverse. Sometimes, these aberrant patterns could be of adaptive value. A case in point could be the increased coherence between frontomesial areas that has been detected in patients with Tourette syndrome when they suppress their tics, as compared with that coherence of control participants (Serrien *et al.*, 2005), thus in this case this pattern may favor the control of the abnormal motor behavior in these patients. In any case, deviant behaviors in some individuals with autism are very slight. Indeed, studies have reported not significantly different behavioral performance between control and autistic persons but, still, there were disparities in the brain activity patterns (for example, Gilbert *et al.* (2009)). Is this due to very subtle variations in brain dynamics resulting in even subtler changes in behaviors? Or, perhaps the individuals with autism use alternative cognitive strategies to achieve certain goals? In this fashion, a neuroimaging study reported that autistic brains seem to process face recognition outside the common fusiform face area of the temporal cortex (Pierce *et al.*, 2001). Perhaps, there is a diversity of brain dynamic strategies that serve almost equally well certain behavioral actions. Again, it should be taken into consideration the wide behavioral range of ASD and that there is not one autism phenotype, but several.

Perhaps, the most common aspect among the broad range of clinical presentations is social interaction impairment, as Kanner already noted in his original description. In this sense, it is conceivable to expect substantial differences in the dynamics in those brain areas related to interpersonal transactions, or "social perception," such as the orbitofrontal cortex, amygdala, and superior temporal sulcus. There have been numerous studies, mostly neuroimaging, that inspected these brain regions in autistic persons involved in a variety of tasks that assess social demands, and the main message from those results, at least for the moment, is that there is an altered functional coordination among these brain networks (reviewed in Zilbovicius *et al.* (2006)). But, since these alterations have been observed in major brain areas that are involved not only in social cognition but in an immense variety of other behaviors (regions like frontal, parietal, occipital, temporal association areas, what more need to be said!), again, one wonders whether there is a general trend in the autistic brain that manifests into a diversity of deviant behaviors.

One question that may emerge in some readers' minds refers to what specific cognitive alterations are described or addressed by the studies aforementioned, is it attention, or cognitive inflexibility? The initial paragraphs in Chapter 1 already described our approach to cognition, following Nieser's views, where it was advised to avoid excessive fractionation. The complexity of the behavioral phenotype in the autism spectrum instigates no few debates in the cognitive/psychological literature (a recent one can be found in Geurts *et al.* (2009)), and at the roots of most of these controversies the reductionism paradigm is to be found, the attempts at the fractionation of cognition and behavior into discrete entities, while many times ignoring the inseparability of the brain and behavior, the brain–behavior continuum whose manifestations scientists measure with diverse techniques and thus the propensity arises to dissect what in reality is very hard to separate.

In summary, we exploit all these observations to highlight once again the distributed nature of cortical information processing, to emphasize the importance of the functional coordination between separate brain networks for sensorimotor transformations, and to suggest that the brain coordination dynamics of individuals with ASD is different from that of the non-autistic population.

This section is concluded with a conjecture, a possible practical result that stems from these studies on brain dynamics aforementioned. Specifically that the altered brain synchrony patterns could inform as to possible interventions that will enhance processing abilities in individuals with autism, depending on where the abnormal patterns occur. This idea, that we have been advocating for the past few years, may sound speculative, but considering that a similar approach has just been implemented in patients with schizophrenia based on initial fMRI changes found in patients and a subsequent cognitive training protocol (Edwards *et al.*, 2010), we think that it can be applied in autism too. There are some interventions currently used to improve attention (Tang and Posner, 2009), such as cognitive training tools like the number–word shift tracing. There are also many sensorimotor interventions, interventions to improve social skills, occupational therapy and more, used in children with autism (Baranek, 2002), but we are thinking about something more specific derived from the possibly altered brain coordinated patterns. Of course, it is too early to speculate which interventions will be more beneficial, because that will depend on the results obtained from the experiments. For example, if frontal lobes are poorly functionally connected, then perhaps an intervention to improve frontal-related cognitive performance may be recommended. Would not be then interesting, if a cognitive intervention is found that facilitates a behavioral improvement, to bring the participants back to the scanner and check their possibly renewed brain dynamics? Would this dynamics be now closer to that found in non-autistic individuals? There is room for creative interventions and experiments here. In fact, we do not envisage an intractable problem in designing experiments and interventional strategies of this kind, rather the main difficulty will be to decide whether or not to change particular behaviors, that, while may be considered deviant, perhaps are not inconvenient for the individuals.

6.2.   Schizophrenia

Schizophrenia is among the most disabling and enigmatic disorders in modern medicine, with very few effective therapeutic interventions available, all of which provide only symptomatic relief. Molecular mechanisms of schizophrenia have been the subject of intense scientific interest, because of the expectation that once a molecular target is identified, treatment and prophylaxis directed toward this specific target can be developed. Decades of research have led to an explosion of information on cellular and molecular mechanisms associated with schizophrenia, yet critical insight into the cause and the pathogenesis of this devastating illness is still lacking. In fact, to this day there is no conceptual framework to link symptomatic manifestations of minds suffering from psychiatric illnesses with their biological underpinnings, "the problem is that psychiatry as a science seems to lack a coherent system linking the mental and molecular domains," in the words of van Beveren and Haan (2008). This is not unexpected, given the complexity of the task and significant technological limitations posed by any attempt to disintegrate the workings of the mind down to discrete modules, molecules, and genes. In all likelihood, schizophrenia may not even have a specific causative molecule, but is rather a phenomenological entity arising from multiple molecular and structural factors that converge onto a particular malfunctioning of neuronal assemblies manifesting as schizophrenia symptomatology. Hence, the understanding of schizophrenia pathology at a more global level would enable us to circumvent (multiple?) causative molecular phenomena and focus instead on the ensuing alterations in collective network functioning, which could be critical for the development of successful therapeutic strategies. In this section, a more global conceptual view of schizophrenia based on clinical observations will be provided, along with current advances in the neuroscience of this syndrome and the application of theoretical frameworks of dynamic systems theory with a hope to advance our understanding of this complex illness. Unfortunately, little experimental evidence is available.

6.2.1. Clinical aspects and neurobiological mechanisms

KS is a young woman suffering from schizophrenia, who has been kept in a secure unit for years because of her tendency to "time travel." She is convinced that "vicious observers" and a "very cruel diagnostic machine" are controlling her thoughts and trying to kill her by suffocating her with her own "phlegm." To make matters worse, she knows for a fact that "others" accuse her of molesting little children, which is excruciating. She hears derogatory voices constantly nagging her with their comments and telling her that she is despicable and deserves to die. She lives a miserable life. However, there is a way out: she "time travels." Her next life is, in fact, very different: she is to be a beautiful, successful, happily married psychologist as soon as she finally succeeds in killing herself on a specific predetermined designated date. For years, her treating team is focused on figuring out the "time-traveling date" well in advance so that the poor soul could be prevented from yet another suicide attempt.

What is happening in her mind as she goes through her agonizing experiences? Why are these minds so different from the rest of ours? Why can she not sustain herself out there in the world without a multitude of antipsychotic medications and a careful supervision to top it up? Besides, why no amount of hard evidence could ever dispel her myth of "cruel machines" watching her every step? The core features of schizophrenia include seemingly unrelated symptoms, such as delusions, hallucinations, thought disorder, affective flattening, and pronounced cognitive deficits. Several features of psychosis are mysterious and fascinating. Delusions are generally classifiable into specific themes (e.g. paranoid, grandiose, religious, etc.), and are time- and culture-dependent. Auditory hallucinations, the predominant perceptual abnormality of schizophrenia, may play into a delusional theme, frequently vocalizing a person's own viewpoints and worst terrors, and their content more often than not is demeaning and derogatory. Is there any way to translate these strange productions of diseased brains into their biological, or perhaps, dynamical correlates?

Until recently, the two leading hypothesis of the pathogenesis of schizophrenia were the dopamine theory and the aberrant neurodevelopment view. The dopamine hypothesis of schizophrenia became the leading theory ever since the advent of antipsychotic medications in the 1960s, based on the fact that antipsychotic medications (typical and later atypical) block dopamine receptors (Seeman *et al.*, 1975). Most evidence for the dopamine theory includes antipsychotic effects of dopamine receptor blockade and the ability of dopaminergic drugs to induce psychosis, although increased mesolimbic dopaminergic signaling in schizophrenic patients has also been demonstrated and even linked to their psychotic symptoms (Laruelle *et al.*, 1996; Brier *et al.*, 1997; Abi-Dargham *et al.*, 1998). Thus, there is reasonable amount of evidence that the acute psychotic state in schizophrenia is associated with heightened dopaminergic transmission. This neurotransmitter has been discussed before, most importantly in Chapter 4 in the context of PD, where it was remarked regarding the importance of this molecule as a general modulator of neural activity. The main postulate of this hypothesis, according to which hyperdopaminergic signaling throughout the brain is responsible for schizophrenia symptomatology, has been recently re-defined. In fact, excessive mesolimbic dopaminergic activation has only been linked to positive symptoms of schizophrenia, such as delusions and hallucinations. Figure 6.2 depicts the major dopaminergic paths.

On the contrary, cognitive abnormalities in schizophrenia have been attributed to decreased dopaminergic tone in mesocortical dopaminergic pathways and prefrontal cortex, and there is some evidence that dopamine agonists improve cognitive function in individuals with schizophrenia (Barch and Carter, 2005). This has led to reintegration of the dopamine hypothesis into a revised theory of schizophrenia as cortical (prefrontal) dopamine hypoactivity and subcortical dopamine hyperactivity. Nevertheless, other neurotransmitter systems, including glutamate, gamma-aminobutyric acid (GABA) and serotonin, have also been implicated in schizophrenia pathophysiology. Currently, it is thought that dopamine dysregulation in schizophrenia is neither exclusive nor primary, and the most recent theories of schizophrenia stipulate that mesolimbic dopaminergic dysregulation is the final common pathway resulting from multiple interacting genetic and environmental risk factors neurochemically converging on mesolimbic dopaminergic hyperactivation (Howes and Kapur, 2009). In any event, this neurotransmitter is a crucial aspect of schizophrenia. How do these neurochemical aberrations translate into KS' "vicious observers" and "cruel machines"?

Sometimes, the best insights into the workings of a diseased brain may come from unexpected sources. Suppose you are falling asleep comfortably when

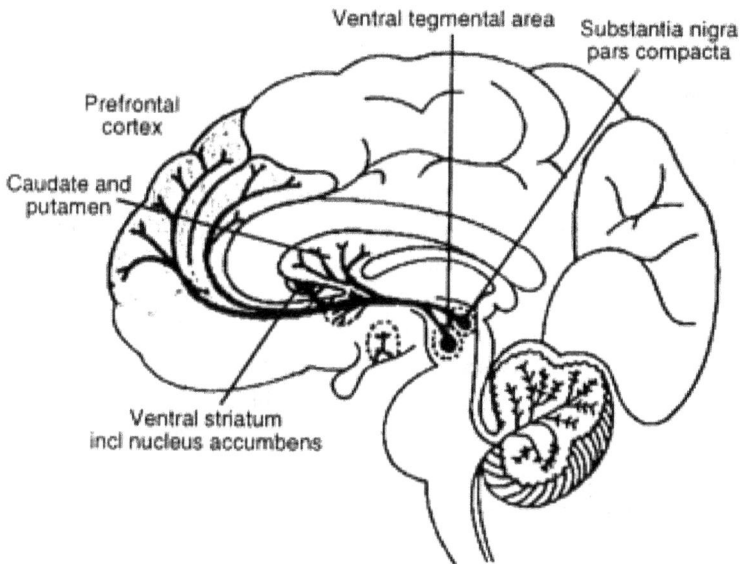

Figure 6.2. Three major divisions of dopaminergic pathways innervate the forebrain and basal ganglia. Mesolimbic and mesocortical pathways originate in the ventral tegmental area and project to most areas of cerebral cortex and the limbic system (e.g. ventral striatum, amygdaloid body). The nigrostriatal dopaminergic system originates in the substantia nigra and terminates in dorsal striatum. There is also the tuberoinfundibular system (projecting from the hypothalamus to the anterior pituitary), which is not discussed here. These dopaminergic neurons are involved in higher motor execution and goal-directed behavior, including reward, motivation, learning, working memory, etc. Reproduced with permission from Schultz (1999).

suddenly from the parking lot near your house you hear an intense and obnoxious sound that you have never heard before (this, incidentally, is a real event that happened many years ago in an era when car alarms were being introduced). As the sound intensifies, you become consumed in the sleep state by a profound fear which is soon transformed into a solid conviction that your parking lot is being invaded by extraterrestrial creatures. At this very point, unexpectedly, a few fluorescent green aliens appear on your parking lot downstairs, exploring it with malicious purposes. Your fear becomes so extreme that you, thankfully, wake up only to make sense of this unusual (and very annoying indeed) sound that you had never heard before, a new type of a car alarm. Here, then, green aliens had invaded a healthy brain's reality, albeit during a dream. Was it that the brain, while sleeping and hence deprived of its normal thinking capacities, did its best to explain irrational fear generated by a frightening sound that crept into a certain stage of sleep? Could this perhaps shed some light on the harsh and distorted reality of psychosis? Could delusions be simply a half-failed attempt of the mind to explain intense and aberrant emotions? The research on split-brain patients demonstrates the powers of the interpreter within the brain, normally assumed to be in the dominant hemisphere. This (these) interpreter(s) is also at work in normally connected brains, in a more integrated fashion and thus challenging to discern. In the transition from wakefulness to sleep, most sensory stimuli cease to reach conscious perception. It has been long thought that brain is shut off from the external world during sleep, as the relay of sensory information to the cortex is extinguished by the slow-oscillating thalamocortical activity (Section 5.1 of Chapter 5). However, recent data indicate that sleep does not make us fully impenetrable to sensory inputs: sounds presented during both slow-wave and rapid eye movement (REM) sleep have been shown to elicit regionally specific responses similar to those observed during wakefulness (Portas *et al.*, 2000; Issa and Wang, 2008); with one important distinction: higher associative or prefrontal cortical regions allowing for conscious perception of the sensory information are less activated during non-REM sleep as compared to wakefulness. In other words, sounds are detected unconsciously while asleep. After revising the discussions in Chapter 5 on altered states of consciousness, hopefully it becomes clear that less widespread distribution of the information, particularly decreased involvement of cortical association areas, is associated with less awareness. Although there seems to be an exception: affectively significant sounds presented during sleep, e.g. presentation of a subject's name in contrast to simple beeps, activates the amygdala (this area is responsible for processing of emotions) and parts of prefrontal cortex to a degree that is as high or even higher than the activation of these areas during awake states (Portas *et al.*, 2000). Moreover, both auditory or direct amygdala stimulation during slow-wave sleep has been shown to elicit waveforms indicative of an REM-sleep episode onset or even awakening, depending on the alerting significance of the stimulus (Sanford *et al.*, 1993; Deboer *et al.*, 1998). This would certainly explain what seems like a supernatural ability of mothers to be aware of the slightest movement of their babies in their sleep, as well as our propensity to dream of an arctic trip if we were to lose a blanket, and other instances of vigilant monitoring that probably serve a protective role. Hence, if one was to translate our aliens in the aforementioned dream episode into the language of neurocircuits, they were most likely produced in the context of a REM-sleep event, triggered by a sound deemed to be too alarming by the amygdala and prefrontal cortex.

Neuroanatomical frameworks of dream generation have been recently studied using functional neuroimaging techniques. Significant increases in regional blood flow during REM sleep were found in limbic and paralimbic areas (such as amygdala, the anterior cingulate cortex, and the insula), as well as in motor and visual associative cortices. By contrast, marked reduction of activity was observed in associative frontal and parietal cortices, including dorsolateral, orbitofrontal, parietal regions, and posterior cingulate. No wonder that there is difficulty in accurately processing information during REM sleep: with the "thinking" parts shut off, the brain is no more than an overactive emotional loop! Meanwhile, secondary associative cortices obediently generate vivid images and bizarre events providing convincing content to our dreamy reality, the one that is seemingly in line with our most genuine feelings and buried insights. Freud labeled our dreams as "the royal road to the unconscious," and sleep neuroimaging certainly seems to provide neurophysiological basis for his controversial yet timeless brilliance.

Despite its intuitive appeal, the persistent idea that hallucinations in schizophrenia are due to intrusions of REM sleep into waking has not been substantiated. Yet, prefrontal cortex deficits have been implicated as one of the major sites of dysfunction in schizophrenia. Imaging studies demonstrate loss of tissue and decreased activation in the frontal lobes, correlating with executive deficits, as well as with other indices of abnormal prefrontal functioning in schizophrenia patients (Goldman-Rakic, 1999; Davidson and Heinrichs, 2003; Berman and Meyer-Lindenberg, 2004). Perhaps, not unlike healthy brains during REM sleep, schizophrenia patients, with their prefrontal cortex "turned off," have difficulty accurately interpreting incoming information. Is there a mechanism that generates an affective surge fueling delusional experiences in psychotic brains? An emerging understanding of the role of abnormal dopaminergic transmission in schizophrenia was posited by S. Kapur (2003) in his influential article "Psychosis as a state of aberrant salience: A framework linking biology, phenomenology, and pharmacology in schizophrenia." Although dopamine has long been considered a neurotransmitter of reward, current leading ideas about its role in behavior suggest that it is rather, or should we say "too," a neurotransmitter of motivation, or emotional salience. Specifically, mesolimbic dopamine is thought to charge affectively neutral stimuli with either attractive or aversive emotional valence. Hence, dopamine contributes to assigning "hedonic valence" to an input (either external or internal), directing the attention and driving an action toward an overall adaptive functioning of the individual. Several lines of evidence support this interpretation. First, the activity of dopaminergic neurons increases not only in response to rewarding stimuli, but also to aversive ones (Salamone *et al.*, 1997; Horvitz, 2002). Moreover, the firing of dopaminergic neurons precedes the consummation of pleasure, regardless of whether or not it is actually consummated (Apicella *et al.*, 1991; Schultz *et al.*, 1992); in addition, dopamine blockers alter the drive to engage in pleasurable activities (Lopez and Ettenberg, 2001). Finally, genetically engineered mice unable to synthesize dopamine, display profound deficits in goal-directed behavior, to the point that they do not engage in reproductive behavior and starve to death unless injected with dopaminergic agents or forced to eat. Yet, these animals have no difficulty producing convincing hedonic responses to sucrose solution delivered directly into their mouths (Cannon and Palmiter, 2003). What a fresh insight these experiments on mice give to our condescending concept of laziness, as those of us, graced with healthy dopaminergic signaling, are fast to judge others far less fortunate with their goal-oriented neurochemistry.

A conceptual framework of dopamine as a mediator of motivational salience has allowed the question about the psychotic mind to be addressed at a neurobiological level. Shitij Kapur, a clinical psychiatrist, astutely uses his clinical observations to resolve a "brain-mind question" of psychosis in schizophrenia. He posits that dopamine, under normal circumstances mediating the process of salience acquisition in response to appropriate environmental (or internal) stimuli, is released out of context, or in a stimulus-independent manner during psychotic states. This leads to aberrant assignment of salience to innocuous external events or internal representations. Endogenous psychosis evolves slowly, "through a series of stages: a stage of heightened awareness and emotionality [. . .], a drive to 'make sense' of the situation, and then usually relief and 'new awareness' as the delusion crystallizes and hallucinations emerge [. . .] most patients report that something in the world around them is changing, leaving them somewhat confused and looking for explanation" (Kapur, 2003). Patients describe their experiences as "I feel that there was some overwhelming significance in this" (McDonald, 1960), or "I felt like I was putting a piece of puzzle together" (Bowers, 1968). These occurrences persist for days to years of prodromal period whereby aberrant salience is experienced without any clear understanding, until patients are finally able to make sense of their mysterious yearnings by arriving at delusional ideas, which provides them with a relief of a "psychotic insight." Delusions are thus conceptualized as a "top-down" cognitive explanation of aberrant salience that an individual eventually conjectures in an effort to make sense of his/her experiences, supposedly fueled by a dopamine surge. This conceptualization, perhaps, explains how a singular neurotransmitter dysregulation can result in a multitude of time- and culture-dependent delusional phenomenological expressions (e.g. whether it is an institution such as the KGB or CIA, or evil witches who are plotting their schemes against you depends on the time and cultural context). Antipsychotics, by dampening dopaminergic transmission reduce aberrant salience and assist in the resolution of the positive symptoms of schizophrenia. The initial response to antipsychotic medications (which rapidly block dopamine receptors) leads to affective dampening of the distressing delusional concerns, while their cognitive reappraisal needs further psychological work by the patient and takes much longer (Kapur, 2003). Although this framework of psychosis remains largely theoretical, several recent studies provide empirical support for a model of aberrant salience in schizophrenia. The disruption of mesolimbic motivational systems in patients suffering from first-episode psychosis has been demonstrated by Murray *et al.* (2008), and significantly greater indices of aberrant salience have also been demonstrated in schizophrenia patients with delusions as compared to those without delusions (Roiser *et al.*, 2009). Does the abnormal salience hypothesis fully answer our question about the pathophysiology of schizophrenia? It is well known that loading a normal brain full of dopamine with a single dose of dopamine-releasing drugs, such as amphetamine, creates an impressive surge of salience, yet it does not result in psychosis. Moreover, people who develop perceptual abnormalities secondary to organic pathology or substances typically preserve their insight. Hence, abnormal salience may be a necessary but not sufficient condition for developing psychosis. Just as well, the abnormal salience hypothesis does not explain the nature of cognitive deficits and other neuropsychiatric and neurophysiological abnormalities characteristic of schizophrenia. Finally, it fails to provide a neurophysiological account for hallucinatory experiences in schizophrenia.

6.2.1.1. *On hallucinations*

Auditory hallucinations, particularly verbal, are a core symptom and the most common perceptual abnormality in schizophrenia. Patients generally believe that the voices are real manifestations of someone talking to them or of someone transmitting a voice. Sometimes they describe that at the start of their illness, voices are heard as unintelligible sounds or noises that eventually progress to whispers and then to clear and understandable words. Often the person hears voices in response to a humming noise of appliances. With time, "experienced" patients may learn to control their voices, sometimes by yelling at them, and sometimes by singing, humming, listening to loud music or even by shifting their posture. Evenings and nights are frequently the times when auditory hallucinations arise or escalate.

How do patients hear voices when nobody speaks? Several neuroimaging (Dierks *et al.*, 1999; Shergill *et al.*, 2000; Lennox *et al.*, 2000) and electrophysiological (Ropohl *et al.*, 2004; Ruelbach *et al.*, 2007) studies report activation in the primary and associative auditory cortices (e.g. planum temporale, an area of receptive language) during auditory hallucinations in patients with schizophrenia. Figure 6.3 shows the activation of brain auditory regions in three patients while hallucinating voices. Auditory hallucinations in schizophrenic subjects have also been associated with activity in higher-order auditory–linguistic associative cortices (temporoparietal), as well as with activation of thalamic, striatal, limbic, and paralimbic structures (Silbersweig *et al.*, 1995). Electrophysiologically, increased beta oscillations was found in actively hallucinating schizophrenia patients, in contrast to their nonhallucinating counterparts, and the source was located to speech-related

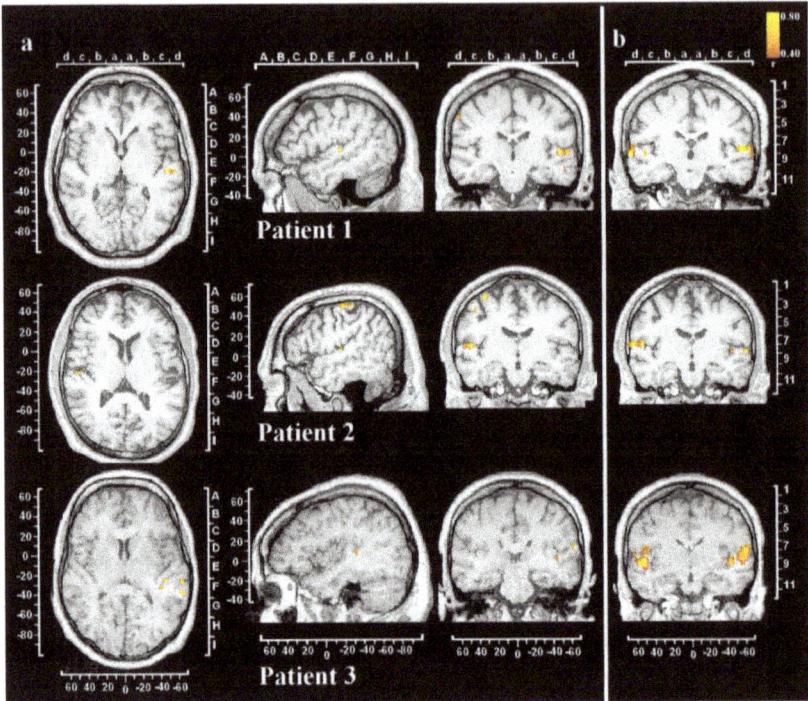

Figure 6.3. Activations of the auditory cortex during hallucinations in three schizophrenic patients. Axial, sagittal, and coronal views of functional data superimposed on magnetic resonance imaging (MRI) scans. Patient 2 has some activation of the sensorimotor cortex, but this could be due to the motor response (button press). Reprinted with permission from Dierks *et al.* (1999).

areas (Lee *et al.*, 2006), and increase of fast MEG activity in the left auditory cortex was observed during auditory hallucinations in schizophrenia (Ropohl *et al.*, 2004; Ruelbach *et al.*, 2007). Enhanced coherence between left and right auditory cortex during auditory hallucinations was derived from scalp EEG recordings (Sritharan *et al.*, 2005). Moreover, medication-resistant auditory hallucinations in schizophrenia responded to PET imaging-guided low-frequency transcranial magnetic stimulation (TMS) targeting the area of excessive neuronal activity within the left auditory cortex (Langguth *et al.*, 2006). What is remarkable is that inner speech, presumed to underlie auditory hallucinations, has not been found to activate primary auditory cortex! No wonder that hallucinations, unlike auditory imagery or inner speech, are experienced as real: the brains of hallucinating patients function as if they were experiencing actual auditory stimulation. Therefore, endogenous activity in auditory areas seems to give rise to auditory hallucinations; in other words, auditory hallucinations reflect abnormal activation of normal auditory pathways. Someone speaks there, in the schizophrenic reality, even though there is no one in the room. Neurophysiological basis of such endogenous activity needs to be established.

6.2.2. The network perspective on the schizophrenia mechanisms

In the past three decades, the field of schizophrenia has moved toward a more comprehensive understanding of the neurophysiological pathology. Current theories and experimental data converge on the notion that a wide range of problems in schizophrenia may result from a failure to integrate the activity of local and distributed neural circuits. The observations include abnormal power and synchronization patterns of induced or evoked EEG rhythmic activity in both medicated and medication-naïve subjects (Clementz *et al.*, 1997; Lee *et al.*, 2003; Spencer *et al.*, 2004; Symond *et al.*, 2005; Uhlhaas *et al.*, 2006), as well as decreased entrainment of oscillatory activity, primarily in high-frequency bands, in response to a steady-state stimulation (Kwon *et al.*, 1999; Light *et al.*, 2006). Impaired ability of distributed neuronal networks to integrate information in schizophrenia has been attributed to abnormalities in some neuronal networks that regulate rhythmic activity, such as inhibitory interneurons. A crucial role of inhibitory deficits in the pathophysiological process of schizophrenia has been suggested by abnormalities in the functional integrity, morphology, and distribution of inhibitory interneurons in schizophrenia patients, and further confirmed by evidence of significant deficits in intracortical inhibition in response to TMS in patients with schizophrenia. Impaired inhibition has also been implicated in abnormal sensory gating, or deficient filtering, considered to be a hallmark of schizophrenia pathology. The deficient gating hypothesis is based on significantly reduced behavioral and electrophysiological indices of sensory gating observed in schizophrenia patients and in animal models of schizophrenia. Human and animal data indicate that similar information-gating deficits may be a corollary to various molecular abnormalities, which include glutamatergic (Adler *et al.*, 1986; Ma *et al.*, 2009), cholinergic (Luntz-Leybman *et al.*, 1992; Bickford and Wear, 1995), dopaminergic (Bickford-Wimer *et al.*, 1990; Light *et al.*, 1999), and GABAergic (Hershman *et al.*, 1995) mechanisms. Interestingly, amphetamine-induced thalamic auditory gating deficits in rats were linked to impairment of bursting activity of the inhibitory thalamic reticular neurons, deficits that were restored by the dopamine D2 antagonist haloperidol and by acetylcholine receptor agonists (an agonist of $\alpha 7$ nicotinic acetylcholine receptor, genetically linked to schizophrenia), supposedly via enhancement of GABAergic neurotransmission (Hajos *et al.*, 2005). Likewise, GABAergic, dopaminergic, and cholinergic mechanisms were implicated in glutamatergic (ketamine-induced) impairment of auditory gating in rat hippocampus (Ma *et al.*, 2009). These studies suggest that multiple molecular factors, perhaps control parameters using the dynamic system terminology, may result in a singular phenotypical expression, a syndrome called "schizophrenia." This notion of multiple causes, same result, has appeared before in other chapters and it is of fundamental importance, so the words of D. Sornette can be brought to mind again: "the richness of out-of-equilibrium systems lies in the multiplicity of mechanisms generating similar behaviors" (Sornette, 2004).

What could be the consequences, at the collective level, of this impaired inhibitory (and/or other filtering parameters) functions? Intricate loops of feedback and feedforward inhibition are known to segregate activated neuronal assemblies into fine spatial and temporal domains specific to the incoming stimulus, which thus generate geometrically discrete rhythmic oscillations in distributed neuronal networks. Binding of multisensory inputs into a coherent cognitive experience seems to be reliant on this inhibitory rhythm-generating "clustering" of activities, as diminution of GABAergic inhibition has been shown to not only distort synchronized brain activity, but also alter perceptual selectivity. Inhibition, then, seems crucial for the integration and segregation of information that was discussed in Chapter 1. Section 1.2.1 provided an account regarding the tendency of inhibitory neurons to synchronize their actions (Whittington *et al.*, 1995). These aspects of inhibitory transmission derive from their hyperpolarizing actions on principal (pyramidal) cells: this limits cell firing thus creating opportunities for the segregation of networks, and, at the same time, a rhythmic interneuronal firing creates rhythms in the pyramidal cell outputs. There is substantial empirical and theoretical evidence indicating that inhibition determines the spread of cortical activation by sculpting oscillatory patterns in time and space. Blocking inhibition has been shown to result in abnormal spread of neuronal activity, resulting in lateral spread of stimulation-induced activation *in vivo* and *in vitro* model systems (Contreras and Llinás, 2001). Spatiotemporal patterns of spreading synaptic activity in response to stimulation in rat frontal brain slices were reported in the presence of both, dopamine and bicuculline, agents supposedly mimicking schizophrenia pathology "in a dish" (Bandyopadhyay *et al.*, 2005). Could deficient inhibitory neurocircuits (and/or any other correlates of aberrant filtering reported in schizophrenia) result in similarly enhanced propagation of neuronal excitation in schizophrenic brains? There is recent TMS–EEG data indicating that this may be the case: TMS induced waves of excitation spreading into remote areas of schizophrenic brains, while in healthy subjects the activation faded away soon after stimulation, remaining circumscribed to the area proximal to stimulation (Frantseva *et al.*, submitted). Of interest, this widespread activation in schizophrenic brains was associated with increased oscillatory activity primarily in gamma- and delta-frequency ranges that was localized not only to the leads proximal to stimulation, but also to fronto-temporo-parietal regions bilaterally. Moreover, contralateral fast oscillatory activity was recorded primarily in temporoparietal regions, most implicated in schizophrenia pathology. These data, although preliminary, suggest fundamental differences in transmission of the information across the cortical mantle in people suffering from schizophrenia.

Now let us try to fill out a missing link between these information-gating deficits in schizophrenic brains with the core symptoms of schizophrenia. What are the possible functional consequences to this "excitation leak"? Could it provide any insight with regards to the mechanisms of schizophrenia symptoms? An indirect answer to this question may be derived from *in vivo* studies in animals utilizing intracortical injections of the $GABA_A$ antagonist bicuculline, which results in localized disinhibition with the ensuing filtering deficits as discussed above. These investigations demonstrated an enlargement of the size of neuronal receptive fields (Alloway and Burton, 1991; Wang *et al.*, 2002) and a change in their spatial selectivity so that affected neurons became responsive to a broader range of stimuli, some which did not resemble the previously preferred inputs (Kyriazi *et al.*, 1996; Wang *et al.*, 2000; Rao *et al.*, 2000). It has thus been concluded that blockade of inhibitory neurotransmission unmasks normally silent excitatory connections, resulting not only in loss of neuronal selectivity, but also in making neurons respond to nonspecific stimuli. It could be said that some disinhibited neurons "hallucinate" by identifying a stimulus that has never been presented. Perhaps, similar mechanisms may be relevant to generating aberrant rhythmic activity responsible for perceptual abnormalities. The signal that "leaks" to the neighboring cortical area due to unfaithful filtering may excite neurons whose activation may be misinterpreted by other connected brain networks, perhaps as a voice when nobody speaks, the emotional/cognitive content fueled by dopaminergic surges. This hypothesis is consistent with aforementioned studies demonstrating increased brain activity during auditory hallucinations in both primary and secondary associative auditory cortices. An interesting interpretation of these findings was suggested by Uhlhaas and Singer (2006), who hypothesized that "hyperconnectivity between higher- and lower-order cortical areas favors back propagation to the respective primary auditory cortices of oscillatory activity generated in higher-sensory areas during visual and auditory imagery, thus generating activation patterns that resemble those induced by sensory stimulation." The "leaky cortex" may be an additional (or primary?) mechanism, underlying spontaneous ectopic activity in areas responsible for hallucinatory experiences. Therefore, auditory hallucinations in response to the sounds of appliances or hallucinations that start as indistinct noises and gradually progress through whisper-like sounds to distinctive voices make perfect sense; as well as the ability of some schizophrenic patients to keep their voices under control by listening to loud music, an intervention that may hypothetically sharpen their auditory receptive fields via thalamic depolarization thus potentially limiting lateral spread of cortical excitation (Llinás et al., 2005). Perhaps, the typical worsening of auditory hallucinations at night time is reflective of the new oscillatory patterns that emerge during drowsiness and sleep, which result in higher synchronous activities as already discussed in Chapters 1 and 5, further impairing the ability of the cortex to maintain crisp localized patterns needed for accurate information processing (of note, the phenomenon of "sundowning" whereby delirious patients worsen at night is well known in geriatric medicine; likewise, most forms of "organic" hallucinations, including those of the Charles Bonnet syndrome, are most prominent by the end of the day).

Thus, from the psychological and neurophysiological perspectives, specified in aberrant salience and abnormal functional conductivity respectively, auditory hallucinations in schizophrenia could be conceptualized as erroneous perceptual awareness resulting from an interaction of anomalous bottom-up (spontaneous oscillatory activity) and top-down (dopamine-fueled interpretation and hypofrontality) processing.

6.2.3.   The dynamical systems perspective on schizophrenia

The application of dynamic system theory to schizophrenia endeavors at the inspection of the possible irregularity at the global, collective level of neural activity, such that a "nonlinear theory of schizophrenia" is increasingly discussed in the literature. Several studies have inspected one or another "nonlinear" characteristic in schizophrenic brains, such as fractal dimensions or Lyapunov exponents, with various results reviewed by Michael Breakspear (2006). While, in general, these investigations of nonlinear properties of brain signals remain debated, as many times commented in this volume, the moral of these nonlinear tales in schizophrenia research seems to suggest that the neural dynamics is different in schizophrenia. In particular, the state space configuration in terms of attractor-like states has been commented in several works to be altered in the schizophrenic brain. Not surprisingly, the attractor metaphor is being used here. The conceptualization of these geometrical entities, the attractors, from a neurobiological standpoint was discussed in previous chapters. Recall, for the present purposes, that the activity of some neuronal assemblies such as the CPGs (e.g. respiratory center) could be described by a limit cycle attractor; or that the attractor reflecting the dynamics in our "sleeping" networks may be considered another limit cycle, the one that, at night, takes over most of the other attractors that keep us busy throughout the day. Furthermore, as you are reading these lines, the synchronized distributed coordinated activity in several areas that likely manifests itself in the comprehension of these lines could be thought of as another attractor (perhaps a synchronization manifold).

Now, let us imagine for a moment that you, while reading this chapter, have not been fed for three whole days. At this point, neuroscience narratives and intellectual resonances would hardly be high on your agenda. Instead, your mind would be consumed with images of the likes of sizzling steaks with pink insides, slowly oozing fresh juices next to a large lump of mashed potatoes (well, unless your attractor is decidedly vegetarian). There would be little you could do to free your brain from culinary imagery and thoughts. Your mind would be entrenched, locked into the workings of your feeding circuitry, easily excluding all other stimuli. The hypothalamus of a starved sufferer would receive "very hungry" signals from the periphery, translating their collective distress via salience generated by the interplay of the ventral tegmental area and orbitofrontal cortex, while committing all higher cortical centers to nothing but food-searching behavior. From a physiological standpoint, this scenario is most likely reflected by a coordinated activity in these distributed systems that control food intake and energy balance. In neurodynamical terms, what happened there is that your "feeding" circuitries, when starved, increased the basin of your feeding attractor-like state (or a synchronization manifold) to the point that nearly any input would switch you into a steak-related activity, be it smell, a word, or any spontaneously generated thought of yours. We suggest that the basin of this attractor (and many others reflecting synchronized coordinated activity or neurophysiological assemblies specialized for a particular behavioral purpose) is determined by the magnitude of your reward system activation, a hedonic valence that your mesolimbic dopaminergic system has assigned to it. Hence, perhaps your salience vector is an important control parameter, the one that under normal circumstances is responsible for building an adaptive hierarchy of actions based on biologically meaningful ranking of rewards (e.g. you would quickly snap out of your gastronomic preoccupations if you were suddenly faced with a rattlesnake). From this metaphorical standpoint, what could be happening in the brain of a schizophrenic person that is producing surges of dopamine out of context, is that its "delusional CPG" (or limit cycle attractor?), the one preoccupied with being followed by the CIA or other "vicious observers," has developed this malignant large basin of attraction, such that it dominates most of other physiological meaningful neuronal activities, not unlike your very objectively real rattlesnake (and indeed, acutely psychotic people are rarely able to concentrate on their physiological and social needs, sometimes requiring hospitalization in order to ensure their survival).

As discussed above, aberrant salience is, perhaps, necessary but unlikely a sufficient condition for developing psychosis, so let us try to add other aberrations of network functioning to our hypothetical dynamical landscape of schizophrenia. As detailed in the previous chapters, thalamocortical rhythmic activity shapes cortical networks into coordinated activity patterns: for instance, high-frequency oscillations (unlike slow ones) result in activation of discrete thalamocortical columns. This particular geometry of activation depends on spatial filtering by inhibitory neurons at multiple (thalamic, cortical etc.) levels, recall that blockade of inhibitory neurons was reported to abolish discrete activation of cortical columns in response to high-frequency stimulation. In other words, being exposed to a humming air conditioner, the sensory cortices would record it with the precision that is allowed by the highly specialized fast-oscillating cortical columns, each of them faithfully reflecting the gadget's shape, color, size, and sound, along with a welcomed change in air temperature. The primary auditory cortex, activated by the humming sound, would faithfully reflect the auditory experience, which can now be correctly classified by higher associative cortical centers as a noisy air conditioner. Now, suppose that the aforementioned impaired filtering characteristics cause a different coordination dynamics such that, for instance, cortical activation spreads nonspecifically over larger areas, to neighboring regions, needing different prefrontal efforts in order to decipher the incoming sound. What is this humming noise then, according to your. . . well, somewhat compromised prefrontal estimation? Perhaps is it whispering? Or maybe, it has been implanted there by police agents so that they can finally transmit their menacing messages directly into your head.

This rather speculative train of thought seems to find some support in models of neural network simulations, designed to simulate schizophrenia pathology using artificial neuronal networks. In these studies, the artificial neuronal network was trained to recognize a number of symbols, and then some of its weak connections were removed in an attempt to reproduce reduction in connectivity, that has been implicated in schizophrenia pathophysiology derived from particular studies (even though it would be more appropriate to say that it is not a matter of just reduced connectivity, but of a different organization of the dynamical interactions across brain regions [Breakspear *et al.*, 2003; Micheloyannis *et al.*, 2006]). At high levels of pruning, this model demonstrated a pattern of activation that did not correspond to any particular memory of previously learned symbols. It manifested autonomous activation that was conjectured to reflect hallucinatory experiences (Hoffman and McGlashan, 1997; Peled, 2000). This pathological activation in the artificial neural network, resulting from excessive pruning, was conceptualized as a "parasitic" attractor, as input did not bring the network into the regions of state space corresponding to symbols that it learned, but rather to a newly emerging attractor. Hence, it can be easily visualized how altered synaptic connectivity underlying (in this particular model) diminished signal-to-noise ratio could result in the appearance of spurious attractors, likely neurodynamic correlates of hallucinations and other neurocognitive disturbances of schizophrenia. This could predispose the system to easily transition to other attractors in response to small perturbations, the transition between attractor-like states described by the notions elaborated in Chapter 2 on chaotic itinerancy, heteroclinic channels or simply a nonhyperbolic attractor (Figures 2.7 and 7.3 of Chapters 2 and 7, respectively), which may ultimately underlie cognitive deficits in schizophrenia (see van Beveren and de Haan (2008) for an elegant visual metaphor addressing these issues). Along similar lines, Loh *et al.* (2007) related the various types of schizophrenia symptoms to instabilities in an attractor neural network that, in their work, consisted of integrate-and-fire model units. With this detailed computational model, the authors were able to ascribe transitions between attractor states in the model to specific ionic conductances, particularly GABA and glutamatergic (N-methyl-D-aspartate, or NMDA), proposing, too, that one reason for the many symptoms of this disease stems from increased statistical fluctuations in brain networks and reduced signal-to-noise ratio. A review of several computational models of schizophrenia and the relation to the dopaminergic modulation can be found in Rolls *et al.* (2008).

Further support for these notions was obtained in an EEG study that analyzed the topographic organization of nonlinear interdependences in schizophrenic patients. Although the rate of occurrence of dynamical interdependences did not statistically differ at any of the sites between the subjects, the topography across the scalp was significantly different between the schizophrenia and normal control groups. Specifically, nonlinear interdependences tended to occur in larger concurrent clusters across the scalp in schizophrenia than in healthy subjects (Breakspear *et al.*, 2003). These findings "do not support a simple 'disconnection' of cortical interactions as implied by the disconnection hypothesis of schizophrenia. Instead, they suggest a loss of the fine-grained organization of cortical interactions and hence can be cautiously interpreted as evidence of an impoverishment of flexibility across hierarchical brain regions in schizophrenia" (Breakspear, 2006). To sum up, a different brain coordination dynamics in schizophrenia is inferred from all these works (Bressler, 2003). Thus, using the language of dynamic system theory, positive symptoms of schizophrenia may be conceptualized as the manifestation of dominant "delusional" attractors (with aberrant salience dictated by a control parameter, however difficult, or impossible, could its quantification be) and the existence of spurious "parasitic" attractors resulting from decreased signal-to-noise ratio and abnormal connectivity.

6.2.4. Conclusions and implications for the treatment of schizophrenia

To sum up in few words, the problems with heightened dopaminergic signaling and inhibitory transmission in schizophrenia here discussed indicate that the brains display altered coordination dynamics, on two accounts based on what is known about inhibition in the nervous system. First, mesolimbic dopaminergic surges result in hyperactive limbic circuitry, which, superimposed on diminished inhibition, promotes higher synchrony in widespread areas, so we find again the familiar scenario of more synchronous activities associated with brain ailment of various natures. Second, inhibitory transmission regulates oscillations and synchronization patterns; hence, a deficient inhibition will alter these oscillations. Therefore, the net result of these events is a more pronounced tendency toward synchrony (particularly in temporoparietal regions) and, in general, different neural coordination dynamics. Although this could be the answer at the collective level of description, finding the core pathology at a more detailed level necessitates the specification of what brain regions are experiencing those altered coordination patterns of activity (not unlike temporoparietal cortex hyperactivity being a correlate of auditory hallucinations).

In the light of the complexity of schizophrenia pathophysiology discussed above, it is not surprising that therapeutic interventions for this devastating disease are strictly symptomatic and are limited to a single class of medications. Indeed, the arduous task of compensating for subtle neuroanatomical abnormalities with an agent targeting specific neurotransmitter system appears almost implausible. As in many other syndromes, so far only relatively nonspecific medications and treatments are useful. Whereas a linear relation between therapeutic concentrations and molecular actions on dopamine D2 receptors of many antipsychotic drugs has been reported, the most effective compounds in schizophrenia (e.g. clozapine) are those that bind to a wide variety of membrane proteins. The nice linear correlation may be a sign of a possible specificity, but with so many other ion channels and receptors targeted by these dugs, one can never be sure. Another very nonspecific method, electroconvulsive therapy (ECT), has provided anecdotal evidence of rapid and almost startling improvements in schizophrenic patients, which offers some hope that the time and space for the right perturbation in schizophrenia treatment is yet to come.

6.3. Stretching the Continuum: Beyond Pathology and Psychiatry

This section covers that part of the continuum form sanity to insanity that leads to psychopathy and extreme violence. The consideration of what is known about the brains of "normal" people and those of the individuals labeled psychopaths provides, perhaps, an adequate illustration of the brain–behavior continuum, the gradation from adaptive to maladaptive behaviors. Psychopathy describes a collection of personality characteristics, started to be described by the seminal work of

H.C. Cleckley in his *"The Mask of Sanity"* (Mosby Co., St. Louis, 1941). Life in our created societies require individuals to overcome ancestral genetic inheritance that drives us to aggression and violence, and not surprisingly waging war has been the major human occupation since antiquity (animals are too disorganized to engage in wars but watch what happens in more organized societies like those of ants). Equally unsurprising is that popular films are those with violence as basic ingredient, or that news and events with bloody images are specially attended to. Trying to repress these instincts may be as fruitless as a bee trying to live without flying. Perhaps, a more fruitful approach is to become fully aware of this tendency (contemplative methods such as meditation can serve this purpose of enhancing awareness) and then to redirect its manifestations to socially accepted actions. Genes may dictate to some extent the functional organization of brain networks, but because brains have evolved in such a manner that, perhaps due to their self-referential recurrent circuitries and/or environmental influences, can modify genetic tendencies in any direction (toward more or toward less, violence in this case here treated), then anomalous behavior arise. Extremes, like excessive altruism or intense aggression, could be in principle considered deviant behaviors, it is just that the former is an accepted one within human societies. Nervous systems are the product of the genes, but now the brain may have reached the capacity to overcome the designs of the gene.

6.3.1. Antisocial behaviors. The mind of a psychopath and the path toward violence

Violence, which can be described as an escalated aggressive behavior expressed out of context, does not have too much adaptive value under normal circumstances and thus it can be declared a deviation. Extreme cases result in psychopathy, serial offenders, among whom serial killers are notorious (these defined as those who intermittently murder three or more persons over a period of more than 30 days and whose motivation for killing is based on psychological gratification). Within this continuum, reactive aggression does not need to be maladaptive, for sometimes it is a proper response to a threat. Not surprisingly, then, aggression is found in animals from insects (Iliadi, 2009) to primates. It can be surmised that the brain circuits for aggressive behavior have been carefully "sculpted" by the genes since ancestral times, and some scholars like the zoologist Konrad Lorenz (1963) emphasized the genetic determinants and talked about aggressive instincts. It is known that, in mammalian brains, areas like the hypothalamus, amygdala, and frontal cortex are fundamental parts of the network that determines aggressive actions, and structural alterations in some of these areas have been observed in brains of psychopaths.

So, stretching the continuum, serial sexual homicides can be thought of as an extreme antisocial variant of phylogenetic predation, and, like L. Miller mentions: "it is pathological only in terms of degree, not the nature of the act, and the brain mechanisms involved are on the same continuum as those related to more "normal" forms of hunting, group combat, romantic pursuit. . ." (Miller, 2000). Hence other authors have remarked that these deviations are part of the continuum in behavior, and where the demarcation is placed could be somewhat arbitrary, nonetheless, these are deviant behaviors.

In clinical terms, aggression is subdivided into two categories: reactive aggression, brought about by a frustrating (or threatening) event without any particular goal in mind, or instrumental aggression, initiated purposefully in order to achieve a specific goal (like taking the victim's money). Reactive and instrumental types of aggression seem to be mediated by separate neurocognitive systems, and there are two populations of antisocial individuals distinguished based on this classification: those presenting with instrumental and reactive aggression and those presenting with predominantly reactive aggression. The neural circuits involved in reactive aggression have been identified in animal studies and are represented by the basic fear circuitry, including amygdala, hippocampus, periaqueductal grey, as well as medial and orbitofrontal cortex that modulate this fear circuit. Genetic and environmental factors (e.g. physical, sexual abuse, and neglect) heighten responsiveness of the fear circuit to threat thus increasing probability of reactive aggression, sometimes leading to a diagnosis of antisocial personality disorder. Likewise, both injuries to prefrontal cortex and dysfunction in the serotonergic system (that uses serotonin as major neurotransmitter) have also been linked to reactive aggression, likely via impaired ability to interpret social cues and to modulate aggressive impulses.

The pathology associated with both instrumental and reactive aggression is indicative of psychopathic traits. Psychopathy, a constellation of personality characteristics that include callousness, lack of empathy, impulsivity, manipulation, and irresponsibility, is thought to stem from a specific biological abnormality present from birth (although multiple other causes contributing to this pathology have also been recently suggested). Deficits in empathy, emotionality, and inability to react to the distress of others (e.g. callousness), are thought to be the core pathology, at the psychological level, of psychopathy. Empathy, defined as the ability to recognize and share another's emotional state, is reliant on intact functioning of limbic (e.g. amygdala) and paralimbic circuitries (insula, anterior cingulate, orbitofrontal, ventromedial cortices, and possibly mirror neurons in general). Based on psychological assessments and neuroimaging data, current theories stipulate that psychopathy is associated with dysfunction of the amygdala (necessary for emotional learning) and prefrontal cortex (required for impulse control, decision making, behavioral adaptation, and emotional learning). The predominant school of thought is that psychopaths are born with temperamental differences such as impulsivity, cortical under-arousal, and fearlessness that lead them to risk-seeking behavior and an inability to comply with social norms (Blair *et al.*, 2005). On the other hand, other, nonpsychopathic sociopaths may have less difficult inborn temperaments, their personality traits being more an effect of negative psychosocial factors like parental neglect, delinquent peers, poverty, and low intelligence.

6.3.1.1. *Behavioral and neural dynamics of antisocial behaviors*

Our main concern here is to try to understand the neurodynamics that psychopathic brains may exhibit, and how it differs from that which manifests as normal, socially accepted behaviors. A look at the literature by organizations like the Federal Beureau of Investigation (FBI), on these subject matters may prove useful. A main conclusion derived from FBI research is that these behaviors tend to develop for a long time, normally since childhood. It was reported that children exposed simultaneously to several of the considered risk factors (like maltreatment, family disruption, parental violence and stress, and low socioeconomic status) were more likely to develop psychological deviations, but exposure to only one of these factors had little impact on children (Appleyard *et al.*, 2005), which has given rise to the multiple risk theory. Perhaps because these antisocial behaviors develop over prolonged periods of time, they become very stable during the life span (Moffitt, 1993). Consequently, using the dynamic terminology introduced in Chapters 2 and 3, these disorders could be considered the manifestation of brain dynamics showing a high degree of structural stability. Very stable behavioral dispositions are not uncommon in psychiatry. For instance, appetite is a drive that is altered in patients with anorexia nervosa, an eating disorder that has one of the highest mortality rate in psychiatry characterized by very restricted food intake and the conscious pursuit of weight loss. Patients with this syndrome, among other things, are anhedonic, and these behavioral characteristics remain even when some patients recover and gain weight to reach normal standards (Kaye *et al.*, 2009). Anhedonia could be a reflection of alterations in dopaminergic transmission and, in general, in the processing of rewards. The point is that the "mental state" persists even when the symptoms disappear. Taken together, these observations in a variety of psychiatric disorders indicate that it is conceivable that the structurally stable dynamics of some brains which manifest in aberrant, nonadaptive behaviors, could be a result of either structural alterations in brain circuitries or subtler dynamical changes in information processing. Obviously, the question of whether the nervous system dynamics or the behaviors are structurally stable from a dynamical (mathematical) perspective cannot be precisely answered in the absence of dynamic models, but to some extent it could be said that the behaviors of all individuals, normal and deviant, tend to exhibit a high degree of stability, notwithstanding the sometimes very profound changes that occur in the transition from adolescence to adulthood.

Slightly different from the continuum view on the interpretation of these psychopathic behaviors, other perspectives propose that these actions result as a consequence of purely pathogenic conditions of the nervous system. In general, both perspectives are very similar: whether anatomically abnormal or not, what seems to occur is a deviant, or "pathological," activation of some brain areas; e.g. some have considered that pathological activations within the limbic system, perhaps due to malformations or misconnections, result in sexual sadism, because the limbic system regulates normal actions related to predation, attack, and drives in general. Even J. Money (1990), one of the proponents of the definite pathogenic view, agrees that the pathological information processing can be due to gross anatomical or biochemical alterations or to minute, undetectable triggers. The modern mammalian brains are complex enough (if we are allowed to mention here complexity without any added quantifier) so that the resulting actions involve a large variety of neural pathways subjected to plasticity: control (frontal cortex), drives, rewards, and addictions (limbic), etc. Therefore, when it is said that there seems to be an impulse that cannot be controlled in serial offenders (clinicians call this lack of control, in general, 'impulse control disorder'), this is a statement of monstrous complexity: the circuitries of addiction may be altered (because patients can be said to be addicted to their harmful behaviors), or the control functions of frontal cortices could be depressed for one or another reason. Having in mind the basic functioning of the nervous system in terms of chains of activations from net to net, then to determine whether that putative "lack of frontal control" derives from altered driving by limbic areas, or because synaptic plasticity (potentiation or depression of synapses) is aberrant in some pathways, or from a myriad convergent inputs from numerous areas forcing the frontal regions to some activity pattern, is virtually improbable to solve. Having said this, one can still indulge in some reflection about possible neurodynamics of the psychopathic brain.

The behavioral dynamics leading to a serial offender has been classified, in the work of Cotter (2009), based on psychological evidence, in the following five stages: (1) emotional problems; (2) initiation; (3) adaptation to murder; (4) a trigger event; and, (5) the act of murder itself, these stages developing over a prolonged time, except the last two, which are precise events. Can this behavioral dynamics be cast into a dynamical system framework? Stages 1 and 2 can be thought of as the original state space configuration, perhaps composed of many attracting sets, as very basically depicted in Figure 7.3 of Chapter 7. Stage 4 may represent an initial condition, a perturbation that leads a behavioral trajectory toward the basin of attraction of one of those attractor-like states, that includes neurophysiological activity in a variety of brain areas processing, among other things, reward, as these violent acts are rewarding and satisfying to these individuals; and stage 5 could be interpreted as the falling into the attractor, the manifestation of the action.

But, how about the neurophysiology underlying these transitions? At the molecular level, serotonergic neurotransmission has been implicated in the regulation of aggression, with sufficient evidence obtained in animal models (Nelson and Chiavegatto, 2001; de Boer *et al.*, 2009), and the circuitry involves a variety of brain regions, among these the amygdala, prefrontal cortex, hypothalamus, and periaqueductal gray. This is the circuitry thought to process emotions, thus it is natural that any slight, or large, abnormality in these networks predispose toward aggression and violence (Davidson *et al.*, 2000). A wide variety of neuropsychological and neuroimaging findings on psychopaths indicate structural abnormalities and dysfunction of frontal and subcortical brain areas especially in the left hemisphere. It is also known that psychosis secondary to traumatic brain injury tend to be associated with lesions in left frontal and temporal areas. Brain alterations observed by neuroimaging performed on serial offenders are reviewed in several works: Miller (2000), Bufkin and Luttrell (2005), Yang and Raine (2009), Alcázar Córcoles *et al.* (2010), and Blair (2010). The general finding from these brain metabolic investigations is that of hypofunction of the frontal cortex relative to hyperactivity of subcortical areas. The role of frontal cortex in maintaining socially accepted behaviors has been well known since the famous case of Phineas Gage first reported in 1868 by John M. Harlow. Hence, the hypoactivity of frontal cortices are interpreted as indicating less "control" directed by these regions. Note that this general pattern of less frontal and more subcortical activity is identical to that found in neuroimaging studies of patients suffering from major depression. It is certain that many specific differences in these patterns exist between depressed and psychopathic individuals, but the observation of a global, common pattern of brain activation derived from these metabolic assessments (fMRI or PET have been the usual studies) in two very dissimilar phenotypes points out a major limitation with these methodologies, in that the contents of those neural networks, the information that is being processed, will never be exactly known. Remarks on this matter appear in the final paragraph of Section 7.1 of Chapter 7. A few details that can be remarked in these neuroimaging studies include amygdala and orbitofrontal cortex hypoactivity, as well as insular, anterior cingulate, and ventromedial prefrontal cortex dysfunctions that have been linked to callous and unemotional traits (Rodrigo *et al.*, 2010). Additionally, dysfunction of limbic and paralimbic circuitries has been shown to impair a person's ability to appreciate emotions of another human, particularly fear. Reduced prefrontal glucose metabolism was found in a group of murderers when compared with normal control subjects (Raine *et al.*, 2000). Interestingly, similarities between the neurocognitive profile of assault/murder criminals (in contrast to drug offenders or theft criminals) and patients with orbitofrontal damage have also been recently demonstrated (Yechiam *et al.*, 2008), implicating prefrontal cortex deficit as the key neurophysiological abnormality in violent criminals.

With regards to more precise neurophysiology, not much has been done. In the mid-1990s, some event related potentials (ERPs) recorded from scalp EEG were assessed for their possible usefulness to identify pedophiles, without much success. One component of ERPs, the P300 waveform, recorded from scalp EEG was reported to be different in some psychopaths (those labeled as "unsuccessful psychopaths"), which represents an indication of distinct patterns of neural activity that may be occurring in these brains, while the ERPs were normal in the "successful" psychopaths (Gao and Raine, 2010). Successful psychopaths are those who are, normally, not incarcerated because they find other, nonviolent more socially acceptable means to achieve their goals; they are a sort of high-functioning psychopaths, according to the original description by H. C. Cleckley in his aforementioned book in the first paragraph of this section. The rest, those convicted and incarcerated, are the unsuccessful psychopaths, these are the typical ones we hear about in the media because of the crimes they tend to commit. The successful types do not show clear structural abnormalities in the brain areas reviewed above, regions that are typically abnormal in the unsuccessful ones. Whereas the alterations reported in those ERPs are suggestive of differences in neural dynamics, this is not precisely known because of the lack of studies. To our knowledge, only one study has evaluated brain synchrony in terms of coherence derived from scalp EEG taken in pedophiles, another set of individuals with antisocial behaviors. This study, that compared EEG patterns in 46 controls and 96 pedophiles, found a reduced interhemispheric and increased intrahemispheric coherence (in verbal tasks) in what the authors call "true pedophiles" (52 of the 96), those men who were exclusively erotically attracted to young children as opposed to others who display more heterogeneous behaviors such as incest offenders or practitioners of adolescent homosexualism (Flor-Henry *et al.*, 1991). The authors hypothesize, based on the observation of the altered coherence pattern only during the verbal task that engaged the dominant hemisphere but neither during rest nor during visuospatial exercises, that there is a dysregulation of left–right hemispheric coordination and an instability of the dominant hemisphere that gives rise to abnormal ideational representations (but which instability, they do not specify). The explosive nature of some forms of violence motivated the idea that these acts may be a manifestation of paroxysmal discharges in brain areas involved in these traits, and indeed epilepsy and sexual sadism are known to coexist in a subset of patients. If this were true, the typical scenario described in the previous chapters emerges once again: an altered coordination dynamics reflected in higher than normal synchronous activities in certain brain networks manifest itself as antisocial behaviors. With such limited data, however, any conclusion on the dynamic patterns of the psychopathic brain could be premature. Some analysis of interest, considering the previously mentioned reports on hyperactivity of subcortical areas found in antisocial individuals, could be the assessment of synchrony derived from electrophysiological signals between subcortical areas and between these and cortical regions, but for these experiments electrodes would have to be implanted into the brain, thus the invasive nature of the procedure may preclude these analyses.

From a theoretical stance, there have been computational models of psychopathology, neural network simulations with the aim to understand the essential characteristics of deviant actions. The famous Hopfield model has been applied to psychiatry since the late 1980s (reviewed by Aakerlund and Hemmingsen (1998)). A model attempting to capture the essence of some, albeit normal, behaviors, was already described in Chapter 4 when discussing a relatively simple sensorimotor task (Section 4.2.1). Further, as discussed there, a fundamental question in this type of studies is the choice of order and control parameters. It was described how the aforementioned sensorimotor study used relative differences in limb position (relative phase) as the order parameter, but now we are talking about very complex behaviors, so, what order parameters can be used for multifaceted behaviors? Behavioral state spaces have been constructed using behavioral intentions as a bifurcation variable to study changes in substance abuse during adolescence (Mazanov and Byrne, 2006). Behavioral intentions have also been used as the variable of choice to predict social behaviors such as voting and criminality, even though sceptics may argue that the term has no clear definition and is ambiguous. Substance abuse has been modeled using the number of drug users, in systems of differential equations of compartmental representations of the consumption of drugs and found, after performing stability analysis (of the type described in Chapter 2), multiple equilibria that determine drug-free or drug-persistent states within a population (Nyabadza and Hove-Musekwa, 2010). The dynamical study of individual behaviors and of societies has not been much explored in general, at least when compared with the immensity of studies on nervous systems' dynamics. Interest, however, may be on the rise. There is even one journal that seems devoted to these queries, the official journal of the Society for Chaos Theory in Psychology & Life Sciences: *Nonlinear Dynamics, Psychology, and Life Sciences*. A relatively early work was that of W. Weidlich (1991), where he advocated the approach of synergetics (introduced in Chapter 2) to study the physics of social science. His framework is probabilistic, starting from decisions of the individuals and introducing probabilistic transition rates between actions and personal attitudes, thus generating equations describing probability distributions of the configurations of societies whose solutions included stable states and bifurcations between social behaviors. The author goes even further and applies this framework to the study of the formation of collective political opinions, formation of settlements, economics, and more. The advice of some seminal thinkers in related fields may suggest other order parameters to be used in sociodynamics. For instance, Adam Smith considered self-interest as an essential drive for economics, and in his view, the interacting components determining society's dynamics were the individual self-interests; but once again, while we are all extremely familiar with the empirical fact of the existence of self-interest, its quantification may not be easy and remain a bit ambiguous. There is certainly room for creativity for those interested in these topics.

A review of the dynamic systems approach to psychopathology, with a focus on development, can be found in Granic and Hollenstein (2003). Here, the authors describe previous proposals to design such an approach to developmental psychology, and comment on the difficulty of selecting order parameters and the even greater intricacy in the identification of valid control parameters. The common framework of these studies exploits the usual metaphorical language: highly stable or persistent behaviors are identified with attractors of a dynamical landscape, bifurcations in behaviors are considered when there is a change in behavioral patterns, which, the authors mention, could be important to target clinical interventions near these critical points. What are the data that can be used in these types of psychodynamic studies? Chapters 2 and 3 described the continuous or discrete (recall the examples using cell spikes or EEG peaks) approaches to construct state spaces. In principle, a first-return plot based on the measurement of one specific behavior at time t and at time $t + 1$ could be constructed. In this manner, the duration of deviant talk (a sort of antisocial talk) measured from videotaped interactions between friends has been used to create a time series, and, from this, a first return, time-delay plot was created. It was observed that, in some pairs of friends, the antisocial talk becomes of longer duration as time advanced, so the investigators conceptualized this as falling into an antisocial-talk attractor (Granic and Hollenstein, 2003). Other means to devise behavioral time series can use questionnaires, which are commonplace in psychology. Another example: using inter-event intervals between successive displays of negative affect in married couples, a first-return graph allowed the scrutiny of marital dynamics (Bakeman and Gottman, 1997). Observations that capture interactions between individuals will probably be useful in psychodynamics. From these studies and current trends, it is felt that researchers are not only using metaphorically dynamical system concepts, but also the methods, and it has been proposed that antisocial trajectories can be anticipated with these analytical tools (Lewis, 2004).

6.3.1.2. *Social dynamics*

While deviations in brain dynamics resulting in diseases and other "insanities" were considered in detail in previous chapters and sections, the study of the emergence of social madness can benefit from analogous dynamical frameworks. In the final analysis, the spatiotemporal patterns of human societies result from the collective (nonlinear and linear) interactions among the constituents. If one cell's activity has meaning only in relation to other cells' activities, it can also be claimed that a brain has meaning in relation to other brains, and thus emergent properties are found in social groupings too. In this sense, there is not much difference between societies, ecosystems with interacting individuals, fluids with patterns emerging from interacting molecules, or brains where the constituents are cells. In the early days of Mediterranean civilization, Greeks already mused on the emergence of social and political order and renowned scholars like Plato provided the basic ingredients: citizens who have to specialize in different skills and organize for cooperative work, exchanging goods and services. In more modern, neuroscientific terms, our philosopher was describing nothing more that the concepts of integration and segregation via interactions that have appeared numerous times in this volume. In further reminiscences of modern ideas in brain research, he also defended the idea of a central control, a government because otherwise instability in the society arises due to conflicting interests of the individuals. However, where the central control in the brain is located, remains unclear, and it may be unclear too the need for a central control in society. Some sort of behavioral vectors are needed to create sociocultural state spaces, to go from the local to the global, from the individual to the collective. The complexity of ecosystems and societies in general calls for a probabilistic approach, for which evolution equations for the groups need to be derived, derivation that can perhaps benefit from the aforementioned Weidlich's studies. It has been explained in previous chapters regarding how to go about capturing neural collective behaviors, and perhaps similar approaches are useful in sociodynamics. The upcoming new edition of Dirk Helbing's *Quantitative Sociodynamics* (Springer) contains stochastic methods that are employed to model social interactions and, as well, the nonlinear dynamic approach.

Synchronization processes in the nervous and other systems have been much discussed in this volume. As well, interpersonal synchronization has been found in experiments on motor coordination (Schmidt *et al.*, 1990, mentioned in Section 4.2.1 of Chapter 4). Equally, there is also synchronization found in the masses. The possibility that the global mechanisms of collective synchronization of cells in nervous systems share parallels with the collective synchronization of minds in social groups should not be underestimated. Indeed, collective synchronization among individual minds seems to occur not only during the well-known phenomenon of spontaneous synchronization during clapping (the applause caused by the hands motioned by the brain, hence a sort of mind synchrony is developed), but also during sports, religious acts, and social revolutions. Perhaps, that is the underlying reason that explains Nietzsche's observation: "In individuals, insanity is rare, but in groups, parties, nations and epochs, it is the rule" (Nietzsche, *Beyond Good and Evil*, 1886). Individuals may be seen as local arrangements of a global phenomenon, there is a continuum here too (as the central doctrine of the Upanishads reads, "our personal awareness of being alive is only a local and imperfect observation of a universal reality," in M.L West, *Early Greek Philosophy and the Orient*, Oxford 1971). This sort of behavioral synchrony in these occasions results, conceivably, in/from specific neural coordinated patterns in each brain of the mass experiencing the collective rapture, the brain–behavior continuum once again manifest (one wonders what brain synchrony can be found between two individuals during a mass social event like the finals of the football world cup when their team scores). Considered as a dynamical system, a society will exhibit the typical stable and unstable states, perhaps bifurcations, critical points very sensitive to perturbations, that may explain the origin of revolutions and the coming to power of certain individuals. The more the dynamics of societies is investigated, the closer it will appear to that of other complex physical systems. Societal dynamics can be affected by large or small perturbations, and history abounds with examples of how slight perturbations completely alter the dynamics of certain societies. A quick example: Abderraman III, first caliph in Europe (with capital in Cordoba), solved in a singular way the social problem that had existed for about two centuries in the Iberian peninsula, by introducing a not-too-drastic perturbation, unlike those of more extreme character popular in medieval times. The three main constituents of his society, Spanish, Arabs, and Berbers, did not get along at all which created innumerable conflicts, so the caliph solved this social dynamical problem by introducing little by little another population, Slaves, from other parts of Europe, perturbation that, apparently and according to the chronicles of the time, solved the social distress. No need for mass executions to alter a complex dynamical system! If adequate order parameters can be found in sociodynamics, could these exhibit growth of their fluctuation amplitudes (Section 2.7, Figure 2.15 of Chapter 2) before, say, a revolution (a societal bifurcation)?

But, to use the advised stochastic approach in these investigations, connoisseurs of probability theory are probably now thinking, the assignment of probabilities is obviously necessary which entails the making of a measurable space where a probability measure (like distribution functions) is applied, and all this requires the usual description in terms of choosing the elementary outcomes of that measurable space and finding a sigma-algebra of the subsets, plus a few more technicalities that can be found in texts on probability theory (Koralov and Sinai, 2007). Whether one is investigating cell networks, brains, or societies, finding the space of realizations of a process where transition functions and initial conditions are defined is the first problem encountered, in some cases being easier than in others. For cells, the firing probabilities can be used; for brains, some synchronization index that represents the likelihood of phase locking; for social groups, perhaps specific individual behaviors can be used to create that space. But before going too far in the realm of speculation, let us now put an end to this digression on social dynamics with the thought that, in the end, neuroscientists, psychologists, and sociologists alike end up studying sums of random variables to capture collective behaviors. Whether membrane potentials, or intensity of pixels in neuroimaging data, or individual tendencies and behaviors, these can be considered random variables and, due to that famous result in complex system theory already mentioned in Section 1.3 of Chapter 1 called the central limit theorem, the individual details are washed out and a final distribution, typically Gaussian, arises as expression of a simple collective phenomenon, regardless of the complexities of the individual constituents.

6.3.1.3. *Between determinism and randomness*

What, then, in a dynamical sense, is different in the mind of a psychopath as compared to any nonpsychopath? The following tale presents in simple terms, that hopefully everybody can understand, the nature of the differences. Let us imagine that a parent and his/her active five year old visit a child's playground on a hot summer day. The parent follows the movements of the little guy lovingly and intently, relishing his loopy laughter and the particular ecstatic expression in his eyes that only happy children can show. The child's happiness is mapped immediately onto the parent's brain and as a result there are now two happy individuals. But, how can the parent be happy in these conditions, one may ask: packed tightly among other sweating parents, having to tolerate yet another hour of this ungodly squealing under the blazing sun, while there is cold beer and a favorite book waiting in a perfectly air-conditioned quiet home? Or else, imagine one of your children smashing his finger, now he is running toward you wailing in distress; pain, hurt, and anger shaking his little body. Your heart sinks. Your mind floats somewhere out there, in anguish of pain, throbbing heart corroborating the very physical nature of your suffering. What has happened here in neural terms? Very much a similar neurodynamics as in the previous example. It is known that certain brain networks get activated when one is in pain or also when pain is observed in others (Craig, 2002). Moreover, the distinction between self and others' painful experiences is encoded primarily by the degree of activation rather than by activation of distinct neuroanatomical structures (Jackson *et al.*, 2006). When the brain responses to pain were compared with the activity obtained in response to an observation of a loved one receiving identical painful stimulus, it turned out that the activity in pain pathways always overlaps, regardless of who received the painful stimulus, you or your loved one (Singer *et al.*, 2004). Activations of the insula and anterior cingulate cortex, the two key areas known to support empathy-related functions, were implicated in this process. Hence, the brain is wired to experience the emotional component of pain or pleasure of others. From this view, the parent's desire to drive the offspring to a playground on a hot summer day is perfectly understandable, as a manner to obtain an actual (albeit vicarious) pleasure. In the final analysis, the parent is experiencing pleasure himself, it cannot be otherwise as the result of a normal brain's activity processing environmental inputs. So much for the high esteem at which we hold highly altruistic people! Indeed, as Richard Bach proclaimed in his 1977 book *Illusions*: "anyone who's ever mattered, [...] anybody who's ever given any gift into the world has been a divinely selfish soul, living for his own best interest." In dynamical terms, the brain is now servicing yet another pleasure-driven attractor-like state, miraculously transforming hot, noisy playgrounds into a joyful experience.

These two tales of joy and grief serve to illustrate what may be happening in the sociopaths' brains, who, as discussed above, are deprived of neural circuitries mediating empathy and other social resonances. They really could not care less about the pain or pleasure of others. Blazing sun and loud children do, perhaps, nothing more than to annoy them. No "pleasure by proxy" rewards there, the dynamical trajectory of their behavior drawing a very different "loveless" picture, antisocial characters being notorious for their reckless child-rearing practices. In fact, if they have to rub someone in order to profit, only the "profit" part registers with them emotionally. In addition to the lack of empathy they are also impulsive, they sooner or later get themselves incarcerated, fulfilling the criteria for the "unsuccessful" psychopaths aforementioned in Section 6.3.1.1. On the contrary, if their intelligence and composure allow for more constructive set of behaviors, they become, perhaps, high-profile "successful" psychopathic chief executive officers (CEOs) (the likes of those who deprive people of their life savings, getting high profits and status, with minimal chances of being jailed). Now, to emphasize the continuum of predatorial functioning, let us think about serial offenders, such as lustful murderers. Some of these folks suffer from what is known as sexual sadism, a paraphilia that involves sadistic fantasies and compulsions leading to rape and homicide. In simple terms, the pleasure centers in these people for some reason are wired to the extreme suffering or even death of others. These fantasies and compulsions are so pervasive and powerful that, in some instances, they lead to most horrible crimes imaginable (these unfortunate instances are supported, as expected, by psychopathic traits and impulsivity). Can they control their antisocial behaviors? Perhaps, no more than one can control an eating frenzy after a prolonged fast. This overly simplistic scenario, yet empirically supported, pictures an interesting phenomenon: the addition of each aberration alters the anatomical architecture of the networks involved in motivating, planning, and execution of certain social behaviors, reflected, in all likelihood, in synchronous patterns of activation of various neuronal subsystems, be it empathic or predatory.

The term "pleasure centers" has appeared above, but perhaps it may not be an accurate expression because, while it is true that there are specific brain areas associated with pleasure, happiness, and reward (recently reviewed by Kringelbach and Berridge (2009)), it is more likely that the general feeling (or "mental state") is the result of the coordinated activity in all those centers (same applies to pain). Other metaphorical terms that have appeared include attractor-like states for reward, pleasure, pain, etc. At this point, some readers could be confused as to whether these states are defined in the anatomy of the brain, the connectivity patterns, or defined as high-level dynamical constructions in a theoretical state space. Section 7.2.1 of Chapter 7 contains more discussion on this matter, and Figure 7.3 presents a hypothetical neural state space. Advancing now what is detailed in Chapter 7, and to try to answer the question posed above, it can be said that the attractor-like states in that state space reflect the dynamics that are not necessarily peculiar to some circuitries: distinct networks can display same dynamics, and it is precisely the nature of these networks that are experiencing one or another type of dynamics that manifest into a behavior. In this regard, we all have powerful "reward attractors," perhaps with enhanced synchrony among areas processing reward-seeking behaviors as shown in Figure 7.2 in Section 7.2.1 of Chapter 7. From this global viewpoint, the dynamics of normal brains is similar to that of psychopathic brains: many of the social offender's actions revolve around reward and the search for pleasure, as we all practise, but for most of us the behaviors revolve around obtaining trivialities of socially acceptable kinds. In the psychopath minds, to their disgrace, those attractors are associated with activity in networks that, as a final result, manifest itself in violent actions. However, the neuronal activity in these brain anatomical networks, most likely, are shared in psychopaths and healthy individuals, for after all it is certain that the amygdala, frontal cortices, and other structures would "light up" in any fMRI scan of someone eating strawberries bathed in chocolate and whipped cream. So, what is the real difference? Same structures are used with similar dynamics, but, we think, with very different coordination dynamics. Here, most probably, lies the main difference, but this field remains very little investigated and the very few studies on brain coordinated activity in psychopathy currently available have been commented above, suggesting, even though perhaps prematurely, altered coordination dynamics.

These views may look very deterministic, in the sense that all behaviors seem to be dictated by the already pre-existing neurocircuits and their dynamics, however the plastic characteristics of nervous systems, something that sets this system apart from most of others (with, perhaps, the exception of the immune system), carry with them the potentiality for transformation. Plasticity, a very common neuroscientific term, normally denotes the activity-driven changes in synaptic transmission, and even a higher-order form, metaplasticity, is being talked about (Mockett and Hulme, 2008). The significance of this phenomenon for the brain and behavior continuum is that sensorimotor transformations performed by brain circuits manifest themselves as behaviors that can feedback onto the same circuitries and alter them. This notion has been repeated often in this volume, and reappears again in Chapter 7: there is no start or end to the chain of neural activations (those synfire chains discussed in Chapter 1), and then, actions create sensory perceptions that recreate more actions. For these reasons, there is always hope for change. It is important to note that some people who derive pleasure from inflicting pain or violating others are able to refrain from committing crimes. Some of them may have intact empathic abilities; others are simply avoiding legal troubles. These people may become hopelessly depressed; they could even seek treatment (even as brutal as castration), and violent suicides of "conscientious" pedophiles are not uncommon in clinical practice. Then, having the "evil" template blueprinted in the nervous system does not automatically sentence one to become a monster. However, one can go further in this loop and consider that even this potential to change is engraved in the brain too: the template already exists in some, perhaps fortunate, brains, a template-changing brain that orchestrate the individuals' functioning patterns of behavior. So, more determinism after all. These discussions already start to sound like the ones revolving around chaotic dynamics, that, as explained in Chapter 2, is deterministic but unpredictable. Well, perhaps, that is it. Since it is not feasible for our limited intellects to be aware of everything, the initial conditions (which are only "initial" depending on the perspective, of course) are never fully known, hence behaviors, while determined by the nervous system's dynamics (plus environmental influences), can be unpredictable. However, does it all mean that, in the words of Dostoyevski (*Notes from the Underground*, 1864), "man will no longer have to answer for his actions"? This was the first quote of four notable ones, at the start of the book, after the preface. Our third quote, B. Spinoza's words, sound equally deterministic, and show the deep understanding of some intellects already in early times. To the Russian's and to the Dutch's words, J.D. Schall has a reply (our second quote) where he posits the question of the constraints in choice making, the boundary conditions (in mathematical terms) within which each individual and the surroundings operate. This is what can be really experimentally addressed and, to our modest opinion, the crux of the matter. Our fourth and last quote on that page by the novelist, poet, playwright and philosopher M. de Unamuno articulated just what has been tried to do in this book. Are we then advising inaction? Do insane or mentally troubled patients have a hope? Certainly, yes, for that indeterminacy of the initial conditions, or of the dynamics in general, invites precisely to experiment with ourselves not in the laboratory but in the theater of life and society. There are always things to win by acting. But action, or will to change, is after all another behavior which should never be assumed to be within the reach of and easily performed by all individuals, and thus we come back to the query of where the cognitive control lies. It is the general acceptance by the immense majority of the population of this belief that change and will to change is a relatively easy behavioral action, that stands at the origin of many misunderstandings, misconceptions, and injustices.

At this point, the Rayleigh-Bénard convection discussed in Chapter 2 can be revisited. Recall the basic phenomenon, the organized geometric arrangement of convective cells, explained in Section 2.3. Would anybody in his/her right mind ask what is in charge of the emergent organized geometric pattern that appears in the oil? Certainly, the temperature gradient is needed, but if the fluid is not oil the patterns may not appear, and let us not forget that gravity is fundamental for the rolling convection of the cells within the fluid: no convection, no pattern. So, what is it in control, gravity, molecules, or temperature? So much for this "control" situation. Then, why do we insist asking this question of brains and behaviors? Nonlinear complex systems do not need a central control. But, in terms of brain dynamics directing behaviors, there are parts that in one or another action remain most crucial and perhaps exert something that can be cautiously labeled as "control." Is there any area that may have more "control" over others? We are not sure.

6.3.2. Potential for therapeutic interventions and the return to "normality"

If, as noted before in Section 6.3.1.1, these deviant brains operate with structurally stable dynamics, is there any hope to modify these socially unacceptable behaviors? From the dynamical systems' perspective, there is very little hope. Likewise, clinicians view treatment of sociopaths with well-deserved pessimism, incarceration remaining to be the most effective intervention (if this can be called intervention) particularly for psychopathy. Basically, the brain state space configuration would have to be completely changed, which means that a new individual would emerge. It is unlikely that there are specific triggers for these conditions, recall the evidence mentioned in Section 6.3.1 that exposure to one "risk factor" was normally not enough to result in a troubled individual. The triggers come later, at the particular time to commit a crime, for example, but the relatively stable and well-formed brain dynamics of these individuals dictates their behaviors, hence we do not think likely that these persons become "ill" due to a very specific trigger, even though if these situations are very prolonged, and always depending upon the person in particular, may influence to a large extent the already existing neurodynamics. Hence, and most likely, a pedophile will always remain a pedophile, and a serial killer will continue to derive pleasure from aggression throughout life, regardless how they pretend to have changed. Nevertheless, within the continuum, some psychiatric deviations will be easier to remedy than others, for instance depression can be successfully treated in some cases.

Therefore, with this idea in mind, perhaps the most efficient treatments will include a variety of approaches so that an important perturbation of the dynamics is produced from different angles, and, we submit as done in the previous chapters, that the more nonspecific the treatment, the better because it will modify many processes. Thus, it may not come as a surprise that effective treatment of conditions like depression, schizophrenia, and a variety of other syndromes like addiction, is direct brain stimulation, using transcranial direct current stimulation (tDCS), TMS, ECT, deep-brain stimulation (DBS), or caloric vestibular stimulation (CVS). These stimulations modulate cellular excitability in a not too specific manner. It is true that the electrodes in DBS may be located very specifically, but once the current is injected, who knows how many brain areas are affected. Further, in the particular case of DBS or other invasive procedures, it is not only the stimulation *per se* but also the damage inflicted by the surgical insertion of the intracranial electrodes, because it is known that diaschisis may occur (this refers to cell damage in areas far from the injured region possibly because of synaptic contacts, neurons tend to die if they lose synaptic inputs). To envisage how nonspecific these methods are, let us take a brief look at one of these techniques. The CVS, which has been used since the very early 20th century and consists in irrigating the auditory canal with cold water, transiently cures phantom limbs, spatial neglect syndrome, anosognosia (denial of deficits), motor neglect, somatoparaphrenia (bizarre beliefs regarding hemiplegic limbs), bipolar disorders, chronic pain disorders, and also modifies binocular rivalry and sound perception. An impressive list resulting from the apparently innocuous procedure of wetting the inner ear! (Been *et al.*, 2007). The basis for all these remarkable effects is that it induces widespread brain activation because of the broad networks that process vestibular information. Similarly, tDCS, that causes broad neural activations too, has been used to ameliorate addiction, depression, and other syndromes.

Another manner to considerably perturb the system is surgery. Thus, treatment of pedophiles include surgeries as varied as hypothalamotomy and orchidectomy (the latter procedure indicating that the problem arising from the brain can some-times be treated by very peripheral perturbations, in these cases delivered to the genitals, for after all it is not only the coordinated activity in the nervous system that is of importance, but also the coordination among all the physiological sys-tems), as well as various medications targeting hormonal pathways, and behavioral therapy including masturbation therapy (reviewed by Hughes (2007)). Apparently, the combination of psychotherapy with medication is successful in the sense that pedophiles show reduction in sexual urges and fantasies. Once again, this com-bination therapy is suggestive of the need for strong perturbations to the brain to ameliorate symptoms. Behavioral interventions, and more specifically the so-called cognitive behavioral therapy, are of interest since these represent a modification of the brain by itself. Cognitive behavioral therapy serves to determine the cogni-tive chain of events lying at the core of a particular pathology, like re-evaluating distorted negative thinking in depression or exaggerated perception of threat in anx-iety for example. This understanding is then used to develop effective cognitive and behavioral strategies that allow for aborting these specific pathological patterns on a day-to-day basis. In simple terms, it teaches the brain its own template-changing strategies. Remarkably, this form of psychotherapy is as effective as (and at times even more effective than) medications for treating illnesses like depression or anx-iety. Mindfulness-based cognitive therapy (Segal *et al.*, 2002) uses contemplative methods such as meditation, which has similarities with hypnosis in terms of brain activations and other physiological features, in order to interfere with pervasive pathological patterns dominating the dynamical landscape of a sufferer. A severely depressed individual spends days, nights, months, and sometimes years ruminating incessantly on morbid thoughts, trapped in a state of emotional torment; suicide being viewed as the only way to end this intolerable pain. Perhaps, switching off this pathological attractor-like state even for the period of a meditation session is enough to alleviate the onerous task of the rest of the "healthy" neural transients to reclaim their space in the dynamical landscape of everyday's living. But in the final analysis, the possible success of these methods rely on the fact that information has been fed into a brain, that then may decide to act on itself, to self-modify its own dynamics.

6.4. Final Notes

The mental and behavioral phenomena commented in this chapter originate certain considerations that affect societal views on brain and behavior, that are treated in-depth in Chapter 7. Presently, only a few words that concern directly the perspectives on ourselves and what we think of others are offered. It was addressed in Section 6.3, the shared neurodynamics between the normal and the psychopathic brain, and underlined that the popular views on the brain misinforms about the mental/behavioral reality of social offenders, in that these characters have very little choice in doing what they do, inasmuch as the autistics cannot help some peculiarities in their behaviors, or the epileptic patients suffer seizures, or the schizophrenic to hear voices. It is not that a general pardon of these antisocial individuals is advocated, for after all some psychopaths are dangerous and thus should be separated from society, but there is no need for any extra judgment. As a rule, these extra judgments comprise hatred, disgust, and hostility. What is advocated here is equanimity, there is no need for those opinions, no need to feel repulsion or hate toward these characters. As the marquis de Sade put it, "One must feel sorry for those who have strange tastes, but never insult them. Their wrong is Nature's too; they are no more responsible for having come into the world with tendencies unlike ours than are we for being born bangy-legged or well-proportioned." Our genetic constitution inherited since time immemorial pushes groups to eliminate competition, individuals to exhibit aggression, and some live with the somehow unrealistic ideal that our intellectual configuration can overcome this hereditary legacy. The majority of people seem to believe that, surely, the frontal cortex, equipped with norms and rules for a civilized living, can exercise control over other, perhaps more limbic areas and thus repress the aggressive, reptilian tendencies. But, we have seen already enough neurodynamics and neurophysiology at work so as to view this notion with scepticism. Chapter 7 focuses on the self-control theme, with the hope of shifting a bit those popular beliefs.

Is there room for shame, then? To what extent do we need to feel responsible for the racial violence, for example, that is still present around the globe? After all, it is the manifestation of a crucial part of the evolutionary landscape through which we emerged and one of the basis of evolution. To some extent, these are pre-existing and stable brain/behavioral dynamics. And how about personal responsibility, a question raised and answered by Michael Gazzaniga in his book *The Ethical Brain* (Dana Press, 2005): "It is one thing to worry about diminished responsibility due to insanity or brain disease, but now the normal person appears to be on the deterministic hook as well. Should we abandon the concept of personal responsibility? [...] Brains are automatic, rule-governed, determined devices, while people are personally responsible agents, free to make their own decision. just as traffic is what happens when physically determined cars interact, responsibility is what happens when people interact. Personal responsibility is a public concept. It exists in a group, not in an individual. If you were the only person on earth, there would be no concept of personal responsibility [...] Brains are determined; people follow rules when they live together, and out of that interaction arises the concept of freedom of action." On the other hand, perhaps full appreciation of how predetermined our individual and global dynamics are could finally enable us to come up with the right perturbation in order to deal with these unfortunate products of our human genes and neural circuitries. A slight sensation of embarrassment comes from the fact that, in spite of our large brains and long sociocultural development, we have not taken all possible opportunities to develop the chance for self-change. The patterns tend to always remain, while the system's constituents may change, but perhaps patterns need not remain at all levels.

CHAPTER 7

SOCIETAL VIEWS ON BRAIN AND BEHAVIOR

In this final chapter, we come back to the popular attitude toward the brain, that special attitude on brain and behavior introduced in the final paragraphs of the first chapter and that reappeared in other sections. This volume has reviewed works that have established some relations between the configuration of the nervous system activity and its manifestations in behaviors, some clearly pathological like seizures or Parkinsonian tremors, others more subtle and hard to classify. Brains determine behaviors, which is what people observe, and thus hold a special position amongst physiological organs. Yet the brain seems to operate, to some extent, in a similar manner as other organs such as liver or kidneys: they all take information from their surroundings, process it, and deliver outputs. In case of kidneys, the information is the materials in the blood and the output is a yellowish liquid. The liver takes up molecules and with its wide biochemical processing machinery metabolizes a multitude of compounds. Brains acquire information through a wider variety of means, from circulating metabolites (hormones, drugs, and so on) to those carried by electrical signals from the sensory terminals. It is the way brains process this latter information that generates the popular special attitude toward this organ, for it entails a relatively fast transmission of information that normally is broadcast to many brain areas to enter awareness, which, combined with the plasticity of its networks that are able to store information and mix it with the incoming stimuli, result in the apprehension of the external world. Owing to this processing of information, basically in terms of perceptions, brains make up realities, create models of the world, classify almost immediately all that goes into them, generate beliefs and conventions, perceive things that are not even out there, and, principally amongst these perceptions, is a sense of unity and continuity in behavior and cognition that is given such names as the "self," agency, or "I" (Taylor, 2007); then brains become certain of their own substantiality and create separations between self and environment, borders between sanity and insanity; they are masters of illusions. And the illusion of control is but one of the many ("The mind's best trick," in the words of D.M. Wegner (2003)), possibly because the nervous system developed not only to perceive with more precision, but also to exert a better control over the surroundings. Control either over the environment or self-control provides obvious evolutionary advantages. The means by which these types of control are implemented are an active area of research, and a characteristic that is becoming apparent from these investigations indicates that some processes are conscious and others unconscious (just think of subliminal perceptions and how these bias responses). But there is one type of control, the self-control, the ability to self-regulate the individual's behavior, that somehow seems to be thought of as a (self-)evident feature of all brains, and from this conception stems that peculiar attitude toward this organ and, in addition, no few misunderstandings and fallacies. These considerations can be encapsulated in the following question, that was posed by J.D. Schall and reproduced in the quotes after our preface: what choices does one really have, and how constrained are those? The neurobiological aspects of choice-making is an active area of research in the cognitive and psychological sciences, but as of today, AD 2011, while a general picture seems to be emerging, crucial features remain to be found for each specific, individual case. May these be found soon because the happiness of many individuals may depend on this knowledge. Nevertheless, the global understanding at a high level of description is already well advanced.

7.1.   The Basics are Known; Details are Missing

Contrary to a popular view, it is our opinion that, at least to a certain level of description, the functioning of the brain and its relation to behavior is relatively clear and well understood. Of course a myriad of details of each particular aspect remain to be discovered, but at the collective level of description, as it has been described throughout this monograph, the basic principles of nervous system function are known and, perhaps, it is time that this knowledge should be widely distributed to the populace (as children we all had to take physical education courses, but, where are the mental education ones? The "mens sana in corpore sano" concept seems to be implemented only in the latter part of the sentence, while the former appears to be left, in most societies, to the individual enterprise).

The basic neural functioning can be condensed into two tendencies: cellular excitability and the tendency to synchronous activity (at the risk of being too repetitious, it is highlighted once again the fact that *tendencies*, rather than definite phenomena, are all we can hope to observe and infer in nature; the fact that significance values are sought and assigned to every possible result obtained in any study should not cloud our vision and create the illusion of the finding of unambiguous phenomena, take a look at J. Lehrer's *"The Truth Wears Off — Is there something wrong with the scientific method?" The New Yorker*, December 13, 2010). It is the degree of these tendencies and where in the nervous system they occur, that leads to spatiotemporal activity patterns that are labeled pathological: if it takes place in the basal ganglia-thalamocortical circuits, then Parkinsonian symptoms may develop, if higher than normal synchrony occurs in limbic areas, then seizures may emerge, or auditory hallucinations in schizophrenia if the auditory cortices are involved, etc. So, one network of synchronously firing cells activates connected cell ensembles, and one of neuroscientists' main occupations is to measure aspects of this chain of activity. Then, they proceed to analyze the measurements with different methods, some more sophisticated than others and terminologies are developed that contain an assortment of words such as coordination, synchronization, amplitudes of recorded brain signals, frequency components of these signals, and metabolic activations or deactivations. These are our descriptions and our conceptualizations of the basic organization of neural activity. Each has importance and reveals some aspects, but becoming wedged in technology or conceptual frameworks should be avoided: these are just the things needed for our most convenient description of phenomena, and truth values should not be derived from them (as many times remarked in Poincaré (1902)). This volume has centered on the analysis and interpretation of the coordinated activity, on the perspective on the collective activity, the dynamical relations amongst the units of the system, for which an assorted multiplicity of experimental techniques and analysis methods have been presented and the interpretations discussed. Conclusions have been derived from the evaluations of measurements that, in one way or another, represent brain activity, of metabolic nature in the case of neuroimaging or electrical potential differences in electrophysiological signals. Of course conclusions can never be drawn with absolute certainty, the only thing one can do is to assign some confidence level. In the final analysis, then, it could be said that what is being sought is the dynamic relations between cell networks. This is something that genomics or proteomics will not be able to answer, with due respect to the importance of those molecular findings. The realization of these phenomena relies more on the study of the dynamic interplay between cell ensembles, between body and surroundings, on recognizing the inseparability of brain, body, and environment that "requires a more poetic, holistic turn of mind and more tolerance for ambiguity and uncertainty that reductionists care for" (Freeman, 1995). Perhaps, and continuing with the "omic" tradition of recent times, a new field called synchronomics or coordinomics can be founded, if only to attract attention in the current "omic" era, where genomics is the monarch.

If, to some extent, the distinction between brain "states" (recall comments on the problematic definition of neural states in Chapters 1 and 2), or the neurophysiological basis of behavior, can be captured by the dynamical relations between networks, one may wonder whether all those measurements of activities and the results from the many analyses aforementioned need to be precisely quantified or they can be left as more qualitative approaches so that greater generalizations can be derived. It has been shown in past chapters the usefulness of constructing dynamical state spaces using a variety of variables, from inter-spike intervals to voltage fluctuations. These can be too specific to the taste of some, but at least these studies go beyond the metaphors and plunge deeper into the technicalities of dynamic system theory applied to brain research. Yet, one may feel that such specificity precludes generalization, and no doubt that some biologists and clinicians would like to see more general approaches that can be applied to various conditions and circumstances. The methods (linear stability analysis, phase space reconstruction, and so on), certainly, are of general applicability, but the collective variables and control parameters will depend on the circumstances, on what can be measured. Using more abstraction may be helpful but dangerous too. For instance, could brain states and the dynamic relations between networks be assessed using a very general dynamical state space by graphing each network as a particular point in this space? This would be a sort of a topological study, the points in space representing some brain or network characteristic. Topology, unlike geometry, is not concerned with precise quantification of lengths and areas or other geometrical features, but with the properties of sets, the topological relations, the analysis of the position of the points (that is why its original name was *analysis situs*), the properties that remain unchanged when figures are distorted. Topology can be described as the geometry of the place, of the position, as distinguished from the metric geometries with which most of us are familiar. In a possible topological "brain dynamical" space, the topological relations amongst the points, which could be cells or networks or whole brains, individuals, organs in a body, etc., could be studied, but a reference frame is needed, possibly constructed with some very general, global observables. Incidentally, this may sound like science fiction but is not, for it is not coincidental that graph theory, a topological method, is currently applied to neuroscience and topological metrics such as clustering coefficients and node connectivity are becoming regular in brain research (recently reviewed in Moreno Vega (2009)). Graphs used in these studies can certainly be abstract enough, and one suggestion for yet pushing further the borders of abstraction is to use the information content of, say, a whole brain, since measures for quantifying information exist and the idea of information distance has been proposed (Bennett *et al.*, 1998). Hence, could information distances between individuals be a metric to assess a sort of topological distance in mental/behavioral space? Once again, to be precise, there are specific properties that a metric of a topological space has to satisfy to become one (Burns and Gidea, 2005), so these matters should not be talked about too hastily. Nonetheless, this hypothetical treatment of brains as points in a topological space based on the information content could facilitate the inclusion of an elusive notion in neuroscience, yet fundamental: the observer is a part in the concept of reality, something the physicists know too well but has been mostly ignored in neuroscience up to recent times (Morowitz, 1980); which is peculiar, because there is nothing more personal and "subjective" than the investigation of the brain by itself. Essentially, the position of points in the space has no intrinsic value, because it all depends on the reference frame, the coordinates of the space. What remains significant, or "real" if an attribution of reality is to be made at the risk of falling into some entangled loops, is the relative position of the points, parameterized by the metrics, the information distance if this measure were chosen. Imagine what a near 0 informational distance between two brains could signify..., well, perhaps there is some exploratory terrain here for matching and dating services. But now it is time to put an end to this speculative digression.

To encapsulate in a simple scheme the essence of phenomena evaluated in this work, Figure 7.1 represents an attempt. To put it in just one sentence, as already expressed at the end of Chapter 3, it could be said that the dynamical repertoire of cells and networks is constrained by the range of microscopic interactions within and outside the system that determine the final dynamical evolution of the nervous system, giving rise to prominent organized collective patterns of activity and, upon reducing the degrees of freedom due to the collective interactions (Hermann Haken's enslaving principle), self-organizing phenomena emerge. Arrows in the diagram are bi-directional because the self-organizing collective patterns themselves will determine (and constrain) further the nature of microscopic interactions (Haken, 2006).

And yet, all this mapping of the various measurements of activity of nervous tissue, whether functional neuroimaging or electrophysiological signals, to behaviors,

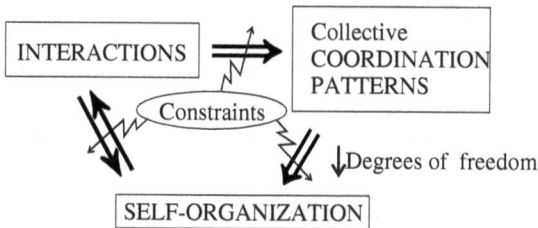

Figure 7.1. Simple scheme that conveys the idea that the dynamical evolution of systems composed by many constituents, and nervous systems in particular, is dictated by the mutual interactions of the components at many levels (micro and macro) and the constraints imposed by the nature of the units themselves and the environment.

is not completely satisfactory, for all that can be done is to associate one pattern of activity with one or another behavioral disposition. Therefore, if, for instance, the amygdala is activated upon some stimuli presentation, conclusions are drawn related to possible emotional aspects of the stimulus, knowing all that is known about the amygdala. Yet, there is no access to the precise and actual items that are being processed by that activity, the symbols that constitute the thoughts if you will, that make up the activity patterns themselves. This is, in our opinion, a major challenge for future brain research, which sounds fiction today (not even science fiction!), for there is no current technique or conceptual framework that can be remotely close to addressing that subject matter. There is plenty of room for creativity here. Aspects such as first-person reports in cognitive science are brought to attention these days, which represent a first, and perhaps not too accurate, step to precisely knowing what the contents of the mind are in relation to some biophysical measurement, steps that, in some future time (provided there is still someone left on this planet to muse about these things), will be looked upon as we do today to those first and primitive attempts at investigating phenomena that today are well characterized and taken for granted. We will probably not witness this future time, but, undoubtedly, it will arrive, with the help of some creative thinking (which is always present) and the financial support of institutions (which is not always present).

7.2. Recapitulation on Aspects of the Brain-Behavior Continuum and Possible Basic General Neural Mechanisms Resulting in Deviations from the "Normal"

To recapitulate briefly the main point, it was proposed in previous chapters that the tendency of cell networks to synchronize their activity results in disease and behavioral deviations, with the corresponding *caveat* in the designation of normal or deviant, as the accepted attributes for "decent" behaviors are relatively arbitrary. While we have been submitting this idea throughout this volume, it is fair to note that it is not novel as it has been around for quite some time in one form or another, and in no way can we attribute ownership to it; however, we consider that it has not permeated the neuroscientific community in general and the clinical in particular.

Aspects of a continuum at several levels have been noted, starting from the book's title to the last chapter on psychiatric deviations. What is the brain-behavior continuum? Is it that transition, sometimes gradual, sometimes abrupt, between wakefulness and sleep, or between sleep stages, or from sanity to insanity? Is it the illusory perception of continuity in the events we perceive, the stream of consciousness? The continuum extends through various levels. It is the incessant changes of molecular and cellular events; the transient collective activities of nervous system networks; the brain-environment loop; the inseparability of the mind, behavior, and surroundings. In behavior, the continuum extends over adaptive and less adaptive actions, ending at one extreme in completely detrimental activities. Where the demarcation of insanity is placed is, as said, a bit arbitrary. The adaptive values of behaviors may not be easy to ascribe either, as this depends on the context, those constraints in the diagram above, the boundary conditions in which individuals live; these should be assessed for each particular case.

Then, there is a continuum at neurophysiological levels, perhaps best represented by the tendency to exhibit synchronous activity which, depending on the networks involved and on the magnitude and the duration of synchrony and on the context too, makes it plausible to create a mapping between this biophysical measure and the adaptability of behaviors. For instance, as noted in previous sections, synchrony may develop transiently in basal ganglia-thalamocortical networks resulting in the normal physiological tremor, or in limbic circuitries during normal information processing; however, if it occurs in widespread regions for long duration and with large magnitude and low variability, then Parkinsonian symptoms in the former case or seizures in the latter may develop. Context is crucial, too: the relatively stable thalamocortical synchrony is beneficial at night as the context requires little sensory processing, that is, sleep, but if the same tendency manifests itself during waking hours, then absence seizures will result. Similarly, the context in which central pattern generators work requires them to produce rhythmic stable outputs, but the neocortex can not operate in this fashion for efficient information processing. In many instances, the so-considered deviations are due to specific, otherwise normal, neural dynamics that appears in the wrong time or in the wrong place.

Thus, a few parameters are of importance: time (in terms of duration), location (networks involved), whole context (internal and external), magnitude, and fluctuations (but this factor can perhaps be accounted for by the duration) of the synchronous activities and neural excitability. The representation would then require a multi-dimensional space, but if only three dimensions were to be chosen (just because a three-dimensional space is convenient to visualize), duration, spread, and magnitude of the excitability/synchrony would already provide an indication as to the health value of the nervous activity: a long duration of high-amplitude and widespread excitability/synchrony, depending on context, would suggest signs of trouble. These considerations are in line with those mentioned in previous chapters on how to quantify, if possible at all, metastability (Section 1.4.2), and the similarity between this problem and that of assigning a value to the fluctuations in coordinated brain activity as relate to health and disease, as well as the comments in Chapter 2 regarding how to assess the stability of equilibria. All these reflections revolve around the same question, and try to address the same phenomenon, which, in essence, refers to the most efficient exchange of information, either among cells, networks, brains, communities, and so on (Perez Velazquez, 2009).

A current field of brain research that exemplifies one of the many aspects of the brain–behavior continuum relates to the fluctuations in brain signals, which represents, we think, the refreshed attitude toward the consideration of the importance of brain–environment interaction, with many investigators turning their attention to it in attempts of finding unifying frameworks of mind–brain–behavior (Fingelkurts *et al.*, 2010). One aspect of this research is the contemporary great interest in the "resting" brain activity, closely associated with the default brain network concept, a field of research that is reaching unprecedented proportions that cannot be covered here due to space constraints (recently extensively reviewed in Buckner *et al.* (2008)). Whether some brain regions constitute a "default network" or not is still debated (the demarcation of what constitutes a network is many times unclear, and will not be discussed here), but the underlying essence of all these investigations is that the incessant, ongoing background activity reflects an internal state that is fundamental for the processing of whatever sensory stimuli arrive to the brain. There is no standard baseline for brain activity; it has to be specified for each experimental condition and many times it is not evident at all that there should be one. Think of the typical cognitive experiment that involves a continuous presentation of a series of many repetitions of stimuli, is not the brain activity overlapping from one stimulus to the next? Is there any unambiguous baseline in between presentations and if so, is not it different from the "baseline" right at the start before the first stimulus? What are the initial conditions to be chosen? These current tendencies in research contrast the old traditional views on background, ongoing EEG activity, traditionally considered to be noise. In fact, the brain has been considered a "noisy" processor from early times (Adey, 1972), which is probably true but nevertheless subject to debate because the distinction of signal from noise in the brain is, we venture, unfeasible in current times. Still, this "noisy" perspective seems to dominate to some extent. For instance, this view underlies the widespread methodology of averaging used in event-related potential (ERP) research: these events represent an average of large number of trials so that the investigators hope they remove any activity unrelated to the event. This deserves a careful consideration, taking into account the research that has been done in the past years on resting neural activity, starting with the early efforts at the characterization of the spontaneous brain activities and how this determines the responses to sensory stimuli (Basar (1980), Arieli *et al.* (1996), and Kenet *et al.* (2003) are three exponents of these initial attempts), studies that motivated a close scrutiny of the variability of neural activity patterns and the interaction between this variability and stimuli (see Northoff *et al.* (2010), for a recent account;

Sadaghiani *et al.* (2010)), such that it is becoming apparent today that what used to be called noise, in the context of neuronal activity and brain function, may have an essential contribution to information processing. The last paragraphs of Chapter 1 remarked the comments by prominent scholars on how sensory-motor processing could be considered a modulation of the background, ongoing brain activity. Especially, the words of Gilles Laurent are worth repeating now, toward the end of the volume: "Our thinking generally ignores the fact that [...] a given neuron is never an end-point or its response an end-product [...] Thinking about sensory integration in these active terms (considering 'responses' not only as products but also as ongoing transformations towards some other goal) may be helpful [...] to understand some brain operations" (Laurent, 2002). We consider that this notion, derived to some extent from the synfire chain concept and the consequences of the serial triggering of activity in connected cell assemblies, is of fundamental importance to the main message of this book, but feel that, while mostly accepted by neuroscientists, its implications have not permeated the scientific community, for what it entails is that there is no start or end to the chain and thus the inseparability of brain and behavior starts to become apparent. The closing paragraphs of Chapter 1 already talked about this aspect, so there will be no more insistence here, just to emphasize that this never-ending chain of collective activity illustrates the continuum between brain and behavior. It is hoped that these investigations will shed light on the interaction between the autonomous and the stimulus-evoked activity of the brain and the extent to which our perceptions and decisions are influenced by the ongoing neural activity. As S. Dehaene appropriately advises: "A major experimental challenge for future years is that autonomous mental activity, too often neglected, must regain its status as a central object of study" (Dehaene, 2007). Indeed, it is estimated, from neuroimaging data mainly, that the responses to environmental stimuli accounts for only 5% of the brain's energy, hence the majority of the energy appears to be allocated to its internal workings. This excursion on the brain autonomous activity is closed with one intriguing possibility, that there may exist predefined dynamical patterns in brain activity that already contain and involve the basic neural networks to perform specific operations/actions. Thus, according to this idea, an external (or internal) input simply reactivates what is already a functional, hard-wired circuitry in the brain. Recall, along these lines, the similarities between the neocortex and central pattern generators that were discussed in Chapter 1.

Can we be more specific in characterizing the continuum? Since the changes in brain activity operate via variations in control parameters, is there a control parameter that can be best qualified as to induce the transitions from the healthy neural function to the dysfunctional? And, which order parameter can best capture this transition? The very general control parameter could be excitability, which depends on so many factors that any of those may be crucial depending on the specific circumstances. The, very general too, order parameter could be a collective variable that captures the coordinated activities of nervous system cells or nets, like the relative phase alluded to in previous chapters. As a rule, variables that describe the interactions between system's constituents should be useful in this regard: whatever captures the relationships, the information exchange; and not only for brain studies, but also in general to investigate behavior and higher-order phenomena, variables that capture interactions are specially indicated. Words that advised attention to the study of relations amongst system's constituents have been pronounced by prominent scholars in different fields: those of Henri Poincaré already reproduced in the first paragraph of Chapter 1, and the psychologist Jean Piaget declaring that: "Concrete reality is the ensemble of the mutual relationships of the environment and the organism, that is to say, the system of interactions which unify them" (Piaget, 1936).

Throughout this volume, subtle and abrupt transitions in brain function and in behavior have been presented, and comments as to the more or less subtlety of these transitions have appeared for specific phenomena, most of the times within the framework of dynamic system theory. Perhaps the most basic of those phenomena, which underlie the aforementioned transitions in excitability and synchrony, is provided by the empirical observations that it is enough that a discrete cell population becomes hyperactive to lead, conditions allowing, to massive changes in collective activity in terms of higher-than-normal synchrony. The dynamics of complex systems does not need strong causes to produce considerable effects, as emphasized in Chapter 2. This nonlinear type of thinking is not too common, yet, amongst biologists. A tiny variation in some factor may be just enough to induce substantial alterations. Nevertheless, the question of subtle or abrupt changes is sort of relative too.

It is hoped that the many examples of this most fundamental characteristic of the nervous system toward synchrony that have been described in this volume associated with pathological conditions did not obscure the finer and more delicate aspects of synchrony with regards to information processing. Next section provides an empirical example and final considerations on the tendencies aforementioned.

7.2.1.   Reward and punishment: A specific example of a neurophysiological tendency

The considerations presented above about the neural tendency to exhibit excitability and synchronization can perhaps be made graphic. Figure 7.2 may serve to exemplify the nature of the phenomenon that has been discussed. The intracerebral

REWARD

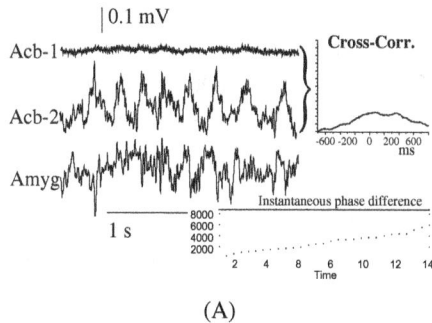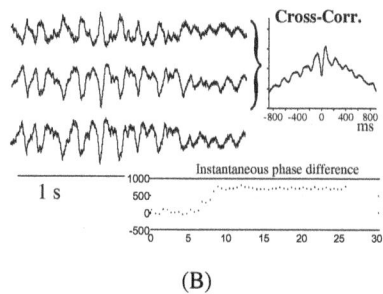

(A)   (B)

STRESS

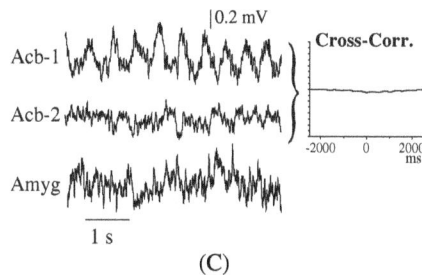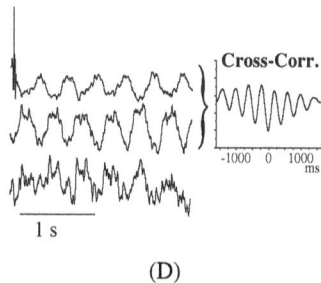

(C)   (D)

Figure 7.2. Intracerebral recordings in rats during rewarding and stressful conditions. Three simultaneous field potential recordings are shown in all panels, the first two traces from electrodes located in the nucleus accumbens (Acb-1 and Acb-2) and separated ∼1.4 mm, and the third trace from an electrode in the lateral nucleus of the amygdala (Amyg). A and B are recordings from one rat, before (A), and during the experience of food reward (B). Visual inspection of the traces, and the cross-correlation graphs on the right-hand side (Cross-Corr., between the two accumbens recordings), demonstrate enhanced synchrony when the rat is reaching and chewing the morsel (actions that took place in the time course of panel B). Some oscillations were present in the accumbens recordings occasionally while the animal was freely moving or standing (for example Acb-2 in A and Acb-1 in C), but normally did not occur for a prolonged time neither were these significantly synchronized. The instantaneous phase difference plots (shown for 14 seconds in panel A and for 26 seconds in B) illustrate the longer-lasting phase synchrony (seen as horizontal plateaus, mostly in B) during the rewarding situation. Panels C and D show recordings in another rat during a "control" condition (rat freely moving, in C) and during a stressful situation (D), as explained in the text. Note the enhanced cross-correlation between the accumbens signals during the stressful circumstances, that would have also been evident if the amygdala recording had been used to compute the correlation. Thus, both rewarding and stressful experiences seem to be associated with long-lasting widespread synchrony amongst several brain areas.

recordings in the figure depict the widespread synchronization between brain areas while a rodent is experiencing reward or stress. The simultaneous local field potential recordings taken from the nucleus accumbens and the amygdala synchronize better (the frequency of the oscillation changes too) during the time when the rat is reaching for and chewing an appetizing pellet (the right-hand side panel, (b)), actions which, after fasting for one day, we have to assume are pleasurable to the animal (even though this no one can know for sure as rodents cannot tell us their subjective experiences). Nevertheless, what can be said without hesitation is that the rat's attention was centered on reaching and eating the morsel, coincident with the periods of more sustained signal synchrony. The instantaneous phase difference between two of these signals shows the tighter phase locking (horizontal segments) for almost 25 seconds, in (B). There were only 3 electrodes implanted into this animal's brain, but, we dare say, if there had been 15 electrodes in different areas, perhaps synchronization in many of these would have been observed during rewarding circumstances. It is not unusual to find oscillatory signals in recordings from the accumbens, perhaps a sign of a considerable propensity to synchronous activity in these deep brain structures, and readers can inspect very similar oscillations to those shown in the figure that have been reported in other studies (such as Goto and O'Donnell, 2001).

But not only rewards capture attention, stressful and damaging situations do a good job at focusing attention too. And a similar widespread synchronization is detected during stress, as shown in another rat's recordings in panels (C) and (D) of Figure 7.2. The stress imposed on this unfortunate animal was the so-called inescapable stress. The reason why the accumbens and amygdala signals may become synchronized in these two situations is that these areas are thought to be involved in processing reward and stress.

As many times mentioned in this book, these field recordings represent most likely synaptic input; therefore, the increased correlation between the accumbens recordings may be due to an enhanced amygdala or orbitofrontal input, as these areas project monosynaptically to the shell of the accumbens. These data, at least, may imply that accumbens cells could be firing in synchrony with amygdala neurons during those specified conditions, even though to be certain of this, a recording using tetrodes (as in Figure 5.3) in these areas would be most adequate. What control parameters may be influencing the transition toward a long-lasting widespread synchrony in these cases? Several, starting with dopamine, fundamental in the basal ganglia circuitry as a messenger of reward (Cools, 2008), and ending with gap junctional communication and excitatory transmission, important too in the accumbens physiology, for it has been shown that mild inhibitors of excitation or gap junctional blockers injected into the nucleus accumbens reduce reward-seeking behavior in rats (Kokarovtseva *et al.*, 2009), however no intracerebral recordings were done in this study so it is not known to what extent the putative higher synchronization was disrupted. As another example of the presence of rhythmic activity in the brains of rats exposed to rewarding conditions, cortical oscillations in primary gustatory

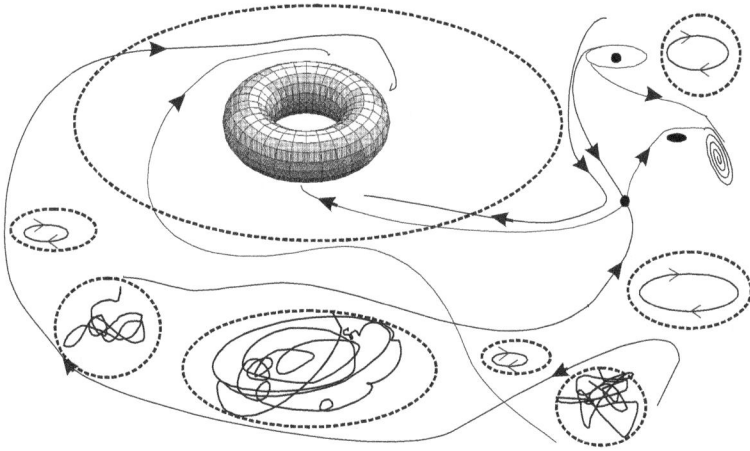

Figure 7.3. A sort of nonhyperbolic attractor, composed by many attracting sets, where the nervous system dynamics may reside. The boundaries of attraction surrounding each attracting set are represented by dotted lines.

cortex, recorded as field potentials, were detected in the range 7–12 Hz while animals sampled a liquid reward (in this experiment rats had been deprived of water before the experiment, so it can be surmised the animals experienced reward (Tort *et al.*, 2010).

The data presented in Figure 7.2 are a particular aspect, a specific solution in a mathematical metaphor, of the general scheme that is presented in Figure 7.3, which is the full "model." This is reminiscent of the concepts detailed in Section 2.2.3 and Figure 2.7. The state space contains many attracting sets, from chaotic to limit cycles and saddle points. Recalling the terminology of Chapter 2, it is then a representation of a nonhyperbolic attractor, containing a variety of attracting sets, each with a basin of attraction, represented by dotted lines in the figure. Note the abundance of limit cycles, signifying the primordial tendency of nervous systems toward producing rhythms, and the most notable torus, the mathematical representation of synchronization, with the largest basin of attraction. The trajectories are the "solutions" of the model, the patterns of brain activity, traveling between neighboring attracting sets, almost never settling on one for long periods, noise and perturbations making them transit from attractor to attractor. The recordings in Figure 7.2 can be thought of a particular trajectory that has fallen in, perhaps, one small torus (because the recordings are synchronized, hence can be represented by tori), or possibly the large one, the great torus, for we all know how important, pervasive and stable pleasure and reward-seeking actions are. Rewards and punishments guide human and animal behavior to a great extent, as expounded in Chapter 10 of E. T. Rolls' volume (Rolls, 1999), hence this theoretical toroidal manifold with the largest basin of attraction can be thought of as the one representing coordinated activity in reward-processing circuits. The many attracting sets portrayed in the figure may represent different anatomical circuitries, but not necessarily, it all depends on the specific interpretation, or how far we want to extend this nonhyperbolic attractor metaphor: distinct networks can display same dynamics, and it is precisely the natures of these anatomical circuitries that are experiencing one or another type of dynamics that manifest themselves into actions. The parameter space that controls the dynamics is large, and what specific bifurcation parameter is chosen depends on the phenomenon assessed. As was described in previous chapters, gap junctional coupling or potassium concentration could be control parameters for epileptiform dynamics, or, dopaminergic transmission for PD and reward. Keep in mind that different parameters can lead to identical dynamics, and be thoughtful about the intricacies in choosing general modulators of activity or more specific bifurcation parameters.

Considering the rodent recordings in Figure 7.2, what could be occurring in the brain when one is absorbed in pleasurable experiences, such as drinking wine (for those who like it)? The attention is captured, the individual becomes absorbed in it. Attention normally requires awareness, thus widespread neural activity, extensive broadcast of the information, the global workspace theory in action (Baars, 1988); at this point, the discussions in Chapter 5 about the spread of neural activity in disorders of consciousness are useful to recall. Perhaps, to broadcast widely the activity, cell networks need to coordinate their actions, and therefore attention in general will be associated with synchronous (and asynchronous too, remember the integration-segregation idea) activities in specified areas. For these reasons, a coordinated activity like that of the rat in the figure may extend throughout our wine drinker's brain, the neural activity of limbic and other networks has fallen into the basin of attraction of the "great torus" of Figure 7.3, pleasure itself, and it is as stable as it can be... as long as the external, boundary conditions, remain the same (until the last wine drop remains in the bottle). But, as many times declared here, cortical dynamics is such that avoids becoming trapped for a long time onto an attractor-like state, so perturbations may result in directing the trajectory away from it. How easily? That depends on the many parameters that participate in the rewarding experience, which will determine how absorbed the individual really is. Minor perturbations may not succeed at disrupting that coordinated activity during the ecstatic experience, but if someone applied a red-hot metal to the fellow's skin, good chances are that the synchronous activity in those areas will be replaced by, yet another activity perhaps equally synchronized, but involving other brain regions (the so-called pain matrix?). The stability of this, arbitrarily labeled, mental state of reward, differs from person to person: normal wine-seeking behavior may not be harmful, but it is well known what may develop in some individuals, in whom there is a need for extremely powerful perturbations to alter their state and modify the behavior, these characters exhibiting such a powerful stability of this particular state that makes them leaving the central range of the Gaussian distribution to become labeled as "ill," alcoholics more specifically.

In a final push to this metaphorical excursion, if we are allowed, it could be of interest to note that there is a mental/brain "state," a perception in the final analysis, that seems to be extremely stable in all individuals, and thus could be regarded as the strongest attractor: that of the sense of self, the perception, more like an illusion, of each individual's continuity/unity in cognition and behavior (we follow the Scottish philosopher and historian David Hume in his view that everything mental consists of perceptions), and the strong belief and attachment to its substantiality. Now, it is true there are "substantial" ion currents, biochemical reactions, and other measurable phenomena, but how substantial mental constructs are is another matter, even though it can be argued that these internal states are "measured" too, but by the individual and not by machines or artifacts. If, after all, everything is a perception, and following Immanuel Kant's concept of apperception (or the perception of a perception), then to label these internal "measurements" as subjective may be controversial because to the individual these are as objective as they can be (the border between objectivity and subjectivity is as nebulous as that between consciousness and unconsciousness). In this regard, the work by J. P. Crutchfield (1994) on the notion of intrinsic emergence, or how some emergent collective properties increase computational capabilities of the system itself and lead to more global information processing so that the system capitalises on those patterns, while a bit technical, may be of relevance for some readers, because of the deep reflections about external or internal "observers" of the emergent patterns.

Current investigations are scrutinizing in great detail the problem of the self and self-awareness, and the published records are already reaching an immense number as it is becoming a popular field of research. Whether this self-construct relies on a unique brain module or on a network of interconnected areas, is still a matter of debate, but, in our opinion, it is more probable that it is the manifestation of a distributed representation in several brain regions giving rise to a certain perception of a sense of unity in behavior and cognition, rather than it being restricted to a single region. It is everybody's experience the extreme stability of this construct, that remains there even in extreme psychiatric deviations, some of them talked about in previous chapters. It stays there even in patients who have lost the capacity to form new memories, like in those typical cases after hippocampal damage or its surgical removal, individuals who live in an eternal present (albeit they can recall events from the time before the injury). To our knowledge, there has been one clear account, the case of Clive Wearing, one of these individuals with hippocampal damage who could not lay down new memories after his injury, where he would lose the sense of the self momentarily, and then would re-appear, only to vanish again (his story documented in his wife's book *Forever Today*, D. Wearing, Random House Press, London, 2005). His self-continuity was distorted, but it can still be appreciated that the "self-attractor" remained there to transiently pull mental trajectories toward its basin of attraction. It is an automatic processing by the brain that the succession of experiences seems continuous, such that a stream of consciousness emerges and constructions such as agencies and selves arise naturally, but recall the comments in Section 4.2.2 on the factual discontinuity of perception. We note on passing that, while this self-construct has some evolutionary advantages, it is also a source of anguish, as someone remarked centuries ago. For those interested, there are remedies to destabilize this "attractor."

As a final note on the cautions that have often appeared throughout this work about dynamic system theory used for neuroscientific purposes, Figure 7.3 is our next-to-last metaphor (Figure 7.4 is the last). It is hoped that the intuitive appeal behind these mathematical concepts has convinced some readers to pursue these investigations in, perhaps, a more rigorous manner, to go beyond the metaphors. Previous sections on epilepsy, Parkinson's disease, and motor coordination have been presented in a more technical fashion trying to transcend the metaphorical use and to demonstrate that a relatively rigorous application of the dynamic system framework is not beyond the reach of brain and behavioral sciences.

7.2.2.   A brief tour through the brain, for the nonspecialist tourist

As a last recollection of the main aspects underscored in this volume, let us take a tour through a normal brain while it is processing information in conscious states. What is sensed in this trip will depend on what neurophysiological feature is recorded. Let us assume cellular activity is recorded. Traveling through the brain networks, activity will be noticed in almost all cells, and not much sense will be derived from it. The trip from net to net will not reveal almost anything of interest, so much activity going on all the time in all places, the traveler will be overwhelmed. Now imagine that a relation between activities is evaluated, like the phase relations so much discussed here, and that the degree of phase locking between cells or networks is color-coded. Now the traveler will see some cells and cell ensembles appearing in different colors, and some patterns may be discerned but only ephemerally, no longer than 200 or 300 milliseconds. This is the normal thalamocortical activity at this level of description during wakefulness.

Visiting other centers in the brain stem or deeper in the spinal cord, more uniform colors would be noticed: perhaps a central pattern generator controlling breathing rhythms, nothing out of the ordinary; all is fine. The same tour during slow wave sleep would detect a more uniform color throughout the cortex and thalamus, which is proper because the context demands it: the organism is at rest. If, instead of the soft matter of neural tissue, the tour takes place on a dynamical state space, then the traveler moves following trajectories that represent the dynamics, and the trip is fast moving amongst regions of that space that look like black holes, the attractor-like states, engulfing activity (trajectories) as soon as it gets closer, but the trajectories almost never get so close, and if they do only temporarily, for noise and fluctuations derived from internal and external sources will push the traveler out of that attractor, the black hole will be left behind and another will be approached: a fast-moving, non-asymptotic journey made up of transients. Again, all is fine.

The same tour through other brains that are a bit unusual for whatever reasons would display a different landscape. The excursion in the neural tissue would reveal more uniformity in the colors that represent the correlations in activity, lasting long times, perhaps seconds or minutes, not milliseconds, and this occurring during conscious states of sensorimotor processing. The traveler may notice this scenario in the striatum and thalamocortical circuits, and perhaps that is a sign of this brain belonging to an individual with a motor disorder such as Parkinson's disease; or it may be present in hippocampal and temporal lobe networks, which would suggest this belongs to an epileptic patient; or it may take place in the auditory sensory cortices, which may indicate a sufferer from auditory hallucinations, perhaps a schizophrenic; or the uniform colors may be seen in some very specific brain regions, perhaps a depressed individual ruminating endlessly about same thoughts. The dynamical trip of these brains would be also different from the previous. Now the traveler would be stuck into a black hole for longer durations, at times wondering whether there is a way out. Only after large perturbations, this attractor-like state will be abandoned, only to come back again after some time. Not everything is fine in these situations. The tour indicates these are no ordinary masses of nervous tissue. The manifestations of these patterns will be noticed in deviant behaviors, but the detection of these will require a different tour through the space of behavioral realizations; tours that are frequented by psychiatrists, neurologists and others. But few of these take the brain tours.

7.3. Epilogue. On Making Choices and the Constraints of This Process: In Search of the Last Ventriloquist

In the examples of the preceding Section 7.2.1, the rats' predicaments and the wine-loving character, is there any brain region in control of those behaviors or of those cellular collective activities? The last section of this work addresses what has been our major central theme: the exercise of self-control, and what, if anything, is in control. This subject matter has important moral implications and is full of legal flavor as it pertains to the individual's responsibility, however here the principal concern is the neuroscientific problem, nicely captured by Theodore Melnechuk poem "Punch and Judy, to their audience (see note 1 at the end)", which let us borrow his terminology and ask, is there any last ventriloquist in the brain? Figure 7.4 depicts a hypothetical control state space, and optimality is ascribed to the shaded region around the center. But, as enquired in the figure legend, who or what is in control in this and other regions of that space? What is controlling what? Is there anybody in charge? Perchance, we know enough of brain function to answer, at least preliminarily and without much specification, these questions.

The main idea behind the pages of this volume is that nervous systems work on the principle of mutual activations, information passed from network to network that depends on potential differences between the intra- and extracellular compartments. This spread of potential differences gives rise to patterns of organized activity, to a coordination of the activities of millions of cells, such that, as repeated often here, one cell's actions has meaning only in relation to other cells' actions. How does this basic neurophysiology enlighten the above posed questions on self-control? S. Frangou considers that "cognitive control describes the ability to optimize processing of contextually relevant information and to select responses in accord to intended goals" (Frangou, 2008). As she then proceeds in her narrative, this apparently "simple" process involves forming associations among perceived stimuli, actions, and outcomes, maintaining in and manipulating goal-relevant stored information (working memory), and monitoring of behavioral performance, to name just a few aspects the brain must achieve in order to gain "cognitive control." In other words, coordination of actions in diverse brain regions seems to be required, once again the concept of integration and segregation of information amongst brain networks, and, it can be surmised, this coordination results from the constantly interacting, chains of self-activating, recurrent circuits in the nervous system and within the brain especially. In Figure 7.4, one cartoon character seems depressed. One can then ask, is any brain region controlling the depressed brain? Naturally, because the individual feels depressed, something must be controlling that in the sense that it is a stable, yet undesirable, behavior! But it does not have to be a tiny, discrete, module here or there. There may not be any last ventriloquist, in the end. It is the combination of the changes in control parameters, the stored information in memory, and the incoming inputs from environmental influences that dictate the emergence of a coordinated pattern of activity, as much as the Rayleigh-Bénard convection (Section 2.3) arose from molecular interactions, and the last paragraph in Section 6.3.1.3 already presented arguments about its similarity with the self-control phenomenon. Emergent properties, self-organized or not, are found at all levels: the calcium concentration dynamics inside a cell is an emergent property of the tiny organism, the patterns of synchronized activity is one of the whole brain, cyclic variations in preys and predators correspond to animal ecosystems, and socio-economic organized activities belong to human societies. All of these emergent properties, from the nonliving to the living, arise from the single theme of (mostly nonlinear) interactions amongst a very large number of constituents. The peculiar aspect of nervous system patterns is that these constitute behaviors that are perceived, judged, and classified by other nervous systems (other people). Patterns of neural activity manifest themselves as behaviors and, it could also be said, behaviors are the neural patterns. Transcending this apparent circular causality requires an extra mental effort that readers are encouraged to attempt. In the final analysis, since the activation of a network depends on their (synaptic) inputs, on the activity of others, so, as Mark Hallett puts it: "At any one time, the activity of the motor cortex [...] will reflect virtually all the activity in the entire brain." This is the key point. In our own experiments trying to identify cortical areas that show differential activity in forced versus willed actions, the differences start to appear merely 60 milliseconds after the command presentation in the occipital (visual) cortex while frontal areas (which were the expected regions to show differential activity) become apparent much later, an indication that the whole brain seems to know what is going to be doing, and not only the executive directors, the frontal and parietal regions (Garcia Dominguez *et al.*, 2011). Then he asks: "Is it necessary that there be anything else? This can be a complete description of the process of movement selection, and even if there is something more — like free will — it would have to operate through such neuronal mechanisms." (Hallett, 2007). A succinct manner to put an end to the free will problem! We could add that this is a complete description not only of movement selection but also of the entire mental life.

There are views that consider that people have little control over their behavior. These views have been criticized in terms of the distinction between conscious or unconscious aspects of control, specifically the apparent need for all control to be conscious (Suhler and Churchland, 2009). Consequently, Suhler and Churchland focus on reviewing some observations about environmental influences on behavior derived from studies demonstrating that, as an example, people are more likely to litter in settings that are already littered as opposed to clean. This reflects a general view from social psychological studies, namely, that choices are affected by various kinds of manipulations. These authors explore the unconscious aspects of decision making from the already extensive neurobiological literature, concluding that some elements of control are unconsciously processed. This is not the time to delve into the very abundant literature of topics such as free will and self-control, and no references will appear here (except for the aforementioned) lest we drive readers become biased toward one or another perspective on these matters. We will limit ourselves to, first, noting that it is true that what enters awareness is a small portion of the computations the brain networks make; second, to observing that, in the end, the discussion on conscious versus unconscious control mechanisms may not be essential to the fact that brain networks are constantly interacting and the final "output" is a specific action/behavior, so to find out what parts of the brain are aware of what, is a very interesting question (subliminal perception, blindsight, split-brain cases, all these remarkable phenomena that underscore the many "ventriloquists" present in one brain and how each of these perceives and computes information) but it is another matter that does not shed too much light onto the bare neurodynamical essentials underlying the phenomenon of action generation; and third, to reproducing one of the figures in that paper, what the authors call a "3D control space," if only because it matches this volume's dynamical approach. Their diagram (Figure 7.4) exemplifies several figures and ideas exposed in previous chapters, in that three control parameters are used (dopamine, serotonin, and norepinephrine) in a state space where the "self-control dynamics" are taking place. The names of possible control parameters are indicated (endorphins, 5-HT, etc.), as

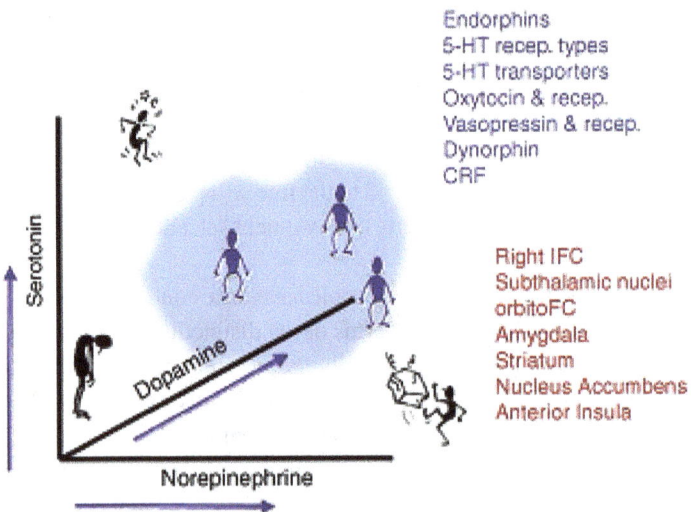

Figure 7.4. 3D "control" space. The optimal, when the individual is exerting what is considered "self-control," is the shaded region. But, what is this self in control? Who is in control in other regions? And what is controlling what? Reproduced, with permission, from Suhler and Churchland (2009).

well as brain areas involved in behavioral control (amygdala, accumbens, etc.). One difficulty in thinking about control parameters in the nervous system, which is not shared with other physical systems, is that their actions may be different in distinct brain regions, hence the spatial, or anatomical, considerations are fundamental. Outside the optimal region in the center (optimality here ascribed to a considered self-control resulting in a happy, content, and decent life, always according to the norms of the current social standards), the individual loses self-control and falls into rage, despair, depression, and so on. In the original figure legend, the authors regret that they "cannot provide for dynamical properties" more specifically, a feat that is beyond the scope of even sophisticated computational models.

It was remarked in other sections that complex systems, which normally exhibit nonlinear dynamics possibly chaotic, do not need any central control for patterns of organized collective activity to emerge. This already provides a hint to the answer that is being sought to the major question of self-control, or whether our brain orchestrates our behavior with the inevitability of an epileptic seizure or a migraine. Most likely, there is nothing specific in control. What, in the end, "controls" the mood of the cartoon character in Figure 7.4 is the resulting organized pattern of activity derived from the modulators (dopamine etc.), control parameters, and the neurocircuits experiencing those activity patterns. But, as said repeatedly in this volume to avoid dichotomies, those patterns are the behaviors themselves.

But wait a moment, Figure 3.12D in Chapter 3 showed a human seizure arrested by cognitive efforts (or by electrical stimulation), so, even ictal events seem to be controlled by the mind itself. Not really, only some, but not the majority (see comments in Section 3.5.5.2). It all depends on how stable that synchrony is within the particular brain networks as extensively commented in that chapter and in the preceding Section 7.2.1. In some cases, a small perturbation will suffice to deviate the brain dynamics toward "normality." Whether in seizures or in mania, tics, depression, antisocial behavior... it is all the same: the more robust the activity, or, using a dynamical metaphor, the stronger the attractor (the larger the basin of attraction), the more difficult to correct whatever needs to be corrected, which here has been reasoned to consist of synchronous activities within specific circuitries for each condition or deviation.

In closing, let us comment on one atypical case study that, we consider, illustrates accurately these points. Wennberg and colleagues reported the behavioral automatisms of an epileptic patient during seizures. One of these automatic actions was the sign of the cross. This patient was unconscious during the seizures (or at least not aware of the actions), as it is the normal case, and, like a devoted Christian, performed (unconsciously) a *signum crucis* towards the end of her seizures (Wennberg *et al.*, 2009). An unusual occurrence, amongst the several automatisms that are observed during seizures. More unusual was the probable reason why this occurred. During her youth, this patient was advised by her parents to cross herself after coming out from the ictus, as these events were seen in the past (and are still by some) as diabolical interventions. She does not do it anymore, well...at least consciously, because now the action has been incorporated into the seizure itself. It seems like the neural activity of her seizures has integrated the neural networks (motor areas and others) that were responsible for the behavioral crossing, so now the action occurs during the ictus and it remains out of consciousness. Now imagine a hypothetical case, where instead of the sign of the cross, the seizures in another, more unlucky, epileptic patient activate a brain network that includes, for example, the memory of one of the myriad of violent scenes seen in the movies. Now the patient, during the seizures and completely unconsciously, starts to hit as hard as (s)he can any surrounding individual. Regardless of how many contusions this behavior has occasioned to the unfortunate caretakers that happen to be around, rest assure that they will not hate the wretched patient and still will empathize with him/her, and will fully understand that the violent behavior was unavoidable, as it was dictated by the activity within the circuitry of the ictal event. On the other hand, let outsiders watch a video of this event, and do not inform them that this is an epileptic patient that is having a seizure. Almost certainly that some of these characters, thinking that the person has committed a violent crime, will dislike the, unknown to them, patient. What is creating this stark difference in the judgments, the verdict being "innocent" from the former set of caretakers, and "guilty" from the anonymous spectators? At the level of the brain function and behavior, nothing is different; at the neurophysiological level, it is the same process: some parts of the brain are activating others to end up in a motor output, whether during the seizure or in normal conditions, the difference lies in the specific networks that are recruited to provide the final input to motor cortices, and there is no need to attribute special "self-control" features to one or another area. The same neurophysiology occurs in "normal" behaviours or mental deviations. It is the interpretation of the behavior in terms of what individuals think they know about brains. If the spectators are explained that they are in reality watching an epileptic attack and that the behavior is an unconscious automatism, their opinion will most likely change dramatically, and now a feeling of sympathy for the patient arises. How about other automatisms, like tics such as those typical of the Tourette syndrome? Why do some individuals get upset with their children that suffer this condition because they cannot control the tics? What if they were told these are mini-seizures? What does make us think that, while it is acknowledged that the epileptic patient could not avoid performing the sign of the cross, other patients suffering from delusions, obsessive-compulsive disorders, pedophilia, and a long *et cetera*, can avoid their offensive or unnatural behaviors? It is that pervasive "control" concept. This is the question that was posed at the end of Chapter 1, the enquiry about self-control, how much control does a brain really have? Or, what parts of the brain, if any, have control over which others? But we already know enough about brains to understand the underlying neurophysiological reasons for all these behaviors.

Consider further one character who has enough will to quit smoking, for example, indicating that this lucky fellow has thinking and executive parts that are powerful enough to exert control over the rest of the brain in pursuit of this lofty, smoke-free long-term goal. This brain seems to be able to change its own template that impelled its owner the pursuit of unhealthy habits, using its rational power resident in, according to current neuroscientific belief, the frontal cortex, the lobe that excels at executive control (Fuster, 2003). Now, this is all very nice. But consider for a moment where this frontal "self-control" comes from. Recall Hallett's words a few paragraphs above; the frontal networks become active thanks to inputs from other areas, where other sensorimotor transformations are being processed. Perhaps this individual has heard and read about the harmful consequences of smoking, about methods to quit, to achieve goals or improve control... so, where is the "self" part, the internal character, in this apparent self-control phenomenon? It seems to be dictated by information gathered from the outside, sensory inputs in the final analysis, that enter the brain and add themselves to the ongoing neural activity (that "resting" activity in Section 7.2), to the chain of activity, those synfire chains of Chapter 1. A cursory and superficial evaluation will attribute the success of smoke-quitting by this subject to self-control, but on a deeper examination, the internal aspect, that "self" part, seems to evanesce the more you scrutinise it. The closer one looks into the brain circuitries, the harder it is to find a last ventriloquist, as there is no origin, no start or end to the chain. But, some will argue, it is the frontal association areas that incorporate the information processed by sensory cortices, combine it with stored information (memories), and finally reach a decision to exert control and act accordingly, the power of the will in effect. Some say that human cognitive abilities depend not so much on external perturbations but on internally generated thoughts. However, nobody said that the sensory inputs occur in a precise moment in time and this represents the absolute start of the "chain reaction"; the sensorium is constantly being incorporated into the brain microstructure; there is a huge literature on the molecular/cellular mechanism of memory formation that cannot be covered here but yet it is very revealing if only because it illustrates how the external world influences the anatomy and neurophysiology at micro and macro levels. Hence, one more dichotomy seems to bite the dust, the distinction between external and internal inputs (see Nachev and Husain (2010), on the distinction between "internal" and "external" stimuli, their view being that "the contrast between internally and externally-generated actions is empirically intractable"), inputs that can make one quit smoking, a pedophile seek children, and a schizophrenic experience hallucinations. The incessant sensory bombardment does not know borders, inputs coming from either the outside or the inside, it does not really matter because all these, perhaps tiny, inputs carry the potentiality of a global and substantial alteration of brain dynamics, by virtue of the multiplicity of associations between the stored and incoming information, by virtue of the dynamics of complex systems. Are we then like Stanislaw Lem's cyber-characters in his *Cyberiada* (1967), sparks of lights traveling from emeralds to sapphires and other metallic parts that constitute his digital personages making them behave deterministically, albeit allowing some stochasticity? (But be careful about what we consider stochastic in these discussions, it could be because the initial conditions can never be known with extreme accuracy and thus it is our own problem, or it may be an inherent part of natural phenomena, but perhaps this musing is of no consequence since it is only what we perceive that we can measure and comprehend.) It is said that the famous physicist Richard Feynman once said that you are today the potatoes you ate yesterday, but when it comes to mental life and the brain, the time frame has to be stretched all the way to the past, for what your brain is today is what it has assimilated since birth (plus, of course, the inherited baggage). Only that those "potatoes" consisted of waves of sensory-motor transformations (Piaget, 1936) that incessantly and through life configure brain circuits, always with the potential to change owing to the omnipresent plasticity at all levels of neural tissue. And this is your brain today. So any novel external (or internal) input enters its circuitries and adds itself to the chains of activity, carrying with it this potentiality for transformation. Can this potentiality be taught, improved, and developed by individuals? Certainly. There are methods that foment mental education and bring about the possibility for the changing of behavioral dispositions. But this is a wide topic that deserves another monograph.

"That since wars begin in the minds of men, it is in the minds of men that defences of peace must be constructed ..."

Preamble of the UNESCO Constitution (1945)

Notes

1. Our puppets strings are hard to see,
   So we perceive ourselves as free,
   Convinced that no mere objects could
   Behave in terms of bad and good.

To you, we mannequins seem less
than live, because our consciousness
is that of dummies, made to sit
on laps of gods and mouth their wit;
Are you, our transcendental gods,
likewise dangled from your rods,
and need, to show spontaneous charm,
some higher god's inserted arm?
We seem to form a nested set,
With each the next one's marionette,
Who, if you asked him, would insist,
that he's the last ventriloquist.

T. Melnechuk, PUNCH AND JUDY, TO THEIR AUDIENCE

Bibliography

Aakerlund, L., Hemmingsen, R. (1998) Neural networks as models of psychopathology. *Biol. Psychiatry* 43, 471–482.

Aarabi, A., Wallois, F., Grebe, R. (2008) Does spatiotemporal synchronization of EEG change prior to absence seizures? *Brain Res.* 1188, 207–221.

Abeles, M. (1982) *Local Cortical Circuits. An Electrophysiological Study.* Springer-Verlag, Berlin.

Abeles, M. (1991) *Corticonics.* Cambridge University Press.

Abi-Dargham, A., Gil, R., Krystal, J., Baldwin, R.M., Seibyl, J.P., Bowers, M., van Dyck, C.H., Charney, D.S., Innis, R.B., Laruelle, M. (1998) Increased striatal dopamine transmission in schizophrenia: confirmation in a second cohort. *Am. J. Psychiatry* 155, 761–767.

Achermann, P., Borbély, A.A. (1998) Coherence analysis of the human sleep electroencephalogram. *Neuroscience* 85, 1195–1208.

Adey, W.R. (1972) Organization of brain tissue: is the brain a noisy processor? *Int. J. Neurosci.* 3, 271–284.

Adrian, E.D., Zotterman, Y. (1926) The impulses produced by sensory nerve endings. Part II: the response of a single end organ. *J. Physiol.* 61, 151–171.

Adleman, N.E., Menon, V., Blasey, C.M., White, C.D., Warsofsky, I.S., Glover, G.H., Reiss, A.L. (2002) A developmental fMRI study of the Stroop color-word task. *Neuroimage* 16, 61–75.

Adler, L.E., Rose, G., Freedman, R. (1986) Neurophysiological studies of sensory gating in rats: effects of amphetamine, phencyclidine and haloperidol. *Biol. Psychiatry* 21, 787–798.

Afraimovich, V.S., Verichev, N.N., Rabinovich, M.I. (1986) Stochastic synchronization of oscillations in dissipative systems. *Izv. Vyssh. Uchebn. Zaved. Radiofiz.* 29, 795–803.

Agid, Y. (1991) Parkinson's disease: pathophysiology. *Lancet* 337, 1321–1323.

Agulhon, C., Fiacco, T.A., McCarthy, K.D. (2010) Hippocampal short and long-term plasticity are not modulated by astrocyte Ca^{2+} signaling. *Science* 327, 1250–1254.

Aitken, P.G., Sauer T., Schiff, S.J. (1995) Looking for chaos in brain slices. *J. Neurosci. Methods* 59, 41–48.

Alberts, W.W. (1972) A simple view of Parkinsonian tremor. Electrical stimulation of cortex adjacent to the rolandic fissure in awake man. *Brain Res.* 44, 357–369.

Alcaro, A., Panksepp, J., Witczak, J., Hayes, D.J., Northoff, G. (2010) Is subcortical-cortical midline activity in depression mediated by glutamate and GABA? A cross-species translational approach. *Neurosci. Biobehav. Rev.* 34, 592–605.

Alcázar Córcoles, M.A., Verdejo Garcia, A., Bouso Saiz, J.C., Bezos Saldana, L. (2010) Neuropsychology of impulsive aggression. *Rev. Neurol.* 50, 291–299.

Allefeld, C., Atmanspacher, H., Wackermann, J. (2009) Mental states as macrostates emerging from brain electrical dynamics. *Chaos* 19, 015102.

Allegrini, P., Fronzoni, L., Pirino, D. (2009) The influence of the astrocyte field on neuronal dynamics and synchronization. *J. Biol. Phys.* 35, 413–423.

Alloway, K.D., Burton, H. (1991) Differential effects of GABA and bicuculline on rapidly- and slowly-adapting neurons in primary somatosensory cortex of primates. *Exp. Brain Res.* 85, 598–610.

Amaral, D.G., Schumann, C.M., Nordahl, C.W. (2008) Neuroanatomy of autism. *Trends Neurosci.* 31, 137–145.

Amit, D.J. (1989) *Modeling Brain Function. The World of Attractor Neural Networks*. Cambridge University Press.

Amor, F., Baillet, S., Navarro, V., Adam, C., Martinerie, J., Le van Quyen, M. (2009) Cortical local and long-range synchronization interplay in human absence seizure initiation. *Neuroimage* 45, 950–962.

Amzica, F., Steriade, M. (2000) Neuronal and glial membrane potentials during sleep and paroxysmal oscillations in the neocortex. *J. Neurosci.* 20, 6648–6665.

Anderson, J.C., Martin, K.A.C. (2009) The synaptic connections between cortical areas V1 and V2 in macaque monkey. *J. Neurosci.* 29, 11283–11293.

Andronov, A. A., Pontryagin, L.S. (1937) Systèmes grossières. *Dokl. Akad. Nauk. SSSR* 14, 247–251.

Andronov, A.A., Vitt, E.A., Khaikin S.E. (1937) *Theory of Oscillators*, Moscow (English translation in Pergamon Press, Oxford, 1966).

Andrzejak, R.G., Widman, G., Lehnertz, K., Rieke, C., David, P., Elger, C.E. (2001) The epileptic process as nonlinear deterministic dynamics in a stochastic environment: an evaluation on mesial temporal lobe epilepsy. *Epilepsy Res.* 44, 129–140.

Anishchenko, V.S., Astakhov, V.V., Neiman, A.B., Vadivasova, T.E., Schimansky-Geier, L. (2002) *Nonlinear Dynamics of Chaotic and Stochastic Systems*. Springer-Verlag, Berlin.

Anishchenko, V.S. Vadivasova, T.E. Strelkova, G.I., Kopeikin, A.S. (1998) Chaotic attractors of two-dimensional invertible maps. *Discrete Dyn. Nature Soc.* 2, 249–256.

Apicella, P., Scarnati, E., Schultz, W. (1991) Tonically discharging neurons of monkey striatum respond to preparatory and rewarding stimuli. *Exp. Brain Res.* 84, 672–675.

Appleyard, K., Egeland, B., van Dulmen, M., Sroufe, A. (2005) When more is not better: the role of cumulative risk in child behaviour outcome. *J. Child Psychol.* 3, 235–245.

Arieli, A., Sterkin, A., Grinvald, A., Aertsen, A. (1996) Dynamics of ongoing activity: explanation of the large variability in evoked cortical responses. *Science* 273, 1868–1871.

Asanuma, K., Tang, C., Ma, Y., Dhawan, V., Mattis, P., Edwards, C., Kaplitt, M.G., Feigin, A., Eidelberg, D. (2006) Network modulation in the treatment of Parkinson's disease. *Brain* 129, 2667–2678.

Aschenbrenner-Scheibe, R., Maiwald, T., Winterhalder, M., Voss, H.U., Timmer, J., Schulze-Bonhage, A. (2003) How well can epileptic seizures be predicted? An evaluation of a nonlinear method. *Brain* 126, 1–11.

Ashtari, M., Bregman, J., Nicholls, S. (2007) Gray matter enlargement in children with high functioning autism and Asperger syndrome using a novel method of diffusion based morphometry. Radiological Society of North America Annual Meeting.

Ashwin, P., Lavric, A. (2010) A low-dimensional model of binocular rivalry using winnerless competition. *Physica D* 239, 529–536.

Assal, F., Schwartz, S., Vuilleumier, P. (2007) Moving with or without will: functional neural correlates of alien hand syndrome. *Ann. Neurol.* 62, 301–306.

Avoli, M. (2001) Do interictal discharges promote or control seizures? Experimental evidence from an *in vitro* model of epileptiform discharge. *Epilepsia* 42 (Suppl. 3), 2–4.

Ayala, G.F., Dichter, M., Gumnit, R.J., Matsumoto, H., Spencer, W.A. (1973) Genesis of epileptic interictal spikes. New knowledge of cortical feedback systems suggests a neurophysiological explanation of brief paroxysms. *Brain Res.* 52, 1–17.

Baars, B.J. (1988) *A Cognitive Theory of Consciousness.* Cambridge University Press.

Baars, B.J., Ramsøy, T.Z., Laureys, S. (2003) Brain, conscious experience and the observing self. *Trends Neurosci.* 26, 671–675.

Babb, T.L., Wilson, C.L., Isokawa-Akesson, M. (1987) Firing patterns of human limbic neurons during stereoencephalography (SEEG) and clinical temporal lobe seizures. *Electroencephalogr. Clin. Neurophysiol.* 66, 467–482.

Babiloni, C., Albertini, G., Onorati, P., Muratori, C., Buffo, P., Condoluci, C., Sarà, M., Pistoia, F., Vecchio, F., Rossini, P.M. (2010) Cortical sources of EEG rhythms are abnormal in Down syndrome. *Clin. Neurophysiol.* 121, 1205–1212.

Babloyantz, A., Destexhe, A. (1986) Low-dimensional chaos in an instance of epilepsy. *Proc. Natl. Acad. Sci. USA* 83, 3513–3517.

Badawy, R., Macdonell, R., Jackson, G., Berkovic, S. (2009) The peri-ictal state: cortical excitability changes within 24 h of a seizure. *Brain* 132, 1013–1021.

Bahar, S. (2006) Pierre Bayle & seizure prediction. *Biol. Physicist* 6, 2–5.

Bakeman, R., Gottman, J.M. (1997) *Observing Interaction: An Introduction to Sequential Analysis.* Cambridge University Press.

Bal, T., McCormick, D.A. (1996) What stops synchronized thalamocortical oscillations? *Neuron* 17, 297–308.

Ballyk, B.A., Quackenbush, S.J., Andrew, R.D. (1991) Osmotic effects on the CA1 neuronal population in hippocampal slices with special reference to glucose. *J. Neurophysiol.* 65, 1055–1066.

Balish, M., Albert, P.S., Theodore, W.H. (1991) Seizure frequency in intractable partial epilepsy: a statistical analysis. *Epilepsia* 32, 642–649.

Bandyopadhyay, S., Gonzalez-Islas, C., Hablitz, J.J. (2005) Dopamine enhances spatiotemporal spread of activity in rat prefrontal cortex. *J. Neurophysiol.* 93, 864–872.

Baranek, G.T. (2002) Efficacy of sensory and motor interventions for children with autism. *J. Autism Dev. Disord.* 32, 397–422.

Barbarosie, M., Avoli, M. (1997) CA3-driven hippocampal-entorhinal loop controls rather than sustains *in vitro* limbic seizures. *J. Neurosci.* 17, 9308–9314.

Barceló, F. (1999) Electrophysiological evidence of two different types of error in the Wisconsin Card Sorting Test. *Neuroreport* 10, 1–5.

Barceló, F., Knight, R.T. (2002) Both random and perseverative errors underlie WCST deficits in prefrontal patients. *Neuropsychologia* 40, 349–356.

Barch, D.M., Carter, C.S. (2005) Amphetamine improves cognitive function in medicated individuals with schizophrenia and in healthy volunteers. *Schizophr. Res.* 77(1), 43–58.

Başar, E. (1980) *EEG Brain Dynamics. Relation between EEG and Brain Evoked Potentials.* Elsevier/North Holland Biomedical Press, Amsterdam.

Baulac, S., *et al.* (2001) First genetic evidence of $GABA_A$ receptor dysfunction in epilepsy: a mutation in the γ2-subunit gene. *Nat. Genet.* 28, 46–48.

Been, G., Ngo, T.T., Miller, S.M., Fitzgerald, P.B. (2007) The use of tDCS and CVS as methods of non-invasive brain stimulation. *Brain Res. Rev.* 56, 346–361.

Behrmann, M., Thomas, C., Humphreys, K. (2006) Seeing it differently: visual processing in autism. *Trends Cogn. Sci.* 10, 258–264.

Belair, J., Glass, L., an der Heiden, U., Milton, J. (1995) Dynamical disease: identification, temporal aspects and treatment strategies of human illness. *Chaos* 5, 1–7.

Belluscio, M.A., Riquelme, L.A., Murer, M.G. (2007) Striatal dysfunction increases basal ganglia output during motor cortex activation in Parkinsonian rats. *Eur. J. Neurosci.* 25, 2791–2804.

Benabid, A.L. (2003) Deep brain stimulation for Parkinson's disease. *Curr. Opin. Neurobiol.* 13, 696–706.

Benini, R., D'Antuono, M., Pralong, E., Avoli, M. (2003) Involvement of amygdala networks in epileptiform synchronization *in vitro*. *Neuroscience* 120, 75–84.

Bennett, C.H., Gacs, P., Li, M., Vitanyi, P.M.B., Zurek, W.H. (1998) Information distance. *IEEE Trans. Inf. Theory* 44, 1407–1423.

Bergé, P., Pomeau, Y., Vidal, C. (1984) *Order within Chaos: Towards a Deterministic Approach to Turbulence.* John Wiley & Sons, New York.

Berger, H. (1929) Uber das Elektroenkephalogram des Menschen. *Archiv für Psychiatrie und Nerven Krankheiten* 87, 527–570.

Bergey, G.K., Franaszczuk, P.J. (2001) Epileptic seizures are characterized by changing signal complexity. *Clin. Neurophysiol.* 112, 241–249.

Berman, K., Meyer-Lindenberg, A. (2004) Functional brain imaging studies in schizophrenia, in: Charney D.S., Nestler E.J. (eds.) *Neurobiology of Mental Illness*, 2nd edition, Oxford University Press.

Bernat, J.L. (1992) How much of the brain must die in brain death? *J. Clin. Ethics* 3, 21–26.

Bertram, E.H., Mangan, P.S., Zhang, D., Scott, C.A., Williamson, J.M. (2001) The midline thalamus: alterations and a potential role in limbic epilepsy. *Epilepsia* 42, 967–978.

Beuter, A., Vasilakos, K. (1995) Tremor: is Parkinson's disease a dynamical disease? *Chaos* 5, 35–42.

Bezruchko, B.P., Dikanev, T.V., Smirnov, D.A. (2001) Role of transient processes for reconstruction of model equations from time series. *Phys. Rev. E* 64, 036210.

Bhattacharya, J. (2000) Complexity analysis of spontaneous EEG. *Acta Neurobiol. Exp.* 60, 495–501.

Bickford-Wimer, P.C., Nagamoto, H., Johnson, R., Adler, L.E., Egan, M., Rose, G.M., Freedman, R. (1990) Auditory sensory gating in hippocampal neurons: a model system in the rat. *Biol. Psychiatry* 27, 183–192.

Bickford, P.C., Wear, K.D. (1995) Restoration of sensory gain of auditory evoked response by nicotine in fimbria-fornix lesioned rats. *Brain Res.* 705, 35–240.

Billings, L., Schwartz, I.B. (2008) Identifying almost invariant sets in stochastic dynamical systems. *Chaos* 18, 023122.

Binnie, C.D., Stefan, H. (1999) Modern electroencephalography: its role in epilepsy management. *Clin. Neurophysiol.* 110, 1671–1697.

Blair, R.J.R. (2010) Neuroimaging of psychopathy and antisocial behaviour: a targeted review. *Curr. Psychiatry Rep.* 12, 76–82.

Blair, J., Mitchell, D., Blair, K. (2005) *The Psychopath: Emotion and the Brain.* Blackwell Publishing, Malden MA.

Blumenfeld, H., Taylor, J. (2003) Why do seizures cause loss of consciousness? *Neuroscientist* 9, 301–310.

Boose, A., Jentgens, C., Spieker, S., Dichgans, J. (1995) Variations on tremor parameters. *Chaos* 5, 52–56.

Boyasrky, A., Góra, P. (2006) Invariant measures in brain dynamics. *Phys. Lett. A* 358, 27–30.

Bowers, M.B. Jr. (1968) Pathogenesis of acute schizophrenic psychosis: an experimental approach. *Arch. Gen. Psychiatry* 19, 348–355.

Braun, H.A., Moss, F., Postnova, S., Mosekilde, E. (2008) Complexity in neurology and psychiatry. *J. Biol. Phys.* 34, 249–252.

Breakspear, M. (2006). The non-linear theory of schizophrenia. *Austr. N.Z. J. Psychiatry* 40, 20–35.

Breakspear, M., Williams, L.M., Stam, C.J. (2004) A novel method for the topographic analysis of neural activity reveals formation and dissolution of 'Dynamic Cell Assemblies'. *J. Comp. Neurosci.* 16, 49–68.

Breakspear, M., Terry, J.R., Friston, K.J., Harris, A.W.F., Williams, L.M., Brown, K., Brennan, J., Gordon, E. (2003) A disturbance of non-linear interdependence in scalp EEG of subjects with first episode of schizophrenia. *Neuroimage* 20, 466–478.

Breakspear, M., Roberts, J.A., Terry, J. R., Rodrigues, S. Mahant, N., Robinson, P. A. (2006) A unifying explanation of primary generalized seizures through nonlinear brain modeling and bifurcation analysis. *Cereb. Cortex* 16, 1296–1313.

Bressler, S.L. (2003) Cortical coordination dynamics and the disorganization syndrome in schizophrenia. *Neuropsychopharmacology* 28 (Suppl. 1), S35–S39.

Bressler, S.L., Kelso, J.A.S. (2001) Cortical coordination dynamics and cognition. *Trends Cogn. Sci.* 5, 26–36.

Brian, J.A., Bryson, S.E. (1996) Disembedding performance and recognition memory in autism/PDD. *J. Child Psychol. Psychiatry* 37, 865–872.

Bricolo, A., Turella, G.S. (1990) Electrophysiology of head injury, in: Braakman, R. (ed.) *Handbook of Clinical Neurology*, Vol. 13(57). Elsevier Science Publishers BV, New York, pp. 181–206.

Brier, A., Su, T.P., Saunders, R., Carson, R.E., Kolachana, B.S., de Bartolomeis, L.S., Weinberger, D.R., Weisenfeld, N., Malhotra, A.K., Eckelman, W.C., Pickar, D. (1997) Schizophrenia is associated with elevated amphetamine-induced synaptic dopamine concentrations: evidence from a novel positron emission tomography method. *Proc. Natl. Acad. Sci. USA* 94, 2569–2574.

Brock, J., Brown, C.C., Boucher, J., Rippon, G. (2002) The temporal binding deficit hypothesis of autism. *Dev. Psychopathol.* 14, 209–224.

Brown, C., Gruber, T., Boucher, J., Rippon, G., Brock, J. (2005) Gamma abnormalities during perception of illusory figures in autism. *Cortex* 41, 364–376.

Bruni, J. (ed.) 1995. *Demystifying Epilepsy.* Grosvernor House Press Inc., Montreal.

Bruni, J., Wilder, B.J. (1979) Valproic acid — review of a new antiepileptic drug. *Arch. Neurol.* 36, 393–398.

Bruno, R.M., Sakmann, B. (2006) Cortex is driven by weak but synchronously active thalamocortical synapses. *Science* 312, 1622–1627.

Bryson, S.E. (1983) Interference effects in autistic children: evidence for the comprehension of single stimuli. *J. Abnorm. Psychol.* 92, 250–254.

Buckner, R.L., Andrews-Hanna, J.R., Schacter, D.L. (2008) The brain's default network. Anatomy, function and relevance to disease. *Ann. N. Y. Acad. Sci.* 1124, 1–38.

Bufkin, J.L., Luttrell, V.R. (2005) Neuroimaging studies of aggressive and violent behavior. *Trauma Violence Abuse* 6(2), 176–191.

Burns, K., Gidea, M. (2005) *Differential Geometry and Topology*. Chapman & Hall/CRC, Boca Raton.

Buzsáki, G. (2006) *Rhythms of the Brain*. Oxford University Press.

Cabrera, J.L., Milton, J.G. (2002) On-off intermittency in a human balancing task. *Phys. Rev. Lett.* 89, 158702.

del Campo, C.M., Pérez Velázquez, J.L., Cortez, M.A. (2009) EEG recording in rodents with a focus on epilepsy, in: *Current Protocols in Neuroscience*, Chapter 6, Unit 6.24, John Wiley & Sons Inc., New Jersey.

Cannon, C.M., Palmiter, R. (2003) Reward without dopamine. *J. Neurosci.* 23, 10827–10831.

Cantero, J.L., Atienza, M., Madsen, J.R., Stickgold, R. (2004) Gamma EEG dynamics in neocortex and hippocampus during human wakefulness and sleep. *NeuroImage* 22, 1271–1280.

Cardon, S.Z., Iberall, A.S. (1970) Oscillations in biological systems. *Curr. Mod. Biol.* 3, 237–249.

Cardoso de Oliveira, S., Thiele, A., Hoffmann, K.P. (1997) Synchronization of neuronal activity during stimulus expectation in a direction discrimination task. *J. Neurosci.* 17, 9248–9260.

Carrington, C.A., Gilby, K.L., McIntyre, D.C. (2007) Effect of focal low-frequency stimulation on amygdala-kindled afterdischarge thresholds and seizure profiles in fast and slow-kindling rat strains. *Epilepsia* 48, 1604–1613.

Castellanos, A., Moleiro, F., Saoudi, N., Interian, A., Myerburg, R.J. (1995) Dynamics of, and alternating Wenckebach periods during, 4:1 and 6:1 atrioventricular block. *Am. J. Cardiol.* 76, 523–525.

Cavada, C., Goldman-Rakic, P.S. (1989) Posterior parietal cortex in rhesus monkey: II. Evidence for segregated corticocortical networks linking sensory and limbic areas with the frontal lobe. *J. Comp. Neurol.* 287, 422–445.

Čenys, A., Anagnostopoulos, A.N., Bleris, G.L. (1997) Distribution of laminar lengths for noisy on-off intermittency. *Phys. Lett. A* 224, 346–352.

Cerf, R., El Ouasad, E.H., Kahane, P. (2004) Criticality and synchrony of fluctuations in rhythmical brain activity: pretransitional effects in epileptic patients. *Biol. Cybern.* 90, 239–255.

Chang, H.-S., Staras, K., Gilbey, M.P. (2000) Multiple oscillators provide metastability in rhythm generation. *J. Neurosci.* 20, 5135–5143.

Chaté, H., Manneille, P. (1987) Transition to turbulence via spatiotemporal intermittency. *Phys. Rev. Lett.* 58, 112–115.

Chavez, M., Adam, C., Navarro, V., Boccaletti, S., Martinerie, J. (2005) On the intrinsic time scales involved in synchronization: a data-driven approach. *Chaos* 15, 023904.

Cherkassky, V.L., Kana, R.K., Keller, T.A., Just, M.A. (2006) Functional connectivity in a baseline resting-state network in autism. *Neuroreport* 17, 1687–1690.

Chow, J.Y., Davids, K., Button, C., Rein, R., Hristovski, R., Koh, M. (2009) Dynamics of multi-articular coordination in neurobiological systems. *Nonlinear Dynamics Psychol. Life Sci.* 13, 27–55.

Christensen, K., Moloney, N.R. (2005) *Complexity and Criticality.* Advanced Physics Texts, Vol. 1. Imperial College Press, London.

Christini, D.J., Collins, J.J. (1997) Control of chaos in excitable physiological systems: a geometric analysis. *Chaos* 7, 544–549.

Christini, D.J., Glass, L. (2002) Introduction: mapping and control of complex cardiac arrhythmias. *Chaos* 12, 732–739.

Christini, D.J., Stein, K.M., Markowitz, S.M., Mittal, S., Slotwiner, D.J., Scheiner, M.A., Iwai, S., Lerman, B.B. (2001) Nonlinear-dynamical arrhythmia control in humans. *Proc. Natl. Acad. Sci. USA* 98, 5827–5832.

Cho, R.Y., Konecky, R.O., Carter, C.S. (2006) Impairment in frontal cortical gamma synchrony and cognitive control in schizophrenia. *PNAS* 103(52), 19878–19883.

Clementz, B.A., Blumenfeld, L.D., Cobb, S. (1997) The gamma band response may account for poor P50 suppression in schizophrenia. *Neuroreport* 8, 3889–3893.

Cohen, I., Navarro, V., Clemenceau, S., Baulac, M., Miles, R. (2002) On the origin of interictal activity in human temporal lobe epilepsy *in vitro. Science* 298, 1418–1421.

Cohn, R., Leader, H.S. (1967) Synchronization characteristics of paroxysmal EEG activity. *Electroencephalogr. Clin. Neurophysiol.* 22, 421–428.

Collet, P., Eckmann, J.P. (1980) *Iterated Maps on the Interval as Dynamical Systems.* Birkhäuser.

Colpan, M.E., Li, Y., Dwyer, J. Mogul, D.J. (2007) Proportional feedback stimulation for seizure control in rats. *Epilepsia* 48, 1594–1603.

Comte, J.C., Ravassard, P., Salin, P.A. (2006) Sleep dynamics: a self-organized critical system. *Phys. Rev. E* 73, 056127.

Connors, B.W. (1984) Initiation of synchronized neuronal bursting in neocortex. *Nature* 310, 685–687.

Contreras, D., Steriade, M. (1995) Cellular basis of EEG slow rhythms: a study of dynamic corticothalamic relationships. *J. Neurosci.* 15, 604–622.

Contreras, D., Llinás, R. (2001) Voltage-sensitive dye imaging of neocortical spatiotemporal dynamics to afferent activation frequency. *J. Neurosci.* 21, 9403–9413.

Cooke, P.M., Snider, R.S. (1955) Some cerebellar influences on electrically induced cerebral seizures. *Epilepsia* 4, 19–28.

Cools, R. (2008) Role of dopamine in the motivational and cognitive control of behaviour. *Neuroscientist* 14, 381–395.

Cooper, R., Winter, A.L., Crow, H.J. *et al.* (1965) Comparison of subcortical, cortical and scalp activity using chronically indwelling electrodes in man. *Electroencephalogr. Clin. Neurophysiol.* 18, 217–228.

Cooper, I.S. (1973) Effect of chronic stimulation of anterior cerebellum on neurological disease. *Lancet* 1(7796), 206.

Cortez, M.A., Perez Velazquez, J.L., Snead III, O.C. (2006) Animal models of epilepsy and progressive effects of seizures, in: Blume, W. *et al.* (eds.) *Intractable Epilepsies* (Advances in Neurology, Vol. 97). Lippincott Williams & Wilkins.

Cossart, R., Aronov, D., Yuste, R. (2003) Attractor dynamics of network UP states in the neocortex. *Nature* 423, 283–288.

Cossart, R., Benard, C., Ben-Ari, Y. (2005) Multiple facets of GABAergic neurons and synapses: multiple fates of GABA signalling in epilepsies. *Trends Neurosci.* 28, 108–115.

Costa, R.M., Lin, S.-C., Sotnikova, T.D., Cyr, M., Gainetdinov, R.R., Caron, M.G., Nicolelis, M.A.L. (2006) Rapid alterations in corticostriatal ensemble coordination during acute dopamine-dependent motor dysfunction. *Neuron* 52, 359–369.

Costalat, R., Chauvet, G. (2008) Basic properties of electrical field coupling between neurons: an analytical approach. *J. Integr. Neurosci.* 7, 225–247.

Cotter, P. (2009) The path to extreme violence: Nazism and serial killers. *Front. Behav. Neurosci.* 3:61. doi: 10.3389/neuro.08.061.2009.

Coulter, D.A., McIntyre, D.C., Loscher, W. (2002) Animal models of limbic epilepsies: what can they tell us? *Brain Pathol.* 12, 240–256.

Courchesne, E., Press, G.A., Yeung-Courchesne, R. (1993) Parietal lobe abnormalities detected with MR in patients with infantile autism. *Am. J. Roentgenol.* 160, 387–393.

Craig, A.D. (2002) How do you feel? Interoception: the sense of the physiological condition of the body. *Nat. Rev. Neurosci.* 8, 655–666.

Crow, T.J. (1986) The continuum of psychosis and its implication for the structure of the gene. *Br. J. Psychiatry* 149, 419–429.

Crutchfield, J.P. (1994) The calculi of emergence: computation, dynamics and induction. *Physica D* 75, 11–54.

Crutchfield, J.P., Kaneko, K. (1988) Are attractors relevant to turbulence? *Phys. Rev. Lett.* 60, 2715–2718.

Cruz, A.V., Mallet, N., Magill, P.J., Brown, P., Averbeck, B.B. (2009) Effects of dopamine depletion on network entropy in the external globus pallidus. *J. Neurophysiol.* 102, 1092–1102.

Curtis, C.E., D'Esposito, M. (2003) Persistent activity in the prefrontal cortex during working memory. *Trends Cogn. Sci.* 7, 415–423.

Damasio, A.R. (1989) The brain binds entities and events by multiregional activation from convergence zones. *Neural Comput.* 1, 123–132.

Davey, M.P., Victor, J.D., Schiff, N.D. (2000) Power spectra and coherence in the EEG of a vegetative patient with severe asymmetric brain damage. *Clin. Neurophysiol.* 111, 1949–1954.

Davidson, R.J., Putnam, K.M., Larson, C.L. (2000) Dysfunction in the neural circuitry of emotion regulation — a possible prelude to violence. *Science* 289, 591–594.

Davidson, L.L., Heirichs, R.W. (2003) Quantification of frontal and temporal lobe brain-imaging findings in schizophrenia: a meta-analysis. *Psychiatry Res.* 122, 69–87.

de Boer, S.F., Caramaschi, D., Natarajan, D., Koolhaas, J.M. (2009) The vicious cycle towards violence: focus on the negative feedback mechanism of brain serotonin neurotransmission. *Front. Behav. Neurosci.* 3:52. doi: 10.3389/neuro.08.052.2009.

Deboer, T., Sanford, L.D., Ross, R., Morrison, A.R. (1998) Effects of stimulation in the amygdale on ponto-geniculo-occipital waves in rats. *Brain Res.* 193, 305–310.

De Clercq, W., Lemmerling, P., Van Huffel, S., Van Paesschen, W. (2003) Anticipation of epileptic seizures from standard EEG recordings. *Lancet* 361, 970–971.

Decroly, O., Goldbeter, A. (1987) From simple to complex oscillatory behaviour: analysis of busting in a multiple regulated biochemical system. *J. Theor. Biol.* 113, 649–671.

De Curtis, M., Manfridi, A., Biella, G. (1998) Activity-dependent pH shifts and periodic recurrence of spontaneous interictal spikes in a model of focal epileptogenesis. *J. Neurosci.* 18, 7543–7551.

DeGiorgio, C.M., Shewmon, A., Murray, D., Whitehurst, T. (2006) Pilot study of trigeminal nerve stimulation (TNS) for epilepsy: a proof-of-concept trial. *Epilepsia* 47, 1213–1215.

Dehaene, S. (2007) A few steps toward a science of mental life. *Mind Brain Educ.* 1(1), 28–47.

Dejean, C., Gross, C.E., Bioulac, B., Boraud, T. (2008) Dynamic changes in the cortex-basal ganglia network after dopamine depletion in the rat. *J. Neurophysiol.* 100, 385–396.

Del Negro, C.A., Wilson, C.G., Butera, R.J., Rigatto, H., Smith, J.C. (2002) Periodicity, mixed mode oscillations and quasiperiodicity in a rhythm-generating neural network. *Biophys. J.* 82, 206–214.

Dement, W., Kleitman, N. (1957) Cyclic variations in EEG during sleep and their relation to eye movements, body motility, and dreaming. *Electroencephalogr. Clin. Neurophysiol.* 9, 673–690.

Deppisch, J., Pawelzik, K., Geisel, T. (1994) Uncovering the synchronization dynamics from correlated neuronal activity quantifies assembly formation. *Biol. Cybern.* 71, 387–399.

Deremble, B., D'Andrea, F., Ghil, M. (2009) Fixed points, stable manifolds, weather regimes, and their predictability. *Chaos* 19, 043109.

Deschenes, M., Paradis, M., Roy, J.P., Steriade, M. (1984) Electrophysiology of neurons of lateral thalamic nuclei in cat: resting properties and burst discharges. *J. Neurophysiol.* 51, 1196–1218.

Deshpande, G., Kerssens, C., Sebel, P.S., Hu, X. (2010) Altered local coherence in the default mode network due to sevoflurane anesthesia. *Brain Res.* 1318, 110–121.

Desmurget, M., Sirigu, A. (2009) A parietal-premotor network for movement intention and motor awareness. *Trends Cogn. Sci.* 13, 411–419.

Destexhe, A. (2009) Self-sustained asynchronous irregular states and up-down states in thalamic, cortical and thalamocortical networks of nonlinear integrate-and-fire neurons. *J. Comput. Neurosci.* 27, 493–506.

Destexhe, A., Contreras, D., Steriade, M. (1999) Spatiotemporal analysis of local field potentials and unit discharges in cat cerebral cortex during natural wake and sleep stages. *J. Neurosci.* 19, 4595–4608.

Destexhe, A., Rudolph, M., Fellous, J.-M., Sejnowski, T.J. (2001) Fluctuating synaptic conductances recreate *in vivo*-like activity in neocortical neurons. *Neuroscience* 107, 13–24.

Destexhe, A., Sejnowski, T.J. (2001) *Thalamocortical Assemblies*, Oxford University Press.

Dichter, M., Spencer, W.A. (1968) Hippocampal penicillin 'spike' discharge: epileptic neuron or epileptic aggregate? *Neurology* 18, 282.

Dierks, T., Linden, D.E., Jandl, M., Formisano, E., Goebel, R., Lanfermann, H., Singer, W. (1999) Activation of Hershl's gyrus during auditory hallucinations. *Neuron* 22, 615–621.

Diesmann, M., Gewaltig, M.O., Aertsen, A. (1999) Stable propagation of synchronous spiking in cortical neural networks. *Nature* 402, 529–532.

Diniz Behn, C.G., Brown, E.N., Scammell, T.E., Kopell, N.J. (2007) Mathematical model of network dynamics governing mouse sleep-wake behavior. *J. Neurophysiol.* 97, 3828–3840.

Dostrovsky, J., Levy, R., Wu, J.P., Hutchison, W.D., Tasker, R.R., Lozano, A.M. (2000) Microstimulation-induced inhibition of neuronal firing in human globus pallidus. *J. Neurophysiol.* 84, 570–574.

Dubois, M., Rubio, M.A., Bergé, P. (1983) Experimental evidence of intermittencies associated with a subharmonic bifurcation. *Phys. Rev. Lett.* 51, 1446–1449.

Ducker, T.B., Thatcher, R.W., Cantor, D.L., Meyer, W., McAlater, R. (1982) Comprehensive assessment of coma in neurotrauma patients, in: *32nd Annual International Congress of Neurological Surgeons*, Toronto, Canada.

Duckrow, R.B., Zaveri, H.P. (2005) Coherence of the electroencephalogram during the first sleep cycle. *Clin. Neurophysiol.* 116, 1088–1095.

Durand, D.M., Bikson, M. (2001) Suppression and control of epileptiform activity by electrical stimulation: a review. *Proc. IEEE* 89, 1065–1081.

Dustman, R., Beck, E. (1965) Phase of alpha brain waves, reaction time and visually evoked potentials. *EEG Clin. Neurophysiol.* 19, 570–575.

Ebersole, J.S. (2005) In search of seizure prediction: a critique. *Clin. Neurophysiol.* 116, 489–492.

Edwards, B.G., Barch, D.M., Braver, T.S. (2010) Improving prefrontal cortex function in schizophrenia through focused training of cognitive control. *Frontiers Human Neurosci.* 4, 32; doi: 10.3389/fnhum.2010.00032.

Einstein, A. (1936) Physics and reality. *J. Franklin Institute* 221(3), 313–347.

El Boustani, S., Destexhe, A. (2010) Brain dynamics at multiple scales: can one reconcile the apparent low-dimensional chaos of macroscopic variables with the seemingly stochastic behavior of single neurons. *Int. J. Bif. Chaos* 20, 1687–1702.

Elger, C.E., Lehnertz, K. (1998) Seizure prediction by non-linear time series analysis of brain electrical activity. *Eur. J. Neurosci.* 10, 786–789.

Engel, A.K., König, P., Gray, C.M., Singer, W. (1990) Stimulus-dependent neuronal oscillations in cat visual cortex: inter-columnar interaction as determined by cross-correlation analysis. *Eur. J. Neurosci.* 2, 588–606.

Engel, J., Wilson, C., Bragin, A. (2003) Advances in understanding the process of epileptogenesis based on patient material: what can the patient tell us? *Epilepsia* 44 (Suppl. 12), 60–71.

Ermentrout, G.B., Galán, R.F., Urban, N.N. (2008) Reliability, synchrony and noise. *Trends Neurosci.* 31, 428–434.

Fanselow, E.E., Reid, A.P., Nicolelis, M.A.L. (2000) Reduction of pentylenetetrazole-induced seizure activity in awake rats by seizure triggered trigeminal nerve stimulation. *J. Neurosci.* 20, 8160–8168.

Fatt, P., Katz, B. (1950) Some observations on biological noise. *Nature* 166, 597–598.

Feigenbaum, M.J. (1983) Universal behaviour in nonlinear systems. *Physica D* 7, 16–39.

Fein, G., Raz, J., Brown, F.F., Merrin, E.L. (1988) Common reference coherence data are confounded by power and phase effects. *Electroencephalogr. Clin. Neurophysiol.* 69, 581–584.

Fell, J., Röschke, J. (1994) Nonlinear dynamical aspects of the human sleep EEG. *Int. J. Neurosci.* 76, 109–129.

Fellous, J.M., Rudolph, M., Destexhe, A., Sejnowski, T.J. (2003) Synaptic background noise controls the input/output characteristics of single cells in an *in vitro* models of *in vivo* activity. *Neuroscience* 122, 811–829.

Ferri, R., Rundo, F., Bruni, O., Terzano, M.G., Stam, C.J. (2005) Dynamics of the EEG slow-wave synchronization during sleep. *Clin. Neurophysiol.* 116, 2783–2795.

Ferri, R., Rundo, F., Bruni, O., Terzano, M.G., Stam, C.J. (2006) Regional scalp EEG slow-wave synchronization during sleep cycle alternating pattern A1 subtypes. *Neurosci. Lett.* 404, 352–357.

Filer, M., Giladi, N. Gruendlinger, L., Yogev-Seligmann, G., Hausdorff, J.M. (2008) Non-rhythmic auditory stimulation prior to walking may improve gait dynamics in patients with moderate Parkinson's disease. Proceedings of the International Congress on Gait and Mental Function.

Fingelkurts, A.A., Fingelkurts, A.A. (2004) Making complexity simpler: multivariability and metastability in the brain. *Int. J. Neurosci.* 114, 843–862.

Fingelkurts, A.A., Fingelkurts, A.A., Rytsäläa, H., Suominen, K., Isometsä, E., Kähkönen, S. (2007) Impaired functional connectivity at EEG alpha and theta frequency bands in major depression. *Hum. Brain Mapp.* 28, 247–261.

Fingelkurts, A.A., Fingelkurts, A.A., Neves, C.F.H. (2010) Natural world physical, brain operational, and mind phenomenal space-time. *Phys. Life Rev.* 7, 195–249.

Fisher, R.S., Webber, W.R., Lesser, R.P., Arroyo, S., Uematsu, S. (1992) High-frequency EEG activity at the start of seizures. *J. Clin. Neurophysiol.* 9, 441–448.

Flor-Henry, P., Lang, R.A., Koles, Z.J., Frenzel, R.R. (1991) Quantitative EEG studies of pedophilia. *Int. J. Psychophysiol.* 10, 253–258.

Florence, G., Dahlem, M.A., Almeida, A.C., Bassani, J.W., Kurths, J. (2009) The role of extracellular potassium dynamics in the different stages of ictal bursting and spreading depression: a computational study. *J. Theor. Biol.* 258, 219–228.

Forster, F.M. (1977) *Reflex Epilepsy, Behavioural Therapy and Conditional Reflexes.* C. C. Thomas, Springfield, USA.

Franceschini, V. (1983) Bifurcations of tori and phase locking in a dissipative system of differential equations. *Physica* 6D, 285–304.

Frangou, S. (2008) Cognitive and emotional control in the major psychoses. *A quem e Alem do Cerebro* (Behind and Beyond the Brain), 7° Simposio da Fundação Bial, Porto, pp. 195–211.

Frank, T.D. (2004) Stochastic feedback, nonlinear families of Markov processes, and nonlinear Fokker-Planck equations. *Physica A* 331, 391–408.

Frank T.D. (2005) *Nonlinear Fokker-Planck Equations.* Springer-Verlag, Berlin.

Frank, T.D., Daffertshofer, A., Peper, C.E., Beek, P.J., Haken, H. (2000) Towards a comprehensive theory of brain activity: coupled oscillator systems under external forces. *Physica D* 144, 62–86.

Frank, T.D., Beek, P.J. (2003) in: Tschacher, W., Dauwalder, J.P. (eds.) *The Dynamical Systems Approach to Cognition*, World Scientific, Singapore, pp. 159–179.

Frank, T.D., Richardson, M.J., Lopresti-Goodman, S.M., Turvey, M.T. (2009) Order parameter dynamics of body-scaled hysteresis and mode transitions in grasping behavior. *J. Biol. Phys.* 35, 127–147.

Frantseva, M.V., Perez Velazquez, J.L., Carlen, P.L. (1998) Changes in membrane and synaptic properties of thalamocortical circuits caused by hydrogen peroxide. *J. Neurophysiol.* 80, 1317–1326.

Frantseva, M.V., Perez Velazquez, J.L., Tsoraklidis G., Mendonca, A.J., Adamchik, Y., Mills, L.R., Carlen, P.L., Burnham, W.M. (2000) Oxidative stress is involved in seizure-induced neurodegeneration in the kindling model of epilepsy. *Neuroscience* 97, 431–435.

Freeman, W.J. (1995) *Societies of Brains.* Lawrence Erlbaum Associates, Hove, U.K.

Freeman, W.J. (2003) Evidence from human scalp electroencephalograms of global chaotic itinerancy. *Chaos* 13, 1067–1077.

Freeman, W.J. (2007a) Definitions of state variables and state space for brain-computer interface. Part 1. Multiple hierarchical levels of brain function. *Cogn. Neurodyn.* 1, 3–14.

Freeman, W.J. (2007b) Definitions of state variables and state space for brain-computer interface. Part 2. Extraction and classification of feature vectors. *Cogn. Neurodyn.* 1, 85–96.

Freeman, W.J., Barrie, J.M. (1994) Chaotic oscillations and the genesis of meaning in cerebral cortex, in: *Temporal Coding in the Brain*, Springer, Heidelberg.

Freyer, F., Aquino, K., Robinson, P.A., Ritter, P., Breakspear, M. (2009) Bistability and non-Gaussian fluctuations in spontaneous cortical activity. *J. Neurosci.* 29, 8512–8524.

Friedrich, R., Uhl, C. (1996) Spatio-temporal analysis of human electroencephalograms: Petit-mal epilepsy. *Physica D* 98, 171–182.

Fries, P. (2005) A mechanism for cognitive dynamics: neuronal communication through neuronal coherence. *Trends Cogn. Sci.* 9, 474–480.

Friston, K.J. (2001) Brain function, nonlinear coupling, and neuronal transients. *Neuroscientist* 7, 406–418.

Frith, U. (1989) *Autism: Explaining the Enigma.* Oxford University Press.

Fuhrmann, G., Markram, H., Tsodyks, M. (2002) Spike frequency adaptation and neocortical rhythms, *J. Neurophysiol.* 88, 761–770.

Fuster, J.M. (1995) *Memory in the Cerebral Cortex.* MIT Press, Cambridge.

Fuster, J.M. (2003) *Cortex and Mind.* Oxford University Press.

Galán, R.F., Ermentrout, G.B., Urban, N.N. (2007) Stochastic dynamics of uncoupled neural oscillators: Fokker-Planck studies with the finite element method. *Phys. Rev. E* 76, 056110.

Gameiro, M., Mischaikow, K., Kalies, W. (2004) Topological characterization of spatio-temporal chaos. *Phys. Rev. E* 70, 035203.

Gao, Y., Raine, A. (2010) Successful and unsuccessful psychopaths: a neurobiological model. *Behav. Sci. Law* 28, 194–210.

Garcia Dominguez, L., Kostelecki, W., Wennberg, R., Perez Velazquez, J.L. (2011) Distinct dynamical patterns that distinguish willed and forced actions. *Cogn. Neurodyn.* 5, 67–76.

García Dominguez, L., Wennberg, R., Gaetz, W., Cheyne, D., Snead, O.C., Perez Velazquez, J.L. (2005) Enhanced synchrony in epileptiform activity? Local versus distant phase synchronization in generalized seizures. *J. Neurosci.* 25, 8077–8084.

García Dominguez, L., Wennberg, R., Perez Velazquez, J.L., Guevara Erra, R. (2007) Enhanced measured synchronization of unsynchronized sources: inspecting the physiological significance of synchronization analysis of whole brain electrophysiological recordings. *Int. J. Phys. Sci.* 2(11), 305–317.

García Domínguez, L., Guevara Erra, R., Wennberg, R., Perez Velazquez, J.L. (2008) On the spatial organization of epileptiform activity. *Int. J. Bif. Chaos* 18, 429–439.

Gastaut, H., Broughton, R. (1972) *Epileptic Seizures: Clinical and Electrographic Features, Diagnosis and Treatments.* Charles C. Thomas, Springfield, pp. 45–47.

Gastaut, H., Zifkin, B.G. (1988) Secondary bilateral synchrony and Lennox-Gastaut syndrome, in: *The Lennox-Gastaut Syndrome*, Liss, New York, pp. 221–242.

Gauthier, D.J., Hall, G.M., Oliver, R.A., Dixon-Tulloch, E.G., Wolf, P.D., Bahar, S. (2002) Progress toward controlling *in vivo* fibrillating sheep atria using a nonlinear dynamics-based closed-loop feedback method. *Chaos* 12, 952–962.

Gerstein, G.L., Perkel, D.H. (1972) Mutual temporal relationships among neuronal spike trains. *Biophys. J.* 12, 453–473.

Getting, P.A. (1989) Emerging principles governing the operation of neural networks. *Annu. Rev. Neurosci.* 12, 185–204.

Geurts, H.M., Corbett, B., Solomon, M. (2009) The paradox of cognitive flexibility in autism. *Trends Cogn. Sci.* 13, 74–82.

Geylin, H.R. (1921) Fasting as a method for treating epilepsy. *Med. Rec.* 99, 1037–1039.

Ghasemi, F., Sahimi, M., Peinke, J., Reza Rahimi Tabar, M. (2006) Analysis of non-stationary data for heart-rate fluctuations in terms of drift and diffusion coefficients. *J. Biol. Phys.* 32, 117–128.

Gibson, J.F., Farmer, J.D., Casdagli, M., Eubank, S. (1992) An analytic approach to practical state space reconstruction. *Physica D* 57, 1–30.

Gilbert, S.J., Meuwese, J.D.I., Towgood, K.J., Frith, C.D., Burgess, P.W. (2009) Abnormal functional specialization within medial prefrontal cortex in high-functioning autism: a multi-voxel similarity analysis. *Brain* 132, 869–878.

Glass, L., Mackey, M. (1979) Pathological conditions resulting from instabilities in physiological control systems. *Ann. N. Y. Acad. Sci.* 316, 214–235.

Glass, M., Dragunow, M. (1995) Neurochemical and morphological changes associated with human epilepsy. *Brain Res. Rev.* 21, 29–41.

Gloor, P. (1986) Consciousness as a neurological concept in epileptology: a critical review. *Epilepsia* 27(Suppl. 2), S14–S26.

Gloor, P. (1997) *The Temporal Lobe and Limbic System.* Oxford University Press, New York.

Gloor, P., Ball, G., Schaul, N. (1977) Brain lesions that produce delta waves in the EEG. *Neurology* 27, 326–333.

Gloor, P., Fariello, R.G. (1988) Generalized epilepsy: some of its cellular mechanisms differ from those of focal epilepsy. *Trends Neurosci.* 11, 63–68.

Gluckman, B.J., Neel, E.J., Netoff, T.I., Ditto, W.L., Spano, M.L., Schiff, S.J. (1996) Electric field suppression of epileptiform activity in hippocampal slices. *J. Neurophysiol.* 76, 4202–4205.

Gluckman, B.J., Nguyen, H., Weinstein, S.L., Schiff, S.J. (2001) Adaptive electric field control of epileptic seizures. *J. Neurosci.* 21, 590–600.

Goldbeter, A. (1996) *Biochemical Oscillations and Cellular Rhythms: The Molecular Bases of Periodic and Chaotic Behaviour.* Cambridge University Press, Cambridge.

Goldman-Rakic, P.S. (1999) The physiological approach: functional architecture of working memory and disordered cognition in schizophrenia. *Biol. Psychiatry* 46, 650–661.

Gomez Mancilla, B., Latulippe, J.F., Boucher, R., Bédard, P.J. (1992) Effect of ethosuximide on rest tremor in the MPTP monkey model. *Mov. Disord.* 7, 137–141.

Gordon, N. (1997) The Landau-Kleffner syndrome: increased understanding. *Brain Dev.* 19, 311–316.

Gotman, J. (1981) Interhemispheric relations during bilateral spike and wave activity. *Epilepsia* 22, 453–466.

Gotman, J. (1983) Measurement of small time differences between EEG channels: method and application to epileptic seizure propagation. *Electroencephalogr. Clin. Neurophysiol.* 56, 501–514.

Goto, Y., O'Donnell, P. (2001) Network synchrony in the nucleus accumbens in vivo. *J. Neurosci.* 21, 4498–4504.

Gouesbet, G., Maquet, J. (1992) Construction of phenomenological models from numerical scalar time series. *Physica D* 58, 202–215.

Goutagny, R., Jackson, J., Williams, S. (2009) Self-generated theta oscillations in the hippocampus. *Nat. Neurosci.* 12, 1491–1493.

Granic, I., Hollenstein, T. (2003) Dynamic systems methods for models of developmental psychopathology. *Dev. Psychopathol.* 15, 641–669.

Greenfield, S.A., Collins, T.F.T. (2005) A neuroscientific approach to consciousness. *Prog. Brain Res.* 150, 11–23.

Greicius, M.D., Flores, B.H., Menon, V., Glover, G.H., Solvason, H.B., Kenna, H., Reiss, A.L., Schatzberg, A.F. (2007) Resting-state functional connectivity in major depression: abnormally increased contributions from subgenual cingulated cortex and thalamus. *Biol. Psychiatry* 62, 429–437.

Grice, S.J., Spratling, M.W., Karmiloff-Smith, A., Halit, H., Csibra, G., de Haan, M., Johnson, M.H., (2001) Disordered visual processing and oscillatory brain activity in autism and Williams syndrome. *NeuroReport* 12, 2697–2700.

Griniasty, M., Tsodyks, M.V., Amit, D.J. (1999) Conversion of temporal correlations between stimuli to spatial correlations between attractors, in: *Neural Codes and Distributed Representations*, MIT Press.

Grinter, E.J., Maybery, M.T., Pellicano, E., Badcock, J.C., Badcock, D.R. (2010) Perception of shapes targeting local and global processes in autism spectrum disorders. *J. Child Psychol. Psychiatry* 51, 717–724.

Grisar, T.M. (1986) Neuron-glia relationships in human and experimental epilepsy: a biochemical point of view. *Adv. Neurol.* 44, 1045–1073.

Gross, J., Timmermann, L., Kujala, J., Dirks, M., Schmitz, F., Salmelin, R., Schnitzler, A. (2002) The neural basis of intermittent motor control in humans. *Proc. Natl. Acad. Sci. USA* 99, 2299–2302.

Guckenheimer, J., Holmes, P. (1983) *Nonlinear Oscillations, Dynamical Systems, and Bifurcations of Vector Fields*. Springer-Verlag.

Guevara Erra, R., Pérez Velazquez, J.L., Nenadovic, V., Wennberg, R., Senjanovic, G., García Dominguez, L. (2005) Phase synchronization measurements using electroencephalographic recordings: what can we really say about neuronal synchrony? *Neuroinformatics* 3, 301–313.

Hajos, M., Hurst, R.S., Hoffmann, W.E., Krause, M., Wall, T.M., Higdon, N.R., Groppi, V.E. (2005) The selective $\alpha 7$ nicotinic acetylcholine receptor agonist PNU-282987 enhances GABAergic synaptic activity in brain slices and restores auditory gating deficits in anesthetized rats. *J. Pharm. Exp. Ther.* 312, 1213–1222.

Haken, H., Kelso, J.A.S., Bunz, H. (1985) A theoretical model of phase transitions in human hand movements. *Biol. Cybern.* 51, 347–356.

Haken, H. (2002) *Brain Dynamics*. Springer-Verlag, Berlin.

Haken, H. (2006) *Information and Self-Organization*, 3rd Edition, Springer, Berlin.

Haken, H. (2007) Towards a unifying model of neural net activity in the visual cortex. *Cogn. Neurodyn.* 1, 15–25.

Hall, K., Christini, D.J., Tremblay, M., Collins, J.J., Glass, L., Billette, J. (1997) Dynamic control of cardiac alternans. *Phys. Rev. Lett.* 78, 4518–4521.

Hallett, M. (2007) Volitional control of movement: the physiology of free will. *Clin. Neurophysiol.* 118, 1179–1192.

Happe, F.G.E., Frith, U. (2006) The weak coherence account: detail-focused cognitive style in autism spectrum disorders. *J. Autism Dev. Disord.* 36, 5–25.

Harris, K.D. (2005) Neural signatures of cell assembly organization. *Nature Rev. Neurosci.* 6, 399–407.

Harrison, M.A., Osorio, I., Frei, M.G., Asuri, S., Lai, Y.C. (2005) Correlation dimension and integral do not predict epileptic seizures. *Chaos* 15, 33106.

Hassler, R., Dalle Ore, G., Bricolo, A., Diecjmann, G., Dolce, G. (1969) Behavioral and EEG arousal induced by stimulation of unspecific projection systems in a patient with post-traumatic appallic syndrome. *Electroencephalogr. Clin. Neurophysiol.* 27, 306–310.

Hauptmann, C., Roulet, J.-C., Niederhauser, J.J., Döll, W., Kirlangic, M.E., Lysyansky, B., Krachkovskyi, V., Bhatti, M.A., Barnikol, U.B., Sasse, L., Bührle, C.P., Speckmann, E.-J., Götz, M., Sturm, V., Freund, H.-J., Schnell, U., Tass, P.A. (2009) External trail deep brain stimulation device for the application of desynchronising stimulation techniques. *J. Neural Eng.* 6, 066003.

Hausdorff, J.M. (2009) Gait dynamics in Parkinson's disease: common and distinct behavior among stride length, gait variability, and fractal-like scaling. *Chaos* 19, 026113.

Hauser, W.A., Hesdorffer, D.C. 1990. *Epilepsy: Frequency, Causes and Consequences.* Demos Publications, New York.

Haut, S.R., Shinnar, S., Moshé, S.L., O'Dell, C., Legatt, A.D. (1999) The association between seizure clustering and convulsive status epilepticus in patients with intractable complex partial seizures. *Epilepsia* 40, 1832–1834.

Henry, T.R. (2002) Therapeutic mechanisms of vagus nerve stimulation. *Neurology* 59 (Suppl. 4), S3–S14.

Herbert, M.R. (2005) Large brains in autism: the challenge of pervasive abnormality. *Neuroscientist* 11, 417–437.

Hernandez, J.L., Valdes, P.A., Vila, P. (1996) EEG spike and wave modelled by a stochastic limit cycle. *Neuroreport* 7, 2246–2250.

Hershkowitz, N. Dretchen, K.L., Raines, A. (1978) Carbamazepine suppression of posttetanic potentiation at the neuromuscular junction. *J. Pharmacol. Exp. Ther.* 207, 810–816.

Hershman, K.M., Freedman, R., Bickford, P.C. (1995) $GABA_B$ antagonists diminish the inhibitory gating of auditory response in the rat hippocampus. *Neurosci. Lett.* 190(2), 133–136.

Higashima, M., Kinoshita, H., Yamaguchi, N., Koshino, Y. (1996) Activation of GABAergic function necessary for afterdischarge generation in rat hippocampal slices. *Neurosci. Lett.* 207, 101–104.

Hill, E.L. (2004) Executive dysfunction in autism. *Trends Cogn. Sci.* 8, 26–32.

Hobson, A. (2004) A model for madness? *Nature* 430, 21.

Hoffman, R.E., McGlashan, T.H. (1997) Synaptic elimination, neurodevelopment, and the mechanisms of hallucinated "voices" in schizophrenia. *Am. J. Psychiatry* 154, 1683–1689.

Hodaie, M., Wennberg, R., Dostrovsky, J.O., Lozano, A.M. (2002) Chronic anterior thalamus stimulation for intractable epilepsy. *Epilepsia* 43, 603–608.

Holmes, G.L., McKeever, M., Saunders, Z. (1981) Epileptiform activity in aphasia of childhood: an epiphenomenon? *Epilepsia* 22, 631–639.

Hopfield, J.J. (1982) Neural networks and physical systems with emergent collective computational abilities. *Proc. Natl. Acad. Sci. USA* 79, 2554–2558.

Hoppensteadt, F.C., Izhikevich, E.M. (1997) *Weakly Connected Neural Networks*. Springer.

Horvitz, J.C. (2002) Dopamine gating of glutamatergic sensorimotor and incentive motivational input signals to the striatum. *Behav. Brain Res.* 137, 65–74.

Howes, O.D., Kapur, S. (2009) The dopamine hypothesis of schizophrenia: version III — the final common pathway. *Schizophr. Bull.* 35, 549–562.

Hramov, A., Koronovskii, A.A., Midzyanovskaya, I.S., Sitnikova, E., van Rijn, C.M. (2006) On-off intermittency in time series of spontaneous paroxysmal activity in rats with genetic absence epilepsy. *Chaos* 16, 043111.

Hubel, D.H., Wiesel, T.N. (1963) Shape and arrangement of columns in cat's striate cortex. *J. Physiol. (Lond.)* 165, 559–568.

Huerta, R., Nowotny, T., Garcia Sanchez, M., Abarbanel, H.D., Rabinovich, M.I. (2004) Learning classification in the olfactory system of insects. *Neural Comput.* 16, 1601–1640.

Hughes, J.R. (2007) Review of medical reports on pedophilia. *Clin. Pediatr.* 46, 667–682.

Hunt, H.W., Antle, J.M., Paustian, K. (2003) False determination of chaos in short noisy time series. *Physica D* 180, 115–127.

Iasemidis, L.D., Sackellares, J.C. (1996) Chaos theory and epilepsy. *Neuroscientist* 2, 118–126.

Iasemidis, L.D., Olson, L.D., Savit, R.S., Sackellares, J.C. (1994) Time dependencies in the occurrences of epileptic seizures. *Epilepsy Res.* 17, 81–94.

Iliadi, K.G. (2009) The genetic basis of emotional behavior: has the time come for a Drosophila model? *J. Neurogenet.* 23, 136–146.

Imas, O.A., Ropella, K.M., Douglas Ward, B., Wood, J.D., Hudetz, A.G. (2005) Volatile anesthetics disrupt frontal-posterior recurrent information transfer at gamma frequencies in rat. *Neurosci. Lett.* 387, 145–150.

Imas, O.A., Ropella, K.M., Wood, J.D., Hudetz, A.G. (2006) Isoflurane disrupts anterior-posterior phase synchronization of flash-induced field potentials in the rat. *Neurosci. Lett.* 402, 216–221.

Issa, E.B., Wang, X. (2008) Sensory responses during sleep in primate primary and secondary auditory cortex. *J. Neurosci.* 28, 14467–14480.

Izhikevich, E.M., Gally, J.A., Edelman, G.M. (2004) Spike-timing dynamics of neuronal groups. *Cereb. Cortex* 14, 933–944.

Jackson, P.L., Brunet, E., Meltzoff, A.N., Decety, J. (2006) Empathy examined through the natural mechanisms involved in imaging how I feel versus how you feel pain. *Neuropsychologia* 44(5), 752–761.

Jahnsen, H., Llinás, R. (1984) Electrophysiological properties of guinea-pig thalamic neurones: an *in vitro* study. *J. Physiol. Lond.* 349, 227–247.

Jarolimek, W., Misgeld, U., Lux, H.D. (1989) Activity dependent alkaline and acid transients in guinea pig hippocampal slices. *Brain Res.* 505, 225–232.

Jefferys, J.G.R. (1990) Basic mechanisms of focal epilepsies. *Exper. Physiol.* 75, 127–162.

Jeffries, C., Perez, J. (1982) Observation of a Pomeau-Manneville intermittent route to chaos in a nonlinear oscillator. *Phys. Rev. A* 26, 2117–2122.

Jensen, M.S., Yaari, Y. (1997) Role of intrinsic burst firing, potassium accumulation, and electrical coupling in the elevated potassium model of hippocampal epilepsy. *J. Neurophysiol.* 77, 1224–1233.

Jenssen, S., Gracely, E.J., Sperling, M.R. (2006) How long do most seizures last? A systematic comparison of seizures recorded in the epilepsy monitoring unit. *Epilepsia* 47, 1499–1503.

Jerger, K., Schiff, S.J. (1995) Periodic pacing an *in vitro* epileptic focus. *J. Neurophysiol.* 73, 876–879.

Jerger, K., Netoff, T.I., Francis, J.T., Sauer, T., Pecora, L., Weinstein, S.L., Schiff, S.J. (2001) Early seizure detection. *J. Clin. Neurophysiol.* 18, 259–268.

Jia, W., Kong, N., Li, F., Gao, X., Gao, S., Zhang, G., Wang, Y., Yang, F. (2005). An epileptic seizure prediction algorithm based on second-order measure. *Physiol. Meas.* 26, 609–625.

Jing, H., Takigawa, M. (2000) Comparison of human ictal, interictal and normal non-linear component analyses. *Clin. Neurophysiol.* 111, 1282–1292.

Jirsa, V.K., Fink, P., Foo, P., Kelso, J.A.S. (2000) Parametric stabilization of biological coordination: a theoretical model. *J. Biol. Phys.* 26, 85–112.

Johanson, M., Revonsuo, A., Chaplin, J., Wedlund, J.-E. (2003) Level and contents of consciousness in connection with partial epileptic seizures. *Epilepsy Behav.* 4, 279–285.

John, E.R., Prichep, L.S. (2005) The anesthetic cascade. *Anesthesiology* 102, 447–471.

Johnston, D., Brown, T.H. (1981) Giant synaptic potential hypothesis for epileptiform activity. *Science* 211, 294–297.

Jouny, C.C., Franaszczuk, P.J., Bergey, G.K. (2005) Signal complexity and synchrony of epileptic seizures: is there an identifiable preictal period? *Clin. Neurophysiol.* 116, 552–558.

Juhász, C., *et al.* (2009) Focal decreases of cortical $GABA_A$ receptor binding remote from the primary seizure focus: what do they indicate? *Epilepsia* 50, 240–250.

Jung, R. (1939) Uber Vegetative Reactionen und Hemmungswirkung von Sinnesreizen im Kleinen epileptischen Anfall. *Nervenartz* 12, 169–185.

Jus, A., Jus, K. (1962) Retrograde amnesia in petit mal. *Arch. Gen. Psychiatry* 6, 163–167.

Just, M.A., Cherkassky, V.L., Keller, T.A., Minshew, N.J. (2004) Cortical activation and synchronization during sentence comprehension in high-functioning autism: evidence of underconnectivity. *Brain* 127, 1811–1821.

Kalinowsky, L.B., Kennedy, F. (1943) Observations in electric shock therapy applied to problems of epilepsy. *J. Nerv. Ment. Dis.* 98, 56–57.

Kana, R.K., Keller, T.A., Minshew, N.J., Just, M.A. (2007) Inhibitory control in high-functioning autism: decreased activation and underconnectivity in inhibition networks. *Biol. Psychiatry* 62, 198–206.

Kaneko, K., Tsuda, I. (2003) Chaotic itinerancy, *Chaos* 13, 926–936.

Kanner, L. (1943) Autistic disturbances of affective contact. *Nerv. Child.* 2, 217–250.

Kantz, H., Schreiber, T. (1997) *Nonlinear Time Series Analysis*. Cambridge University Press.

Kapur, S. (2003) Psychosis as a state of aberrant salience: a framework linking biology, phenomenology, and pharmacology of schizophrenia. *Am. J. Psychiatry* 160(1), 13–25.

Kaye, W.H., Fudge, J.L., Paulus, M. (2009) New insights into symptoms and neurocircuit function of anorexia nervosa. *Nat. Rev. Neurosci.* 10, 573–584.

Kelso, J.A.S. (1981) On the oscillatory basis of movement. *Bull. Psychon. Soc.* 18, 63.

Kelso, J.A.S. (1984) Phase transitions and critical behaviour in human bimanual coordination. *Am. J. Physiol.* 15, R1000–R1004.

Kelso, J.A.S. (1995) *Dynamic Patterns: The Self-Organization of Brain and Behaviour*. MIT Press, Cambridge.

Kelso, J.A.S. (2008) An essay on understanding the mind. *Ecol. Psychol.* 20, 180–208.

Kelso, J.A.S., Scholz, J.P., Schöner, G. (1986) Nonequilibrium phase transitions in coordinated biological motion: critical fluctuations. *Phys. Lett. A* 118, 279–284.

Kelso, J.A.S., Fuchs, A. (1995) Self-organizing dynamics of the human brain: critical instabilitites and Šil'nikov chaos. *Chaos* 5, 64–69.

Kelso, J.A.S., Engstrøm, D.A. (2006) *The Complementary Nature*. MIT Press, Cambridge.

Kelso, J.A.S., Tognoli, E. (2007) Toward a complementary neuroscience: metastable coordination dynamics of the brain, in: Perlovsky, L.I., Kozma, R. (eds.) *Neurodynamics of Cognition and Consciousness*, Springer-Verlag, Berlin-Heidelberg.

Kenet, T., Bibitchkov, D., Tsodyks, M., Grinvald, A., Arieli, A. (2003) Spontaneously emerging cortical representations of visual attributes. *Nature* 425, 954–956.

Kestler, J., Kopelowitz, E., Kanter, I., Kinzel, W. (2008) Patterns of chaos synchronization. *Phys. Rev. E* 77, 046209.

Khalilov, I., Holmes, G.L., Ben-Ari, Y. (2003) *In vitro* formation of a secondary epileptogenic mirror focus by interhippocampal propagation of seizures. *Nat. Neurosci.* 6, 1079–1083.

Khosravani, H., Carlen, P.L., Perez Velazquez, J.L. (2003) The control of seizure-like activity in the rat hippocampal slice. *Biophys. J.* 84, 1–9.

Kim, J.-M., Jung, K.-Y., Choi, C.-M. (2002). Changes in brain complexity during valproate treatment in patients with partial epilepsy. *Neuropsychobiology* 45, 106–112.

Kim, J.W., Robinson, P.A. (2008) Controlling limit-cycle behaviors of brain activity. *Phys. Rev. E* 77, 051914.

Kimpo, R.R., Theunissen, F.E., Doupe, A.J. (2003) Propagation of correlated activity through multiple stages of a neural circuit. *J. Neurosci.* 23, 5750–5761.

Kiss, I.Z., Hudson, J.L. (2001) Phase synchronization and suppression of chaos through intermittency in forcing of an electrochemical oscillator. *Phys. Rev. E* 64, 1–8.

Kiviniemi, V.J., Haanpää, H., Kantola, J.H., Jauhiainen, J., Vainionpää, V., Alahuhta, S., Tervonen, O. (2005) Midazolam sedation increases fluctuation and synchrony of the resting brain BOLD signal. *Magn. Reson. Imaging* 23, 531–537.

Koerner, E. (1996) Comparative reduction of theories — or over-simplification? *Behav. Brain Sci.* 19, 301–302.

Kokarovtseva, L., Jaciw-Zurakiwsky, T., Arbocco, R.M., Frantseva, M.V., Perez Velazquez, J.L. (2009) Excitability and gap junction-mediated mechanisms in nucleus accumbens regulate self-stimulation reward in rats. *Neuroscience* 159, 1257–1263.

Kolmogorov, A.N. (1957) General theory of dynamical systems and classical mechanics, in: *Proceedings of the* 1954 *International Congress of Mathematics*, North Holland, Amsterdam.

Komiyama, T., Sato, T.R., O'Connor, D.H., Zhang, Y.-X., Huber, D., Hooks, B.M., Gabitto, M., Svoboda, K. (2010) Learning-related fine-scale specificity imaged in motor cortex circuits of behaving mice. *Nature* 464, 1182–1186.

Koralov, L.B., Sinai, Y.G. (2007) *Theory of Probability and Random Processes*. Springer-Verlag, Berlin.

Kowalik, Z.J. (2001) The noise of chaos. *Behav. Brain Sci.* 24, 820.

Kringelbach, M.L., Jenkinson, N., Owen, S.L.F., Aziz, T.Z. (2007) Translational principles of deep brain stimulation. *Nat. Rev. Neurosci.* 8, 623–635.

Kringelbach, M.L., Berridge, K.C. (2009) Towards a functional neuroanatomy of pleasure and happiness. *Trends Cogn. Sci.* 13, 479–487.

Krupa, M. (1997) Robust heteroclinic cycles. *J. Nonlinear Sci.* 7, 129–176.

Kryukov, V.I., Borisyuk, G.N., Borisyuk, R.M., Kirillov, A.B., Kovalenko, Y.I. (1990) Metastable and unstable states in the brain, in: *Stochastic Cellular Systems: Ergodicity, Memory, Morphogenesis*, Manchester University Press.

Kuizenga, K., Wierda, J.M., Kalkman, C.J. (2001) Biphasic EEG changes in relation to loss of consciousness during induction with thiopental, propofol, etomidate, midazolam or sevoflurane. *Br. J. Anaesth.* 86, 354–360.

Kulisek, R., Hrncir, Z., Hrdlicka, M., Faladov, L., Sterbova, K., Krsek, P., Vymlatilova, E., Palus, M., Zumrová, A., Komárek, V. (2008) Nonlinear analysis of the sleep EEG in children with pervasive developmental disorder. *Neuro Endocrinol. Lett.* 29, 512–517.

Kuncel, A.M., Grill, W.M. (2004) Selection of stimulus parameters for deep brain stimulation. *Clin. Neurophysiol.* 115, 2431–2441.

Kwon, J.C., O'Donnell, B.F., Wallenstein, G.V., Greene, R.W., Hirayasu, Y., Nestor, P.G., Hasselmo, M.E., Potts, G.F., Shenton, M.E., McCarley, R.W. (1999) Gamma frequency-range abnormalities to auditory stimulation in schizophrenia. *Arch. Gen. Psychiatry* 56, 1001–1005.

Kyriazi, H.T., Carvell, G.E., Brumberg, J.C., Simons, D.J. (1996) Quantitative effects of GABA and bicuculline methiodide on receptive field properties of neurons in real and simulated whisker barrels. *J. Neurophysiol.* 75, 547–560.

Lachaux, J.P., Rodriguez, E., Martinerie, J., Varela, F.J. (1999) Measuring phase synchrony in brain signals. *Hum. Brain Mapp.* 8, 194–208.

Lado, F.A., Moshé, S.L. (2008) How do seizures stop? *Epilepsia* 49, 1651–1664.

Lai, Y.-C., Grebogi, C., Yorke, J.A., Kan, I. (1993) How often are chaotic saddles nonhyperbolic? *Nonlinearity* 6, 779–797.

Lamme, V.A.F. (2006) Towards a true neural stance on consciousness. *Trends Cogn. Sci.* 10, 494–501.

Landau, W.M., Kleffner, F.R. (1957) Syndrome of acquired aphasia with convulsive disorder in childhood. *Neurology* 7, 523–530.

Landmesser, L.T., O'Donovan, M.J. (1984) Activation patterns of embryonic chick hind limb muscles recorded *in ovo* and in an isolated spinal cord preparation. *J. Physiol.* 347, 189–204.

Langguth, B., Eichhammer, P., Zowe, M., Marienhagen, J., Spiessl, H., Hajak, G. (2006) Neuronavigated transcranial magnetic stimulation and auditory hallucinations in schizophrenic patient: monitoring of neurobiological effects. *Schizophr. Res.* 84, 185–186.

Laruelle, M., Abi-Dargham, A., van Dyck, C.H., Gil, R., D'Souza, C.D., Erdos, J., McCance, E., Rosenblatt, W., Fingado, C., Zoghbi, S.S., Baldwin, R.M., Seibyl, J.R., Krystal, J.H., Charney, D.S., Innis, R.B. (1996) Single photon emission computerized tomography imaging of amphetamine-induced dopamine release in drug-free schizophrenia subjects. *Proc. Natl. Acad. Sci. USA* 93, 9235–9240.

Lasztóczi, B., Antal, K., Nyikos, L., Emri, Z., Kardos, J. (2004) High-frequency synaptic input contributes to seiure initiation in the low-[Mg^{2+}] model of epilepsy. *Eur. J. Neurosci.* 19, 1361–1372.

Laurent, G. (2000) What does 'understanding' mean? *Nat. Neurosci.* 3, 1211.

Laurent, G. (2002) Olfactory network dynamics and the coding of multidimensional signals. *Nat. Rev. Neurosci.* 3, 884–895.

Laureys, S. (2005) Death, unconsciousness and the brain. *Nat. Rev. Neurosci.* 6, 899–909.

Laureys, S., Lemaire, C., Maquet, P., Phillips, C., Franck, G. (1999) Cerebral metabolism during vegetative state and after recovery to consciousness. *J. Neurol. Neurosurg. Psychiatry* 67, 121–122.

Laureys, S., Owen, A.M., Schiff, N.D. (2004) Brain function in coma, vegetative state, and related disorders. *Lancet Neurol.* 3, 537–546.

Le Bon-Jego, M., Yuste, R. (2007) Persistently active, pacemaker-like neurons in neocortex. *Front. Neurosci.* 1, 123–129.

Lee, K.H., Williams, L.M., Breakspear, M., Gordon, E. (2003) Synchronous gamma activity: a review and contribution to an integrative neuroscience model of schizophrenia. *Brain Res. Rev.* 41(1), 57–78.

Lee, K.H., Hitti, F.L., Shalinsky, M.H., Kim, U., Leiter, J.C., Roberts, D.W. (2005) Abolition of spindle oscillations and 3-Hz absence seizurelike activity in the thalamus by using high-frequency stimulation: potential mechanism of action. *J. Neurosurg.* 103, 538–545.

Lee, S.H., Wynn, J.K., Green, M.F., Kim, H., Lee, K.J., Nam, M., Park, J.K., Chung, Y.C. (2006) Quantitative EEG and low resolution electromagnetic tomography (LORETA) imaging of patients with persistent auditory hallucinations. *Schiz. Res.* 83, 111–119.

Lee, J.S., Yang, B.H., Lee, J.H., Choi, J.H., Choi, I.G., Kim, S.B. (2007) Detrended fluctuation analysis of resting EEG in depressed outpatients and healthy controls. *Clin. Neurophysiol.* 118, 2489–2496.

Lee, U., Kim, S., Noh, G.-J., Choi, B.-M., Hwang, E., Mashour, G.A. (2009) The directionality and functional organization of frontoparietal connectivity during consciousness and anesthesia in humans. *Conscious. Cogn.* 18, 1069–1078.

Lefevre, F., Aronson, N. (2000) Ketogenic diet for the treatment of refractory epilepsy in children: a systemic review of efficacy. *Pediatrics* 105(4), E46.

Lehnertz, K. (2008) Epilepsy and nonlinear dynamics. *J. Biol. Phys.* 34, 253–266.

Lehnertz, K., et al. (eds.) (2000) *Chaos in Brain?* World Scientific, Singapore.

Lehmann, D. (1984) EEG assessment of brain activity: spatial aspects, segmentation and imaging. *Int. J. Psychophysiol.* 1, 267–276.

Lehmann, D., Ozaki, H., Pal, I. (1987) EEG alpha map series: brain microstates by space-oriented adaptive segmentation. *Electroencephalogr. Clin. Neurophysiol.* 67, 271–288.

Lehmann, D. (1989) From mapping to the analysis and interpretation of EEG/EP maps, in: *Topographic Brain Mapping of EEG and Evoked Potentials*, Springer (Berlin), pp. 53–75.

Lehmann, D., Strick, W.K., Henggeler, B., Koenig, T., Koukkou, M. (1998) Brain electric microstates and momentary conscious mind states as building blocks of spontaneous thinking: I. Visual imagery and abstract thoughts. *Int. J. Psychophysiol.* 29, 1–11.

Leise, E.M. (1990) Modular reconstruction of nervous systems: a basic principle of design for invertebrates and vertebrates. *Brain Res. Rev.* 15, 1–23.

Lennox, B.R., Park, S.B., Medley, I., Morris, P.G., Jones, P.B. (2000). The functional anatomy of auditory hallucinations in schizophrenia. *Psychiatry Res.* 100, 13–20.

León-Carrión, J., van Eeckhout, P., Domínguez Morales, M.R. (2002) The locked-in syndrome: a syndrome looking for a therapy. *Brain Inj.* 16, 555–569.

León-Carrión, J., Martin-Rodriguez, J.F., Damas-Lopez, J., Barroso y Martin, J.M., Dominguez-Morales, M.R. (2008) Brain function in the minimally conscious state: a quantitative neurophysiological study. *Clin. Neurophysiol.* 119, 1506–1514.

Lerner, D.E. (1996) Monitoring changing dynamics with correlation integrals: case study of an epileptic seizure. *Physica D* 97, 563–576.

Leslie, S.W., Friedman, M.B., Coleman, R.R. (1980) Effects of chlordiazepoxide on depolarisation-induced calcium influx into synaptosomes. *Biochem. Pharmacol.* 29, 2439–2443.

Letellier, C., Aguirre, L.A. (2002) Investigating nonlinear dynamics from time series: the influence of symmetries and the choice of observables. *Chaos* 12, 549–558.

Letellier, C., Moroz, I.M., Gilmore, R. (2008) Comparison of tests for embeddings. *Phys. Rev. E* 78, 026203.

Le van Quyen, M., Martinerie, J., Adam, C., Varela, F.J. (1997) Unstable periodic orbits in human epileptic activity. *Phys. Rev. E* 56, 3401–3411.

Le van Quyen, M., Martinerie, J., Navarro, V., Baulac, M., Varela, F.J. (2001) Characterizing neurodynamic changes before seizures. *J. Clin. Neurophysiol.* 18, 191–208.

Le van Quyen, M., Navarro, V., Baulac, M., Renault, B., Martinerie, J. (2003a) Anticipation of epileptic seizures from standard EEG recordings. *Lancet* 361, 970–971.

Le van Quyen, M., Navarro, V., Martinerie, J., Baulac, M., Varela, F.J. (2003b) Toward a neurodynamical understanding of ictogenesis. *Epilepsia* 44 (Suppl. 12), 30–43.

Le van Quyen, M., Amor, F., Rudrauf, D. (2006) Exploring the dynamics of collective synchronizations in large ensembles of brain signals. *J. Physiol. (Paris)* 100, 194–200.

Lewis, M.D. (2004) Trouble ahead: predicting antisocial trajectories with dynamic systems concepts and methods. *J. Abnorm. Child Psychol.* 32(6), 665–671.

Li, C.-L. (1959) Synchronization of unit activity in the cerebral cortex. *Science* 129, 783–784.

Librizzi, L., de Curtis, M. (2003) Epileptiform ictal discharges are prevented by periodic interictal spiking in the olfactory cortex. *Ann. Neurol.* 53, 382–389.

Light, G.A., Malaspina, D., Geyer, M.A., Luber, B.M., Coleman, E.A., Sackeim, H.A., Braff, D.L. (1999) Amphetamine disrupts P50 suppression in normal subjects. *Biol. Psychiatry* 46, 990–996.

Light, G.A., Hsu, J.L., Hsieh, M.H., Meyer-Gomes, K., Sprock, J., Swerdlow, N.R., Braff, D.L. (2006) Gamma band oscillations reveal neural network cortical coherence dysfunction in schizophrenia patients. *Biol. Psychiatry* 60, 1231–1240.

Lipsitz, L.A., Goldberger, A.L. (1992) Loss of "complexity" and aging: potential applications of fractals and chaos theory to senescence. *JAMA* 267, 1806–1809.

Litt, B., Esteller, R., Echauz, J., *et al.* (2001) Epileptic seizures may begin hours in advance of clinical onset: a report of five patients. *Neuron* 30, 51–64.

Litt, B., Lehnertz, K. (2002) Seizure prediction and the preseizure period. *Curr. Opin. Neurol.* 15, 173–177.

Little, W.A. (1974) The existence of persistent states in the brain. *Math. Biosci.* 19, 101–120.

Liu, J., She, Z.-S., Guo, H., Li, L., Ouyang, Q. (2004) Hierarchical structure description of spatiotemporal chaos. *Phys. Rev. E* 70, 036215.

Livanov, M.N. (1977) *Spatial Organization of Cerebral Processes*. Wiley, New York.

Llinás, R., Ribary, U. (1993) Coherent 40-Hz oscillation characterizes dream states in humans. *Proc. Natl. Acad. Sci. USA* 90, 2078–2081.

Llinás, R., Urbano, F.J., Leznik, E., Ramirez, R.R., van Marle, H.J. (2005) Rhythmic and dysrhythmic thalamocortical dynamics: GABA systems and the edge effect. *Trends Neurosci.* 28, 325–333.

Logothetis, N.K. (2008) What we can do and what we cannot do with fMRI. *Nature* 453, 869–878.

Loh, M., Rolls, E.T., Deco, G. (2007) A dynamical systems hypothesis of schizophrenia. *PLoS Comput. Biol.* 3(11), e228.

Lopantsev, V., Avoli, M. (1998) Participation of $GABA_A$-mediated inhibition in ictal-like discharges in the rat entorhinal cortex. *J. Neurophysiol.* 79, 352–360.

Lopes da Silva, F.H., Pijn, J.P., Wadman, W.J. (1994) Dynamics of local neuronal networks: control parameters and state bifurcations in epileptogenesis. *Prog. Brain Res.* 102, 359–370.

Lopes da Silva, F.H., Pijn, J.P.M., Gorter, J.A., van Vliet, E., Daalman, E.W., Blanes, W. (2000) Rhythms of the brain: between randomness and determinism, in: Lehnertz, K., *et al.* (eds.) *Chaos in Brain?* World Scientific, Singapore, pp. 63–76.

Lopes da Silva, F.H., Blanes, W., Kalitzin, S.N., Parra, J., Suffczynski, P., Velis, D.N. (2003) Epilepsies as dynamical diseases of brain systems: basic models of the transition between normal and epileptic activity. *Epilepsia* 44 (Suppl. 12), 72–83.

Lopez, H.H., Ettenberg, A. (2001) Dopamine antagonism attenuates the unconditioned incentive value of estrous female cues. *Pharmacol. Biochem. Behav.* 68, 411–416.

Lorente de Nó, R. (1938) Cerebral cortex: architecture, intracortical connections, motor projections, in: Fulton, J.F. (ed.) *Physiology of the Nervous System*, Oxford University Press.

Lorenz, E.N. (1963) Deterministic nonperiodic flow. *J. Atmos. Sci.* 20, 130–141.

Lorenz, K. (1963) *Das sogenannte Böse zur Naturgeschichte der Aggression*. Verlag Dr. G. Borotha-Schoeler, Vienna. Translated as *On Aggression*, Mehtuen & Co., London, 1966.

Löscher, W. (1998) New visions in the pharmacology of anticonvulsion. *Eur. J. Pharmacol.* 342, 1–13.

Löscher, W., Poulter, M.O., Padjen, A.L. (2006) Major targets and mechanisms of antiepileptic drugs and major reasons for failure. *Adv. Neurol.* 97, 417–427.

Luntz-Leybman, V., Bickford, P.C., Freedman, R. (1992) Cholinergic gating of response to auditory stimuli in rat hippocampus. *Brain Res.* 587(1), 130–136.

Luria, A.R. (1966) *Higher Cortical Functions in Man*. Basic Books, New York.

Ma, J., Tai, S.K., Leung, L.S. (2009) Ketamine-induced deficit of auditory gating in the hippocampus of rats is alleviated by medial septal inactivation and antipsychotic drugs. *Psychopharmacology* 206, 457–467.

MacDonald, R.L., McLean, M.J. (1986) Anticonvulsant drugs: mechanisms of action. *Adv. Neurol.* 44, 713–736.

McDonald, N. (1960) Living with schizophrenia. *Can. Med. Assoc. J.* 82, 218–221.

Mackey, M., Milton, J. (1987) Dynamical diseases. *Ann. N. Y. Acad. Sci.* 504, 16–32.

MacLeod, K., Bäcker, A., Laurent, G. (1998) Who reads temporal information contained across synchronized and oscillatory spike trains? *Nature* 395, 693–698.

Maiwald, T., Winterhalder, M., Aschenbrenner-Scheibe, R., Voss, H.U., Schulze-Bonhage, A., Timmer, J. (2004) Comparison of three nonlinear seizure prediction methods by means of the seizure prediction characteristic. *Physica D* 194, 357–368.

Majumdar, K.K. (2007) A structural and a functional aspect of stable information processing by the brain. *Cogn. Neurodyn.* 1, 295–303.

Makarov, V.A., Nekorkin, V.I., Velarde, M.G. (2001) Spiking behaviour in a noise-driven system combining oscillatory and excitatory properties. *Physiol. Rev. Lett.* 86, 3431–3434.

Malach, R. (1994) Cortical columns as devices for maximizing neuronal diversity. *Trends Neurosci.* 17, 101–104.

Mandelbrot, B. (1977) *Fractals: Form, Chance and Dimension.* W.H. Freeman & Co., San Francisco.

Mandell, A.J., Selz, K.A. (1993) Brain stem neuronal noise and neocortical "resonance". *J. Stat. Phys.* 70, 355–373.

Mantovani, J.F., Landau, W.F. (1980) Acquired aphasia with convulsive disorder: course and prognosis. *Neurology* 30, 524–529.

Mañé, R. (1983) *Ergodic Theory and Differentiable Dynamics.* Springer-Verlag, Berlin.

Marder, E., Calabrese, R.L. (1996) Principles of rhythmic motor pattern generation. *Physiol. Rev.* 76, 687–717.

Martineau, J., Cochin, S., Magne, R., Barthelemy, C. (2008) Inpaired cortical activation in autistic children: is the mirror neuron system involved? *Int. J. Psychophysiol.* 68, 35–40.

Masquelier, T., Hugues, E., Deco, G., Thorpe, S.J. (2009) Oscillations, phase-of-firing coding, and spike timing-dependent plasticity: an efficient learning scheme. *J. Neurosci.* 29, 13484–13493.

May, R.M. (1976) One-dimensional first return maps in ecological models. *Nature* 261, 459–467.

Mazanov, J., Byrne, D.G. (2006) A cusp catastrophe model analysis of changes in adolescent substance use: assessment of behavioural intention as a bifurcation variable. *Nonlin. Dyn. Psychol. Life Sci.* 10(4), 445–470.

McCollum, G. (2000) Social barriers to a theoretical neuroscience. *Trends Neurosci.* 23, 334–336.

McCormick, D.A., Contreras, D. (2001) On the cellular and network bases of epileptic seizures. *Annu. Rev. Physiol.* 63, 815–846.

McGinn, C. (1989) Can we solve the mind-body problem? *Mind* 98(391), 349–366.

McIntosh, A.R. (2004) Contexts and catalysts: a resolution of the localization and integration of function in the brain. *Neuroinformatics* 2, 175–182.

McIntosh, A.R., Kovacevic, N., Itier, R.J. (2008) Increased brain signal variability accompanies lower behavioural variability in development. *PLoS Comput. Biol.* 4(7), e1000106.

McIntosh, G.C., Brown, S.H., Rice, R.R., Thaut, M.H. (1997) Rhythmic auditory-motor facilitation of gait patterns in patients with Parkinson's disease. *J. Neurol. Neurosurg. Psychiatry* 62, 22–26.

McAlonan, G.M., Cheung, V., Cheung, C., Suckling, J., Lam, G.Y., Tai, K.S., Yip, L., Murphy, D.G.M., Chu, S.E. (2005) Mapping the brain in autism. A voxel-based MRI study of volumetric differences and intercorrelations in autism. *Brain* 128, 268–276.

McLean, M.J., MacDonald, R.L. (1983) Multiple actions of phenytoin on mouse spinal cord neurons in cell culture. *J. Pharmacol. Exp. Ther.* 227, 779–789.

McMillen, D., Kopell, N. (2003) Noise-stabilized long-distance synchronization in populations of model neurons. *J. Comp. Neurosci.* 15, 143–157.

McSharry, P.E., Smith, L.A., Tarassenko, L. (2003) Prediction of epileptic seizures: are nonlinear methods relevant? *Nat. Medicine* 9, 241–242.

Meeren, H.K.M., Pijn, J.P.M., van Luijtelaar, E.L.J.M., Coenen, A.M.L., Lopes da Silva, F.H. (2002) Cortical focus drives widespread corticothalamic networks during spontaneous absence seizures in rats. *J. Neurosci.* 22, 1480–1495.

Menendez de la Prida, L., Sanchez-Andres, J.V. (1999) Nonlinear transfer function encodes synchronization in a neural network from the mammalian brain. *Phys. Rev. E* 60, 3239–3243.

Meyer-Lindenberg, A., Ziemann, U., Hajak, G., Cohen, L., Faith Berman, K. (2002) Transitions between dynamical states of differing stability in the human brain. *Proc. Natl. Acad. Sci. USA* 99, 10948–10953.

Micheloyannis, S., Pachou, E., Stam, C.J., Breakspear, M., Bitsios, P., Vourkas, M., Erimaki, S., Zervakis, M. (2006) Small-world networks and disturbed functional connectivity in schizophrenia. *Schizophr. Res.* 87, 60–66.

Miller, L. (2000) The predator's brain: neuropsychodynamics of serial killing, in: Schlesinger, L.B. (ed.) *Serial Offenders: Current Thought, Recent Findings.* CRC Press, pp. 135–166.

Miller, E.K., Cohen, J.D. (2001) An integrative theory of prefrontal cortex function. *Annu. Rev. Neurosci.* 24, 167–202.

Milner, B. (1963) Effects of different brain lesions on card sorting: the role of the frontal lobes. *Arch. Neurol.* 9, 100–110.

Milner, P.M. (1996) Neural representations: some old problems revisited. *J. Cogn. Neurosci.* 8, 69–77.

Milnor, J. (1985) On the concept of attractor. *Commun. Math. Phys.* 99, 177–195.

Milton, J.G., Gotman, J., Remillard, G.M., Andermann, F. (1987) Timing seizure recurrence in adult epileptic patients: a statistical analysis. *Epilepsia* 28, 471–478.

Milton, J., Black, B. (1995) Dynamic diseases in neurology and psychiatry. *Chaos* 5, 8–13.

Mirowski, P., Madhavan, D., Lecun, Y., Kuzniecky, R. (2009) Classification of patterns of EEG synchronization for seizure prediction. *Clin. Neurophysiol.* 120, 1927–1940.

Mizuno, A., Villalobos, M.E., Davies, M.M., Dahl, B.C., Müller, R.-A. (2006) Partially enhanced thalamocortical functional connectivity in autism. *Brain Res.* 1104, 160–174.

Mockett, B.G., Hulme, S.R. (2008) Metaplasticity: new insights through electrophysiological investigations. *J. Integr. Neurosci.* 7, 315–336.

Modolo, J., Henry, J., Beuter, A. (2008) Dynamics of the subthalamo-pallidal complex in Parkinson's disease during deep brain stimulation. *J. Biol. Phys.* 34, 351–366.

Moffitt, T.E. (1993) Adolescence-limited and life-course persistent antisocial behavior: a developmental taxonomy. *Psychol. Bull.* 100, 674–701.

Money, J. (1990) Forensic sexology: paraphilic serial rape (biastophilia) and lust murder (erotophonophilia). *Am. J. Psychother.* 44(1), 26–36.

Monk, C.S., Peltier, S.J., Wiggins, J.L., Weng, S.J., Carrasco, M., Risi, S., Lord, C. (2009) Abnormalities of intrinsic functional connectivity in autism spectrum disorders. *NeuroImage* 47, 764–772.

Monteiro, C., Lima, D., Galhardo, V. (2006) Switching-on and switching-off of bistable spontaneous discharges in rat spinal deep dorsal horn neurons. *Neurosci. Lett.* 398, 258–263.

Moreno Vega, Y. (2009) Complex network modeling: a new approach to neurosciences, in: Perez Velazquez, J.L., Wennberg, R. (eds.) *Coordinated Activity in the Brain: Measurements and Relevance to Brain Function and Behaviour*, Springer, Heidelberg.

Mormann, F., Lehnertz, K., David, P., Elger, C.E. (2000) Mean phase coherence as a measure for phase synchronization and its application to the EEG of epilepsy patients. *Physica D* 144, 358–369.

Mormann, F., Kreuz, T., Andrzejak, R.G., David, P., Lehnertz, K., Elger, C.E. (2003) Epileptic seizures are preceded by a decrease in synchronization. *Epilepsy Res.* 53, 173–185.

Morowitz, H.J. (1980) Rediscovering the mind. *Psychology Today*, August 1980.

Morrell, F., Whisler, W.W., Smith, M.C., *et al.* (1995) Landau-Kleffner syndrome: treatment with subpial intracortical transection. *Brain* 118, 1529–1546.

Morrison, R.S., Dempsey, E.W. (1943) Mechanisms of thalamocortical augmentation and repetition. *Am. J. Physiol.* 138, 297–308.

Mountcastle, V.B. (1957) Modality and topographic properties of single neurons of cat's somatic sensory cortex. *J. Neurophysiol.* 20, 408–434.

Moussaid, M., Garnier, S., Theraulaz, G., Helbing, D. (2009) Collective information processing and pattern formation in swarms, flocks, and crowds. *Top. Cogn. Sci.* 1, 469–497.

Murias, M., Webb, S.J., Greenson, J., Dawson, J. (2007) Resting state cortical connectivity reflected in EEG coherence in individuals with autism. *Biol. Psychiatry* 62, 270–273.

Murray, G.K., Corlett, P.R., Clark, L., Pessiglione, M., Blackwell, A.D., Honey, G., Jones, P.B., Bullmore, E.T., Robbins, T.W., Fletcher, P.C. (2008) Substantia nigra/ventral tegmental reward prediction error disruption in psychosis. *Mol. Psychiatry* 13, 267–276.

Murthy, V.N., Fetz, E.E. (1996) Oscillatory activity in sensorimotor cortex of awake monkeys: synchronization of local field potentials and relation to behavior. *J. Neurophysiol.* 76, 3949–3967.

Nachev, P., Hacker, P.M.S. (2010) Covert cognition in the persistent vegetative state. *Prog. Neurobiol.* 91, 68–76.

Nachev, P., Husain, M. (2010) Action and the fallacy of the 'internal'. *Trends Cogn. Sci.* 14, 192–193.

Nandrino, J.L., Pezard, L., Martinerie, J., el Massioui, F., Renault, B., Jouvent, R., Allilaire, J.F., Widlocher, D. (1994) Decrease of complexity in EEG as a symptom of depression. *Neuroreport* 5, 528–530.

Navarro, V., Martinerie, J., Le Van Quyen, M., Clemenceau, S., Adam, C., Baulac, M., Varela, F. (2002) Seizure anticipation in human neocortical partial epilepsy. *Brain* 125, 640–655.

Nelson, R.J., Chiavegatto, S. (2001) Molecular basis of aggression. *Trends Neurosci.* 24, 713–718.

Nenadovic, V., Hutchison, J.S., Garcia Dominguez, L., Otsubo, H., Gray, M.P., Belkas, J., Perez Velazquez, J.L. (2008) Fluctuations in cortical synchronization may predict paediatric traumatic brain injury outcome. *J. Neurotrauma* 25, 615–627.

Nenadovic, V., Topjian, A., Abend, N., Garcia Dominguez, L., Dlugos, D., Berg, R., Nadkarni, V., Ichord, R., Clancy, R., Hutchison, J., Perez Velazquez, J.L. (2009) Reduced variability in electroencephalographic spatial synchrony patterns during therapeutic hypothermia following pediatric cardiac arrest is associated with unfavourable short-term neurologic outcomes. *Resuscitation Science Symposium*, Orlando, USA.

Netoff, T.I., Schiff, S.J. (2002) Decreased neuronal synchronization during experimental seizures. *J. Neurosci.* 22, 7297–7307.

Netoff, T.I., Clewley, R., Arno, S., Keck, T., White, J.A. (2004a) Epilepsy in small-world networks. *J. Neurosci.* 24, 8075–8083.

Netoff, T.I., Pecora, L.M., Schiff, S.J. (2004b) Analytical coupling detection in the presence of noise and nonlinearity. *Phys. Rev. E* 69, 017201.

Newell, K.M., Gao, F., Sprague, R.L. (1995) The dynamical structure of tremor in tardive dyskinesia. *Chaos* 5, 43–47.

Nicolelis, M.A.L., Baccala, L.A., Lin, R.C.S., Chapin, J.K. (1995) Sensorimotor encoding by synchronous neural ensemble activity at multiple levels of the somatosensory system. *Science* 268, 1353–1358.

Nicoll, R.A., Madison, D.V. (1982) General anesthetics hyperpolarize neurons in the vertebrate central nervous system. *Science* 217, 1055–1057.

Nieser, U. (1967) *Cognitive Psychology*, Appleton, New York.

Niessing, J., Friedrich, R.W. (2010) Olfactory pattern classification by discrete neuronal network states. *Nature* 465, 47–52.

Nishitani, N., Avikainene, S., Hari, R. (2004) Abnormal imitation-related cortical activation sequences in Asperger's syndrome. *Ann. Neurol.* 55, 558–562.

Noebels, J.L. (2003) The biology of epilepsy genes. *Annu. Rev. Neurosci.* 26, 599–625.

Norden, A.D., Blumenfeld, H. (2002) The role of subcortical structures in human epilepsy. *Epilepsy Behav.* 3, 219–231.

Northoff, G., Qin, P., Nakao, T. (2010) Rest-stimulus interaction in the brain: a review. *Trends Neurosci.* 33, 277–284.

Nunez, P.L. (2000) Toward a quantitative description of large-scale neocortical dynamic function and EEG. *Behav. Brain Sci.* 23, 371–437.

Nyabadza, F., Hove-Musekwa, S.D. (2010) From heroin epidemics to methamphetamine epidemics: modelling substance abuse in a South African province. *Math. Biosci.* 225, 132–140.

Oberheim, N.A., Tian, G.-F., Han, X., Peng, W., Takano, T., Ransom, B., Nedergaard, M. (2008) Loss of astrocytic domain organization in the epileptic brain. *J. Neurosci.* 28, 3264–3276.

O'Donnell, P. (2003) Dopamine gating of forebrain neural ensembles. *Eur. J. Neurosci.* 17, 429–435.

O'Donovan, M.J. (1999) The origin of spontaneous activity in developing networks of the vertebrate nervous system. *Curr. Opin. Neurobiol.* 9, 94–104.

Omaya, A.K., Gennarelli, T.A. (1974) Cerebral concussion and traumatic unconsciousness. Correlation of experimental and clinical observations of blunt head injuries. *Brain* 97, 633–654.

Osorio, I., Frei, M.G., Manly, B.F., Sunderam, S., Bhavaraju, N.C., Wilkinson, S.B. (2001) An introduction to contingent (closed-loop) brain electrical stimulation for seizure blockage, to ultra-short-term clinical trials, and to multidimensional statistical analysis of therapeutic efficacy. *J. Clin. Neurophysiol.* 18, 533–544.

Osorio, I., Overman, J., Giftakis, J., Wilkinson, S.B. (2007) High frequency thalamic stimulation for inoperable mesial temporal epilepsy. *Epilepsia* 48, 1561–1571.

Osorio, I., Frei, M.G. (2009) Seizure abatement with single DC pulses: is phase resetting at play? *Int. J. Neural Systems* 19, 149–156.

Ott, E., Grebogi, C., Yorke, J.A. (1990) Controlling chaos. *Phys. Rev. Lett.* 64, 1196–1199.

Oullier, O., Lagarde, J., Jantzen, K.J., Kelso, J.A.S. (2006) Coordination dynamics: (in)stability and metastability in the behavioural and neural systems. *J. Soc. Biol.* 200, 145–167.

Owen, A.M. (2008) Disorders of consciousness. *Ann. N.Y. Acad. Sci.* 1124, 225–238.

Ozen, L.J., Teskey, G.C. (2009) One Hertz stimulation to the corpus callosum quenches seizure development and attenuates motor map expansion. *Neuroscience* 160, 567–575.

Ozonoff, S. (1995) Reliability and validity of the Wisconsin Card Sorting Test in studies of autism. *Neuropsychology* 9, 491–500.

Ozonoff, S., Strayer, D.L., McMahon, W.M., Filloux, F. (1994) Executive function abilities in autism and Tourette syndrome: an information processing approach. *J. Child Psychol. Psychiatry* 35, 1015–1032.

Ozonoff, S., Jensen, J. (1999) Brief report: specific executive function profiles in three neurodevelopmental disorders. *J. Autism Dev. Disord.* 29, 171–177.

Packard, N.H., Crutchfield, J.P., Farmer, J.D., Shaw, R.S. (1980) Geometry from a time series. *Phys. Rev. Lett.* 45, 712–716.

Palva, S., Palva, J.M. (2007) New vistas for alpha-frequency band oscillations. *Trends Neurosci.* 30(4), 150–158.

Pandarinath, C., Bomash, I., Victor, J., Prusky, G., Tschetter, W., Nirenberg, S. (2010) A novel mechanism for switching a neural system from one state to another. *Front. Comput. Neurosci.* 4:2; doi: 10.3389/fncom.2010.00002.

Paré, D., Curro Dossi, R., Steriade, M. (1990) Neuronal basis of the Parkinsonian resting tremor: a hypothesis and its implications for treatment. *Neuroscience* 35, 217–226.

Parra, J., Kalitzin, S.N., Iriarte, J., Blanes, W., Velis, D.N., Lopes da Silva, F.H. (2003) Gamma-band phase clustering and photosensitivity: is there an underlying mechanism common to photosensitive epilepsy and visual perception? *Brain* 126, 1164–1172.

Parrent, A., Serrano Almeida, C. (2006) Deep brain stimulation and cortical stimulation in the treatment of epilepsy. *Adv. Neurol.* 97, 563–572.

Pavlides, C., Greenstein, Y.J., Grudman, M., Winson, J. (1988) Long-term potentiation in the dentate gyrus is induced preferentially on the positive phase of the θ rhythm. *Brain Res.* 439, 383–387.

Pecora, L.M., Carroll, T.L., Heagy, J.F. (1995) Statistics for mathematical properties of maps between time series embeddings. *Phys. Rev. E* 52, 3420–3439.

Pecora, L.M., Carroll, T.L., Johnson, G.A., Mar, D.J., Heagy, J.F. (1997) Fundamentals of synchronization in chaotic systems, concepts, and applications. *Chaos* 7, 520–543.

Pecora, L.M., Moniz, L., Nichols, J., Carroll, T.L. (2007) A unified approach to attractor reconstruction. *Chaos* 17, 013110.

Peled, A. (2000) A new diagnostic system for psychiatry. *Med. Hypotheses* 54, 367–380.

Penfield, W., Jasper, H. (1954) *Epilepsy and the Functional Anatomy of the Human Brain*. Little, Brown and Co., Boston.

Perea, G., Navarrete, M., Araque, A. (2009) Tripartite synapses: astrocytes process and control synaptic information. *Trends Neurosci.* 32, 421–431.

Perez Velazquez, J.L. (2003a) Mathematics and the gap junctions: in-phase synchronization of identical neurons. *Int. J. Neurosci.* 113, 1095–1101.

Perez Velazquez, J.L. (2003b) Bicarbonate-dependent depolarizing potentials in pyramidal cells and interneurons during epileptiform activity. *Eur. J. Neurosci.* 18, 1337–1342.

Perez Velazquez, J.L. (2005) Brain, behaviour and mathematics: are we using the right approaches? *Physica D* 212, 161–182.

Perez Velazquez, J.L. (2006) Brain research: a perspective from the coupled oscillators field. *NeuroQuantology* 4, 155–165.

Perez Velazquez, J.L. (2009) Finding simplicity in complexity: general principles of biological and nonbiological organization. *J. Biol. Phys.* 35, 209–221.

Perez Velazquez, J.L., Garcia Dominguez, L., Nenadovic, V., Wennberg, R. (2011) Experimental observation of increased fluctuations in an order parameter before epochs of extended brain synchronization. *J. Biol. Phys.* 37, 141–152.

Perez Velazquez, J.L., Valiante, T.A., Carlen, P.L. (1994) Modulation of gap junctional mechanisms during calcium-free induced field burst activity: a possible role for electrotonic coupling in epileptogenesis. *J. Neurosci.* 14(7), 4308–4317.

Perez Velazquez, J.L., Carlen, P.L. (1999) Synchronization of GABAergic interneuronal networks in seizure-like activity in the rat horizontal hippocampal slice. *Eur. J. Neurosci.* 11(11), 4110–4118.

Perez Velazquez, J.L., Carlen, P.L. (2000) Gap junctions, synchrony and seizures. *Trends Neurosci.* 23, 68–74.

Perez Velazquez, J.L., Khosravani, H., Lozano, A., Carlen, P.L., Bardakjian, B.B., Wennberg, R. (1999) Type III intermittency in human partial epilepsy. *Eur. J. Neurosci.* 11, 2571–2576.

Perez Velazquez, J.L., Carlen, P.L., Skinner, F. (2001) Artificial electrotonic coupling affects neuronal firing patterns depending upon intrinsic cellular characteristics. *Neuroscience* 103, 841–849.

Perez Velazquez, J.L., Cortez, M.A., Snead, O.C., Wennberg, R. (2003). Dynamical regimes underlying epileptiform events: role of instabilities and bifurcations in brain activity, *Physica D* 186, 205–220.

Perez Velazquez, J.L., Khosravani, H. (2004) A subharmonic dynamical bifurcation during *in vitro* epileptiform activity. *Chaos* 14, 333–342; *Virtual J. Biol. Phys. Res.* 7(11), 2004 (www.vjbio.org).

Perez Velazquez, J.L., Wennberg, R. (2004) Metastability of brain states and the many routes to seizures: numerous causes, same result, in: *Recent Research Developments in Biophysics* (Vol. 3, pp. 25–59), Transworld Research Network.

Perez Velazquez, J.L., Garcia Dominguez, L., Guevara Erra, R. (2007a) Fluctuations in neuronal synchronization in brain activity correlate with the subjective experience of visual recognition. *J. Biol. Phys.* 33, 49–59.

Perez Velazquez, J.L., Galán, R.F., Garcia Dominguez, L., Leshchenko, Y., Lo, S., Belkas, J., Guevara Erra, R. (2007b) Phase response curves in the characterization of epileptiform activity. *Phys. Rev. E* 76, 061912.

Perez Velazquez, J.L., Garcia Dominguez, L., Guevara Erra, R., Wennberg, R. (2007c) The fluctuating brain: dynamics of neuronal activity, in: Wang, C.W. (ed.) *Nonlinear Phenomena Research Perspectives*, Nova Science Publishers, New York.

Perez Velazquez, J.L., Garcia Dominguez, L., Wennberg, R. (2007d) Complex phase synchronization in epileptic seizures: evidence for a devil's staircase. *Phys. Rev. E* 75, 011922; also in the *Virtual J. Biol. Phys. Res.* 13(3), 2007.

Perez Velazquez, J.L., Huo, J., Garcia Dominguez, L., Leshchenko, Y., Snead, O.C. (2007e) Typical versus atypical absence seizures: network mechanisms of the spread of paroxysms. *Epilepsia* 48, 1585–1593.

Perez Velazquez, J.L., Wennberg, R. (eds.) (2009) *Coordinated Activity in the Brain: Measurements and Relevance to Brain Function and Behaviour.* Springer, Heidelberg.

Perez Velazquez, J.L., Guevara Erra, R., Wennberg, R., Garcia Dominguez, L. (2009a) Correlations of cellular activities in the nervous system: physiological and methodological considerations, in: Perez Velazquez, J.L., Wennberg, R. (eds.) *Coordinated Activity in the Brain: Measurements and Relevance to Brain Function and Behaviour*, Springer, Heidelberg.

Perez Velazquez, J.L., Barceló, F., Hung, Y., Leshchenko, Y., Nenadovic, V., Belkas, J., Raghavan, V., Brian, J., Garcia Dominguez, L. (2009b) Decreased brain coordinated activity in autism spectrum disorders during executive tasks: reduced long-range synchronization in the fronto-parietal networks. *Int. J. Psychophysiol.* 73, 341–349.

Perko, L. (1991) *Differential Equations and Dynamical Systems.* Springer-Verlag, Berlin.

Petsche, H., Brazier, M.A.B. (1972) *Synchronization of EEG Activity in Epilepsies.* Springer-Verlag, Berlin.

Petsche, H., Pockberger, H., Rappelsberger, P. (1984) On the search for the sources of the EEG. *Neuroscience* 11, 1–27.

Piaget, J. (1936) La Naissance de l'intelligence chez l'enfant, ©The Estate of Jean Piaget. Translated as '*The Origin of Intelligence in the Child*', Penguin Books, Middlesex, U.K., 1977.

Pierce, K., Müller, R.A., Ambrose, J., Allen, G., Courchesne, E. (2001) Face processing occurs outside the fusiform 'face area' in autism: evidence from functional MRI. *Brain* 124, 205–2073.

Pikovsky, A., Rosemblum, M., Kurths, J. (2001) *Synchronization: A Universal Concept in Nonlinear Sciences.* Cambridge University Press.

Pinault, D. (2003) Cellular interactions in the rat somatosensory thalamocortical system during normal and epileptic 5–9 Hz oscillations. *J. Physiol.* 552.3, 881–905.

Pincus, S.M. (2001) Assessing serial irregularity and its implications for health. *Ann. N.Y. Acad. Sci.* 954, 245–267.

Pinto, R.D., Sartorelli, J.C., Gonçalves, W.M. (2001) Homoclinic tangencies and routes to chaos in a dripping faucet experiment. *Physica A* 291, 244–254.

Pittson, S., Himmel, A.M., MacIver, M.B. (2004) Multiple synaptic and membrane sites of anesthetic action in the CA1 region of rat hippocampal slices. *BMC Neurosci.* 5, 52 (www.biomedcentral.com/1471-2202/5/52).

Platt, N., Spiegel, E.A., Tresser, C. (1993) On-off intermittency: a mechanism for bursting. *Phys. Rev. Lett.* 70, 279–282.

Plum, F., Schiff, N.D., Ribary, U., Llinás, R. (1998) Coordinated expression in chronically unconscious persons. *Philos. Trans. R. Soc. Lond. B Biol. Sci.* 353, 1929–1933.

Poincaré, H. (1885) Sur les courbes definies par une èquation diffèrentielle. *J. de Math. Pures et Appl. Serie IV* 1, 167–277.

Poincaré, H. (1899) *Les Mèthodes Nouvelles de la Mèchanique Celeste.* Gauthiers-Villars, Paris.

Poincaré, H. (1902) *La Science et l'hypothèse.* Transl. *Science and Hypothesis,* Walter Scott Publishing, London (1905).

Pomeau, Y., Manneville, P. (1980) Intermittent transition to turbulence in dissipative dynamical systems. *Commun. Math. Phys.* 74, 189–197.

Popovych, O.V., Hauptmann, C., Tass, P.A. (2008) Impact of nonlinear delayed feedback on synchronized oscillators. *J. Biol. Phys.* 34, 367–379.

Portas, C.M., Krakow, K., Allen, P., Josephs, O., Armony, J.L., Frith, C.D. (2000) Auditory processing across the sleep-wake cycle: simultaneous EEG and fMRI monitoring in humans. *Neuron* 28, 991–999.

Poulet, J.F.A., Petersen, C.C.H. (2008) Internal brain state regulates membrane potential synchrony in barrel cortex of behaving mice. *Nature* 454, 881–885.

Prasad, A., Biswal, B., Ramaswamy, R. (2003) Strange nonchaotic attractors in driven excitable systems. *Phys. Rev. E* 68, 037201.

Prigogine, I., Stengers, I. (1980) *Order Out of Chaos.* Bantam Books, Toronto.

Prince, D.A., Connors, B.W. (1986) Mechanisms of interictal epileptogenesis. *Adv. Neurol.* 44, 275–299.

Prinz, A.A., Bucher, D., Marder, E. (2004) Similar network activity from disparate circuit parameters. *Nat. Neurosci.* 7, 1345–1352.

Proctor, J., Holmes, P. (2010) Reflexes and preflexes: on the role of sensory feedback on rhythmic patterns in insect locomotion. *Biol. Cybern.* 102, 513–531.

Proulx, E., Leshchenko, Y., Kokarovtseva, L., Khokhotva, V., El-Beheiry, M., Carter Snead, O., Perez Velazquez, J.L. (2006) Functional contribution of specific brain areas to absence seizures: role of thalamic gap junctional coupling. *Eur. J. Neurosci.* 23, 489–496.

Pulvermüller, F., Shtyrov, Y. (2009) Spatiotemporal signatures of large-scale synfire chains for speech processing as revealed by MEG. *Cereb. Cortex* 19, 79–88.

Pyragas, K. (1992) Continuous cntor of chaos by self-controlling feedback. *Phys. Lett. A* 170, 421–428.

Quiroga, R.Q., Krastkov, A., Kreuz, T., Grassberger, P. (2002) Performance of different synchronization measures in real data: a case study on electroencephalographic signals. *Phys. Rev. E* 65, 041903.

Raabe, W., Gumnit, R.J. (1977) Anticonvulsant action of diazepam: increase of cortical postsynaptic inhibition. *Epilepsia* 18, 117–120.

Rabinovich, M.I., Volkovskii, A., Lecanda, P., Abarbanel, H.D., Laurent, G. (2001) Dynamical encoding by networks of competing neuron groups: winnerless competition. *Phys. Rev. Lett.* 87, 068102.

Rabinovich, M.I., Huerta, R., Varona, P., Afraimovich, V.S. (2008) Transient cognitive dynamics, metastability, and decision making. *PLoS Comput. Biol.* 4(5), e1000072; doi: 10.1371/journal.pcbi.1000072.

Rae-Grant, A.D., Kim, Y.W. (1994) Type III intermittency: a nonlinear dynamic model of EEG burst suppression. *Electroencephalogr. Clin. Neurophysiol.* 90, 17–23.

Raethjen, J., Pawlas, F., Lindemann, M., Wenzelburger, R., Deuschl, G. (2000a) Determinants of physiologic tremor in a large normal population. *Clin. Neurophysiol.* 111, 1825–1837.

Raethjen, J., Lindemann, M., Schmaljohann, H., Wenzelburger, R., Pfister, G., Deuschl, G. (2000b) Multiple oscillators are causing Parkinsonian and essential tremor. *Mov. Disord.* 15, 84–94.

Raethjen, J., Lemke, M.R., Lindemann, M., Wenzelburger, R., Krack, P., Deuschl, G. (2001) Amitriptyline enhances the central component of physiological tremor. *J. Neurol. Neurosurg. Psychiatry* 70, 78–52.

Raine, A., Lencz, T., Bihrle, S., LaCasse, L. Coletti, P. (2000) Reduced prefrontal gray matter volume and reduced autonomic activity in antisocial personality disorder. *Arch. General Psychiatry* 57, 1190127.

Ramocki, M.B., Zoghbi, H.Y. (2008) Failure of neuronal homeostasis results in common neuropsychiatric phenotypes. *Nature* 455, 912–918.

Rao, S.G., Williams, G.V., Goldman-Rakic, P.S. (2000) Destruction and creation of spatial tuning by disinhibition: $GABA_A$ blockade of prefrontal cortical neurons engaged by working memory. *J. Neurosci.* 20, 485–494.

Rapp, P.E. (1994) A guide to dynamical analysis. *Integrat. Physiol. Behav. Sci.* 29, 311–327.

Rapp, P.E., Zimmerman, I.D., Vining, E.P., Cohen, N., Albano, A.M., Jimenez Montano, M.A. (1994) The algorithmic complexity of neural spike trains increases during focal seizures. *J. Neurosci.* 14, 4731–4739.

Regesta, G., Tanganelli, P. (1999) Clinical aspects and biological bases of drug-resistant epilepsies. *Epilepsy Res.* 34, 109–122.

Reimann, H.A. (1963) *Periodic Diseases*. F.A. Davis, Philadelphia.

Riffell, J.A., Lei, H., Hildebrand, J.G. (2009) Neural correlates of behavior in the moth *Manduca sexta* in response to complex odors. *Proc. Natl. Acad. Sci. USA* 106, 19219–19226.

Ring, H.A., Baron-Cohen, S., Wheelwright, S., Williams, S.C.R., Brammer, M., Andrew, C., Bullmore, E.T. (1999) Cerebral correlates of preserved cognitive skills in autism. *Brain* 122, 1305–1315.

Rippon, G., Brock, J., Brown, C., Boucher, J. (2007) Disordered connectivity in the autistic brain: challenges for the 'new psychophysiology'. *Int. J. Psychophysiol.* 63, 164–172.

Robinson, P.A., Rennie, C.J., Rowe, D.L. (2002) Dynamics of large-scale brain activity in normal arousal states and epileptic seizures. *Phys. Rev. E* 65, 041924.

Rodrigo, C., Rajapakse, S., Jayananda, G. (2010) The "antisocial" person: an insight in to biology, classification and current evidence on treatment. *Ann. Gen. Psychiatry* 9, 31.

Rodrigues, S., Goncalves, J., Terry, J.R. (2007) Existence and stability of limit cycles in a macroscopic neuonal population model. *Physica D* 233, 39–65.

Rogawski, M.A. (2000) KCNQ2/KCNQ3 K^+ channels and the molecular pathogenesis of epilepsy: implications for therapy. *Trends Neurosci.* 23, 393–398.

Roiser, J.P., Stephan, K.E., den Ouden, H.E.M., Barnes, T.R.E., Friston, K.J., Joyce, E.M. (2009) Do patients with schizophrenia exhibit aberrant salience? *Psychol. Med.* 39, 199–209.

Rolls, E.T. (1999) *The Brain and Emotion.* Oxford University Press.

Rolls, E.T., Loh, M., Deco, G., Winterer, G. (2008) Computational models of schizophrenia and dopamine modulation in the prefrontal cortex. *Nat. Rev. Neurosci.* 9, 696–709.

Ropohl, A., Sperling, W., Elstner, S., Tomandl, B., Ruelbach, U., Kaltenhauser, M., Kornhuber, J., Maihofner, C. (2004) Cortical activity associated with auditory hallucinations. *Brain Imaging* 15, 523–526.

Rosanova, M., Casali, A., Bellina, V., Resta, F., Mariotti, M., Massimini, M. (2009) Natural frequencies of human corticothalamic circuits. *J. Neurosci.* 29, 7679–7685.

Rosso, O.A., Martin, M.T., Figliola, A., Keller, K., Plastino, A. (2006) EEG analysis using wavelet-based information tools. *J. Neurosci. Methods* 153, 163–182.

Roux, J.C. (1983) Experimental studies of bifurcations leading to chaos in the Beousof-Zhabotinsky reaction. *Physica D* 7, 57–68.

Ruelbach, U., Bleich, S., Maihöfner, C., Kornhuber, J., Sperling, W. (2007) Specific and unspecific auditory hallucinations in patients with schizophrenia. *Neuropsychobiology* 55, 89–95.

Ruelle, D., Takens, F. (1971) On the nature of turbulence. *Commun. Math. Phys.* 20, 167–192.

Rulkov, N.F. (2002) Modelling of spiking-bursting neural behavior using two-dimensional map. *Phys. Rev. E* 65, 041922.

Sadaghiani, S., Hesselmann, G., Friston, K.J., Kleinschmidt, A. (2010) The relation of ongoing brain activity, evoked neural responses, and cognition. *Front. Syst. Neurosci.* 4:20, doi: 10.3389/fnsys.2010.00020.

Salamone, J.D., Cousins, M.S., Snyder, B.J. (1997) Behavioural functions of nucleus accumbens dopamine: empirical and conceptual problems with the anhedonia hypothesis. *Neurosci. Biobehav. Rev.* 21, 341–359.

Sanford, L.D., Morrison, A.R., Ball, W.A., Ross, R.J., Mann, G.L. (1993) The amplitude of elicited PGO waves: a correlate of orienting. *Electroencephalogr. Clin. Neurophysiol.* 86, 438–445.

Sarà, M., Pistoia, F. (2010) Complexity loss in physiological time series of patients in a vegetative state. *Nonlin. Dynamics Psychol. Life Sci.* 14, 1–13.

Sarkisian, M.R. (2001) Overview of the current animal models for human seizure and epileptic disorders. *Epilepsy Behav.* 2, 201–216.

Sauer, T. (1994) Reconstruction of dynamical systems from interspike intervals. *Phys. Rev. Lett.* 72, 3811–3814.

Sauer, T. (1995) Interspike interval embedding of chaotic signals. *Chaos* 5, 127–132.

Sauer, T., Yorke, J.A., Casdagli, M. (1991) Embedology. *J. Stat. Phys.* 65, 579–616.

Schäffer, C., Rosemblum, M.G., Kurths, J., Abel, H.-H. (1998) Heartbeat Synchronized with ventilation. *Nature* 392, 239–240.

Schevon, C.A., Cappell, J., Emerson, R., Isler, J., Grieve, P., Goodman, R., McKhann, G. Jr., Weiner, H., Doyle, W., Kuzniecky, R., Devinsky, O., Gilliam, F. (2007) Cortical abnormalities in epilepsy revealed by local EEG synchrony. *Neuroimage* 35, 140–148.

Schiff, N.D. (2010) Recovery of consciousness after brain injury: a mesocircuit hypothesis. *Trends Neurosci.* 33, 1–9.

Schiff, N.D., Labar, D.R., Victor, J.D. (1999) Common dynamics in temporal lobe seizures and absence seizures. *Neuroscience* 91, 417–428.

Schiff, N.D., Rodriguez Moreno, D., Kamal, A., Kim, K.H.S., Giacino, J.T., Plum, F., Hirsch, J. (2005) fMRI reveals large-scale network activation in minimally conscious patients. *Neurology* 64, 514–523.

Schiff, N.D., Giacino, J.T., Kalmar, K., Victor, J.D., Baker, K., Gerber, M., Fritz, B., Eisenberg, B., O'Connor, J., Kobylarz, E.J., Farris, S., Machado, A., McCagg, C., Plum, F., Fins, J.J., Rezai, A.R. (2007) Behavioural improvements with thalamic stimulation after severe traumatic brain injury. *Nature* 448, 600–603.

Schiff, S.J., Jerger, K., Duong, D.H., Chang, T., Spano, M.L., Ditto, W.L. (1994) Controlling chaos in the brain. *Nature* 370, 615–620.

Schiff, S.J. (2005). Dangerous phase. *Neuroinformatics* 3, 315–318.

Schiff, S.J., Sauer, T., Kumar, R., Weinstein, S.L. (2005) Neuronal spatiotemporal pattern discrimination: the dynamical evolution of seizures. *Neuroimage* 28, 1043–1055.

Schild, D. (1984) Coordination of neuronal signals as structures in state space. *Int. J. Neurosci.* 22, 283–297.

Schmid-Schönbein, C. (1998) Improvement of seizure control by psychological methods in patients with intractable epilepsies. *Seizure* 7, 261–270.

Schmidt, R.C., Carello, C., Turvey, M.T. (1990) Phase transitions and critical fluctuations in the visual coordination of rhythmic movements between people. *J. Exp. Psychol. Hum. Percept. Perform.* 16, 227–247.

Schneidman, E., Freedman, B., Segev, I. (1998) Ion channel stochasticity may be critical in determining the reliability and precision of spike timing. *Neural Comput.* 10, 1679–1703.

Schöner, G., Haken, H., Kelso, J.A.S. (1986) A stochastic theory of phase transitions in human hand movements. *Biol. Cybernet.* 53, 247–257.

Schroeder, C.E., Lakatos, P. (2008) Low-frequency neuronal oscillations as instruments of sensory selection. *Trends Neurosci.* 32, 9–18.

Schultz, W. (1999) The reward signal of midbrain dopamine neurons. *News Physiol. Sci.* 14, 249–255.

Schultz, W., Apicella, P., Scarnati, E., Ljinberg, T. (1992) Neuronal activity in monkey ventral striatum related to expectation of reward. *J. Neurosci.* 12, 4595–4610.

de Schutter, E. (2008) Reviewing multi-disciplinary papers: a challenge in neuroscience? *Neuroinformatics* 6, 253–255.

Schwarz, J., Spielmann, R.P. (1983) Flurazepam: effects on sodium and potassium currents in myelinated nerve fibres. *Eur. J. Pharmacol.* 90, 359–366.

Seeman, P. (1972) The membrane action of anaesthetics and tranquilizers. *Pharmacol. Rev.* 24, 583–655.

Seeman, P., Chau-Wong, M., Tadesco, J., Wong, K. (1975) Brain receptors for antipsychotic drugs and dopamine: direct binding assays. *Proc. Natl. Acad. Sci. USA* 72, 4376–4380.

Segal, Z.V., Williams, J.M.G., Teasdale, J.D. (2002) *Mindfulness-Based Cognitive Therapy for Depression: A New Approach to Preventing Relapse.* The Guilford Press, New York.

Seidenbecher, T., Staak, R., Pape, H.-C. (1998) Relations between cortical and thalamic cellualr activities during absence seizures in rats. *Eur. J. Neurosci.* 10, 1103–1112.

Seri, S., Cerquiglini, A., Pisani, F. (1998) Spike-induced interference in auditory sensory processing in Landau-Kleffner syndrome. *Electroencephalogr. Clin. Neurophysiol.* 108, 506–510.

Serquina, R., Lai, Y.-C., Chen, Q. (2008) Characterization of nonstationary chaotic systems. *Phys. Rev. E* 77, 026208.

Serrien, D.J., Orth, M., Evans, A.H., Lees, A.J., Brown, P. (2005) Motor inhibition in patients with Gilles de la Tourette syndrome: functional activation patterns as revealed by EEG coherence. *Brain* 128, 116–125.

Shah, A., Frith, U. (1983) An islet of ability in autistic children: a research note. *J. Child Psychol. Psychiatry* 24, 613–620.

Shaw, N.A. (2002) The neurophysiology of concussion. *Prog. Neurobiol.* 67, 281–344.

Shergill, S.S., Brammer, M.J., Williams, S.C., Murray, R.M., McGuire, P.K. (2000) Mapping auditory hallucinations in schizophrenia using functional magnetic resonance imaging. *Arch. Gen. Psychiatry* 57(11), 1033–1038.

Shields, D.C., Leiphart, J.W., McArthur, D.L., Vespa, P.M., Le van Quyen, M., Martinerie, J., Soss, J.R. (2007) Cortical synchrony changes detected by scalp electrode electroencephalograph as traumatic brain injury patients emerge from coma. *Surgical Neurol.* 67, 354–359.

Shinbrot, T., Grebogi, C., Ott, E., Yorke, J.A. (1993) Using small perturbations to control chaos. *Nature* 363, 411–417.

Shinnar, S., Berg, A.T., Moshé, S.L., Shinnar, R. (2001) How long do new-onset seizures in children last? *Ann. Neurol.* 49, 659–664.

Shuai, J., Bikson, M., Hahn, P.J., Lian, J., Durand, D.M. (2003) Ionic mechanisms underlying spontaneous CA1 neuronal firing in Ca^{2+}-free solution. *Biophys. J.* 84, 2099–2111.

Silbersweig, D.A., Stern, E., Frith, C., Cahil, C., Holmes, A., Grootoonk, S., Seaward, J., McKenna, P., Chua, S.E., Schnorr, L., Jones, T., Frackowiak, R.S.J. (1995) A functional neuroanatomy of hallucinations in schizophrenia. *Nature* 378, 176–179.

Šil'nikov, L.P. (1965) A case of the existence of a countable number of periodic motions. *Soviet Math. Dokl.* 6, 163–166.

Silva, S., Alacoque, X., Fourcade, O., Samii, K., Marque, P., Woods, R., Mazziotta, J., Chollet, F., Loubinoux, I. (2010) Wakefulness and loss of awareness: brain and brainstem interaction in the vegetative state. *Neurology* 74, 313–320.

Singer, T., Seymur, B., O'Doherty, J., Kaube, H., Dolan, R.J., Frith, C.D. (2004) Empathy for pain involves the affective but not sensory components of pain. *Science* 303, 1157–1162.

Singer, W. (2006) Phenomenal awareness and consciousness from a neurobiological perspective. *NeuroQuantology* 4, 134–154.

Singer, W., Gray, C.M. (1995) Visual feature integration and the temporal correlation hypothesis. *Ann. Rev. Neurosci.* 18, 555–586.

Skerritt, J.H., Davies, L.P., Johnston, G.A.R. (1982) A purinergic component in the anticonvulsant action of carbamazepine. *Eur. J. Pharmacol.* 82, 195–197.

Skinner, F.K., Kopell, N., Marder, E. (1994) Mechanisms for oscillation and frequency control in reciprocal inhibitory model neural networks. *J. Comput. Neurosci.* 1, 69–87.

Sleigh, J.W., Steyn-Roos, D.A., Steyn-Ross, M.L., Grant, C., Ludbrook, G. (2004) Cortical entropy changes with general anaesthesia: theory and experiment. *Physiol. Meas.* 25, 921–934.

Slutzky, M.W., Cvitanovic, P., Mogul, D.J. (2003) Manipulating epileptiform bursting in the rat hippocampus using chaos control and adaptive techniques. *IEEE Trans. Biomed. Eng.* 50, 559–570.

Smirnov, D.A., Bezruchko, B.P., Seleznev, Y.P. (2002) Choice of dynamical variables for global reconstruction of model equations from time series. *Phys. Rev. E* 65, 026205.

Smith, K.A., Bierkamper, G.G. (1990) Paradoxical role of GABA in a chronic model of petit mal (absence)-like epilepsy in the rat. *Eur. J. Pharmacol.* 176, 45–55.

Snead, O.C., Gibson, K.M. (2005) γ-Hydroxybutyric acid. *N. Engl. J. Med.* 352, 2721–2732.

So, P., Francis, J.T., Netoff, T.I., Gluckman, B.J., Schiff, S.J. (1998) Periodic orbits: a new language for neuronal dynamics. *Biophys. J.* 74, 2776–2785.

Sornette, D. (2004) *Critical Phenomena in Natural Sciences.* Springer-Verlag, Berlin.

Spach, M.S., Josephson, M.E. (1994) Initiating reentry: the role of nonuniform anisotropy in small circuits. *J. Cardiovasc. Electrophysiol.* 5, 182–209.

Spencer, K.M., Nestor, P.G., Perlmutter, R., Niznikiewicz, M.A., Klump, M.C., Frumin, M., Shenton, M.E., McCarley, R.W. (2004) Neural synchrony indexes disordered perception and cognition in schizophrenia. *Proc. Natl. Acad. Sci. USA* 101, 17288–17293.

Sperry, R.W. (1952) Neurology and the mind-brain problem. *Am. Sci.* 40, 291–312.

Sritharan, A., Line, P., Sergejew, A., Silberstein, R., Egan, G., Copolov, D. (2005) EEG coherence measures during auditory hallucinations in schizophrenia. *Psychiatry Res.* 136, 189–200.

Staley, K., Hellier, J.L., Dudek, F.E. (2005) Do interictal spikes drive epileptogenesis? *Neuroscientist* 11, 272–276.

Stanton, P.K., Sejnowski, T.J. (1989) Associative long-term depression in the hippocampus induced by hebbian covariance. *Nature* 339, 215–217.

Steinlein, O.K. (2004) Genetic mechanisms that underlie epilepsy. *Nat. Rev. Neurosci.* 5, 400–408.

Steriade, M. (2001) Impact of network activities on neuronal properties in corticothalamic systems. *J. Neurophysiol.* 86, 1–39.

Steriade, M., Domich, L., Dakson, G., Deschenes, M. (1987) The deafferented reticular thalamic nucleus generates spindle rhythmicity. *J. Neurophysiol.* 57, 260–273.

Steriade, M., Contreras, D., Curro Dossi, R., Nunez, A. (1993) The slow (<1 Hz) oscillation in reticular thalamic and thalamocortical neurons: scenario of sleep rhythm generation in interacting thalamic and neocortical networks. *J. Neurosci.* 13, 3284–3299.

Steriade, M., Contreras, D., Amzica, F., Timofeev, I. (1996) Synchronization of fast (30–40 Hz) spontaneous oscillations in intrathalamic and thalamocortical networks. *J. Neurosci.* 16, 2788–2808.

Stevens, C.F. (2000) Models are common; good theories are scarce. *Nat. Neurosci.* 3, 1177.

Stewart, M., Fox, S.E. (1990) Do septal neurons pace the hippocampal theta rhythm? *Trends Neurosci.* 13, 163–168.

Steyn-Ross, M.L., Steyn-Ross, D.A., Sleigh, J.W. (2004) Modelling general anaesthesia as a first-order phase transition in the cortex. *Prog. Biophys. Mol. Biol.* 85, 369–385.

Steyn-Ross, D.A., Steyn-Ross, M.L., Sleigh, J.W., Wilson, M.T., Gillies, I.P., Wright, J.J. (2005) The sleep cycle modelled as a critical phase transition. *J. Biol. Phys.* 31, 547–569.

Stopfer, M., Bhagavan, S., Smith, B.H., Laurent, G. (1997) Impaired odour discrimination on desynchronization of odour-encoding neural assemblies. *Nature* 390, 70–74.

Strata, F., Atzori, M., Molnar, M., Ugolini, G., Tempia, F., Cherubini, E. (1997) A pacemaker current in dye-coupled hilar interneurons contributes to the generation of giant GABAergic potentials in developing hippocampus. *J. Neurosci.* 17, 1435–1446.

Stratonovich, R.L. (1963, 1967) *Selected Topics in the Theory of Random Noise*, Vols. 1 & 2, Gordon & Breach, New York.

Stroop, J.R. (1935) Studies of interference in serial verbal reactions. *J. Exp. Psychol.* 18, 643–662.

Sturman, M.M., Vaillancourt, D.E., Metman, L.V., Bakay, R.A., Corcos, D.M. (2004) Effects of subthalamic nucleus stimulation and medication on resting and postural tremor in Parkinson's disease. *Brain* 127, 2131–2143.

Suffczynski, P., Kalitzin, S., Lopes da Silva, F.H. (2004) Dynamics of non-convulsive epileptic phenomena modeled by a bistable neuronal network. *Neuroscience* 126, 467–484.

Suhler, C.L., Churchland, P.S. (2009) Control: conscious and otherwise. *Trends Cogn. Sci.* 13, 341–347.

Swartzwelder, H.S., Lewis, D.V., Anderson, W.W., Wilson, W.A. (1987) Seizure-like events in brain slices: suppression by interictal activity. *Brain Res.* 410, 362–366.

Symond, M.P., Harris, A.W., Gordon, E., Williams, L.M. (2005) "Gamma synchrony" in first-episode schizophrenia: a disorder of temporal connectivity? *Am. J. Psych.* 162, 459–465.

Takens, F. (1981) *Detecting Strange Attractors in Turbulence*. Lecture Notes in Mathematics, Vol. 898, Springer, New York.

Takeshita, D., Sato, Y.D., Bahar, S. (2007) Transitions between multistable states as a model of epileptic seizure dynamics. *Phys. Rev. E* 75, 051925.

Tang, Y.Y., Posner, M.I. (2009) Attention training and attention state training. *Trends Cogn. Sci.* 13, 222–227.

Tass, P., Rosenblum, M.G., Weule, J., Kurths, J., Pikovsky, A., Volkmann, J., Schnitzler, A., Freund, H.-J. (1998) Detection of $n{:}m$ phase locking from noisy data: application to magnetoencephalography. *Phys. Rev. Lett.* 81, 3291–3294.

Taylor, J.G. (2007) On the neurodynamics of the creation of consciousness. *Cogn. Neurodyn.* 1, 97–118.

Taylor, C.P., Dudek, F.E. (1984) Synchronization without active chemical synapses during hippocampal afterdischarges. *J. Neurophysiol.* 52, 145–155.

Tegnér, J., Compte, A., Wang, X.-J. (2002) The dynamical stability of reverberatory neural circuits. *Biol. Cybern.* 87, 471–481.

Teitelbaum, A., Nenadovic, V., Barceló, F., Garcia Dominguez, L., Kostelecki, W., Wennberg, R., Brian, J., Perez Velazquez, J.L. (2009) Patterns and fluctuations of brain coordinated activity in health and disease. International Workshop on Trends in Complex Systems-Synchronization and Multiscale Complex Dynamics in the Brain, Dresden, Germany.

Tél, T., lai, Y.-C., Gruiz, M. (2008) Noise-induced chaos: a consequence of long deterministic transients. *Int. J. Bif. Chaos* 18, 509–520.

Teramae, J.-N., Tanaka, D. (2004) Robustness of the noise-induced phase synchronization in a general class of limit cycle oscillators. *Phys. Rev. Lett.* 93, 204103.

Tergau, F., Naumann, U., Paulus, W., Steinhoff, B.J. (1999) Low-frequency repetitive transcranial magnetic stimulation improves intractable epilepsy. *Lancet* 353, 2209.

Thatcher, R.W., Walker, R.A., Gerson, I., Geisler, F.H. (1989) EEG discriminant analyses of mild head trauma. *Electroencephalogr. Clin. Neurophysiol.* 73, 94–106.

Thelen, E., Ulrich, B.D. (1991) Hidden skills: a dynamic systems analysis of treadmill stepping during the first year. *Monographs of the Society for Research in Child Development* 56 (1, Serial No. 223).

Thomasson, N., Pezard, L. (1999) Dynamical systems and depression: a framework for theoretical perspectives. *Acta Biotheor.* 47, 209–218.

Timofeev, I., Bazhenov, M., Avramescu, S., Nita, D.A. (2010) Posttraumatic epilepsy: the roles of synaptic plasticity. *Neuroscientist* 16, 19–27.

Titcombe, M.S., Glass, L., Guehl, D., Beuter, A. (2001) Dynamics of Parkinsonian tremor during deep brain stimulation. *Chaos* 11, 766–773.

Tononi, G., Edelman, G.M., Sporns, O. (1998) Complexity and coherency: integrating information in the brain. *Trends Cogn. Sci.* 2, 474–484.

Tort, A.B.L., Fontanini, A., Kramer, M.A., Jones-Lush, L.M., Kopell, N.J., Katz, D.B. (2010) Cortical networks produce three distinct 7–12 Hz rhythms during single sensory responses in the awake rat. *J. Neurosci.* 30, 4315–4324.

Träff, J., Petrovic, P., Ingvar, M. (2000) Thalamic activation in photic myoclonus. *Acta Neurol. Scand.* 101, 339–343.

Trujillo, L.T., Peterson, M.A., Kaszniak, A.W., Allen, J.J.B. (2005) EEG phase synchrony differences across visual perception conditions may depend on recording and analysis methods. *Clin. Neurophysiol.* 116, 172–189.

Tsacopoulos, M., Magistretti, P.J. (1996) Metabolic coupling between glia and neurons. *J. Neurosci.* 16, 877–885.

Tsuda, I. (2001) Toward an interpretation of dynamic neural activity in terms of chaotic dynamical systems. *Behav. Brain Sci.* 24, 793–847.

Tsuda, I., Unemura, T. (2003) Chaotic itinerancy generated by coupling of Milnor attractors. *Chaos* 13, 937–946.

Tsuda, I., Fuji, H., Tadokoro, S., Yasuoka, T., Yamaguti, Y. (2004) Chaotic itinerancy as a mechanism of irregular changes between synchronization and desynchronization in a neural network. *J. Integr. Neurosci.* 3, 159–182.

Tsuda, I. (2009) Hypotheses on the functional roles of chaotic transitory dynamics. *Chaos* 19, 015113.

Tucker, W. (2002) Computing accurate Poincarè maps. *Physica D* 171, 127–137.

Tukhlina, N., Rosenblum, M., Pikovsky, A., Kurths, J. (2007) Feedback suppression of neural synchrony by vanishing stimulation. *Phys. Rev. E* 75, 011918.

Tukhlina, N., Rosenblum, M. (2008) Feedback suppression of neural synchromny in two interacting populations by vanishing stimulation. *J. Biol. Phys.* 34, 301–314.

Turing, A.M. (1952) The chemical basis of morphogenesis. *Phil. Trans. R. Soc. B.* 237, 37–72.

Uhlhaas, P.J., Singer, W. (2006) Neural synchrony in brain disorders: relevance for cognitive dysfunctions and pathophysiology. *Neuron* 52, 155–168.

Uhlhaas, P.J., Singer, W. (2007) What do disturbances in neural synchrony tell us about autism? *Biol. Psychiatry* 62, 190–191.

Uhlhaas, P.J., Linden, D.E., Singer, W., Haenschel C., Lindner, M., Maurer, K., Rodriguez, E. (2006) Dysfunctional long-range coordination of neural activity during Gestalt perception in schizophrenia. *J. Neurosci.* 26, 8168–8175.

Vallbo, A.B., Wessberg, J. (1993) Organization of motor output in slow finger movements in man. *J. Physiol.* 469, 673–691.

van Beveren, N.J.M., de Haan, L. (2008) A visual metaphor describing neural dynamics in schizophrenia. *PLoS ONE* 3(7), e2577.

van der Heyden, M.J., Velis, D.N., Hoekstra, B.P.T., Pijn, J.P.M., van Emde Boas, W., van Veelen, C.W.M., van Rijen, P.C., Lopes da Silva, F.H., DeGoede, J. (1999) Non-linear analysis of intracranial human EEG in temporal lobe epilepsy. *Clin. Neurophysiol.* 110, 1726–1740.

van Elswijk, G., Maij, F., Schoffelen, J.M., Overeem, S., Stegeman, D.F., Fries, P. (2010) Corticospinal beta-band synchronization entails rhythmic gain modulation. *J. Neurosci.* 30, 4481–4488.

van Putten, M.J. (2003) Nearest neighbor phase synchronization as a measure to detect seizure activity from scalp EEG recordings. *J. Clin. Neurophysiol.* 5, 320–325.

VanRullen, R., Koch, C. (2003) Is perception discrete or continuous? *Trends Cogn. Sci.* 7, 207–213.

van Vreeswijk, C., Abbott, L.F., Ermentrout, G.B. (1994) When inhibition not excitation synchronizes neural firing. *J. Comput. Neurosci.* 1, 313–321.

Velasco, F., Velasco, M., Velasco, A.L., Jiménez, F. (1993) Effect of chronic electrical stimulation of the centromedian thalamic nuclei on various intractable seizure patterns: I. Clinical seizures and paroxysmal EEG activity. *Epilepsia* 34, 1052–1064.

Velišek, L., Velkišková, J., Stanton, P.K. (2002) Low-frequency stimulation of the kindling focus delays basolateral amygdala kindling in immature rats. *Neurosci. Lett.* 326, 61–63.

Venkataramani, S.C., Antonsen, T.M., Ott, E., Sommerer, J.C. (1996) On-off intermittency: power spectrum and fractal properties of time series. *Physica D* 96, 66–99.

Viglione, S.S., Walsch, G.O. (1975) Proceedings: epileptic seizure prediction. *Electroencephalogr. Clin. Neurophysiol.* 39, 435–436.

Villalobos, M.E., Mizuno, A., Dahl, B.C., Kemmotsu, N., Muller, R. (2005) Reduced functional connectivity between V1 and inferior frontal cortex associated with visuomotor performance in autism. *NeuroImage* 25, 916–925.

Volgushev, M., Chistiakova, M., Singer, W. (1998) Modification of discharge patterns of neocortical neurons by induced oscillations of the membrane potential. *Neuroscience* 83, 15–25.

Volkmann, J., Joliot, M., Mogilner, A., Ioannides, A.A., Lado, F., Fazzini, E., Ribary, U., Llinás, R. (1996) Central motor loop oscillations in Parkinsonian retsing tremor revealed by magnetoencephalography. *Neurology* 46, 1359–1370.

von der Malsburg, C. (1981) The correlation theory of brain function. Internal report, Max-Planck Institute, Göttingen, Germany.

von Holst, E. (1939) The behavioural physiology of animals and man, in: *The Selected Papers of Erich von Holst*. Methuen, London.

von Krosigk, M., Bal, T., McCormick, D.A. (1993) Cellular mechanisms of a synchronized oscillation in the thalamus. *Science* 261, 361–364.

Voss, H.U., Uluç, A.M., Dyke, J.P., Watts, R., Kobylarz, E.J., McCandliss, B.D., Heier, L.A., Beattie, B.J., Hamacher, K.A., Vallabhajosula, S., Goldsmith, S.J., Ballon, D., Giacino, J.T., Schiff, N.D. (2006) Possible axonal regrowth in late recovery from the minimally conscious state. *J. Clin. Invest.* 116, 2005–2011.

Wackermann, J. (1994) Segmentation of EEG map series in n-dimensional state space. *Brain Topogr.* 6, 246.

Wagenaar, D.A., Nadasdy, Z., Potter, S.M. (2006) Persistent dynamic attractors in activity patterns of cultured neuronal networks. *Phys. Rev. E* 73, 051907.

Wallenstein, G.V., Kelso, J.A.S., Bressler, S.L. (1995). Phase transitions in spatiotemporal patterns of brain activity and behavior. *Physica D* 84, 626–634.

Wang, X.-J. (2001) Synaptic reverberation underlying mnemonic persistent activity. *Trends Neurosci.* 24, 455–463.

Wang, X.-J., Rinzel, J. (1992) Alternating and synchronous rhythms in reciprocally inhibitory model neurons. *Neural Comp.* 4, 84–97.

Wang, Y., Fujita, I., Murayama, Y. (2000) Neuronal mechanisms of selectivity for object features revealed by blocking inhibition in inferotemporal cortex. *Nature Neurosci.* 3, 807–813.

Wang, Y., Fujita, I., Tamura, H., Murayama, Y. (2002) Contribution of GABAergic inhibition to receptive field structures of monkey inferior temporal neurons. *Cereb. Cortex* 12, 62–74.

Warren, W.H. (2006) The dynamics of perception and action. *Psychol. Rev.* 113, 358–389.

Weaver, D.F. (2006) Designing future drugs for the treatment of intractable epilepsy. *Adv. Neurol.* 97, 429–434.

Weaver, A.L., Roffman, R.C., Norris, B.J., Calabrese, R.L. (2010) A role for compromise: synaptic inhibition and electrical coupling interact to control phasing in the leech heartbeat CPG. *Front. Behav. Neurosci.* 4, 38, doi: 10.3389/fnbeh.2010.00038.

Wegner, D.M. (2003) The mind's best trick: how we experience conscious will. *Trends Cogn. Sci.* 7, 65–69.

Weidlich, W. (1991) Physics and social science — the approach of synergetics. *Phys. Rep.* 204(1), 1–163.

Weiss, S.R.B., Li, X.-L., Rosen, J.B., Li, H., Heynen, T., Post, R.M. (1995) Quenching: inhibition of development and expression of amygdala kindled seizures with low frequency stimulation. *NeuroReport* 6, 2171–2176.

Wendling, F., Bartolomei, F., Bellanger, J.J., Chauvet, P. (2002) Epileptic fast activity can be explained by a model of impaired GABAergic dendritic inhibition. *Eur. J. Neurosci.* 15, 1499–1508.

Wennberg, R. (2004) Short term benefit of battery depletion in vagus nerve stimulation for epilepsy. *J. Neurol. Neurosurg. Psychiatry* 75, 939.

Wennberg, R., Arruda, F., Quesney, L.F., Olivier, A. (2002) Preeminence of extrahippocampal structures in the generation of mesial temporal seizures: evidence from human depth electrode recordings. *Epilepsia* 43, 716–726.

Wennberg, R., McAndrews, M.-P., Zumsteng, D., Perez Velazquez, J.L. (2009) The sign of the cross as a learned automatism? *Epilepsy Behav.* 15(3), 394–398.

Weyl, H. (1932) *The Open World, Three Lectures on the Metaphysical Implications of Science.* Yale University Press.

Wheal, H.V., Bernard, C., Chad, J.E., Cannon, R.C. (1998) Pro-epileptic changes in synaptic function can be accompanied by pro-epileptic changes in neuronal excitability. *Trends Neurosci.* 21, 167–174.

Whittington, M.A., Traub, R.D., Jefferys, J.G.R. (1995) Synchronized oscillations in interenuron networks driven by metabotropic glutamate receptor activation. *Nature* 373, 612–615.

Wiener, N. (1961) *Cybernetics*. MIT Press, Cambridge.

Wiggins, S. (1994) *Normally Hyperbolic Invariant Manifolds in Dynamical Systems*. Springer-Verlag, New York.

Williams, S.R., Turner, J.P., Crunelli, V. (1995) Gamma-hydroxybutyrate promotes oscillatory activity of rat and cat thalamocortical neurons by a tonic $GABA_B$ receptor-mediated hyperpolarization. *Neuroscience* 66, 133–141.

Williams, E.R., Soteropoulos, D.S., Baker, S.N. (2009) Coherence between motot cortical activity and peripheral discontinuities during slow finger movements. *J. Neurophysiol.* 102, 1296–1309.

Wilson, M.T., Steyn-Ross, D.A., Sleigh, J.W., Steyn-Ross, M.L., Wilcocks, L.C., Gillies, I.P. (2006) The K-complex and slow oscillation in terms of a mean-field cortical model. *J. Comput. Neurosci.* 21, 243–257.

Wilson, M.T., Barry, M., Reynolds, J.N.J., Hutchison, E.J.W., Steyn-Ross, D.A. (2008) Characteristics of temporal fluctuations in the hyperpolarized state of the cortical slow oscillation. *Phys. Rev. E* 77, 061908.

Wilson, M.T., Barry, M., Reynolds, J.N.J., Crump, W.P., Steyn-Ross, D.A., Steyn-Ross, M.L., Sleigh, J.W. (2010) An analysis of the transitions between down and up states of the cortical slow oscillation under urethane anaesthesia. *J. Biol. Phys.* 36, 245–259.

Winfree, A.T. (2001) *The Geometry of Biological Time*. Springer-Verlag, Berlin.

Winterhalder, M., Maiwald, T., Voss, H.U., Aschenbrenner-Scheibe, R., Timmer, J., Schulze-Bonhage, A. (2003) The seizure prediction characteristic: a general framework to assess and compare seizure prediction methods. *Epilepsy Behav.* 4, 318–325.

Winterhalder, M., Schelter, B., Maiwald, T., Brandt, A., Schad, A., Schulze-Bonhage, A., Timmer, J. (2006) Spatio-temporal patient-individual assessment of synchronization changes for epileptic seizure prediction. *Clin. Neurophysiol.* 117, 2399–2413.

Wolfart, J., Debay, D., Le Masson, G., Destexhe, A., Bal, T. (2005) Synaptic background activity controls spike transfer from thalamus to cortex. *Nature Neurosci.* 12, 1760–1767.

Womelsdorf, T., Schoffelen, J.-M., Oostenveld, R., Singer, W., Desimone, R., Engel, A.K., Fries, P. (2007) Modulation of neuronal interactions through neuronal synchronization. *Science* 316, 1609–1612.

Worrell, G.A., Cranstoun, S.D., Echauz, J., Litt, B. (2002) Evidence for self-organized criticality in human epileptic hippocampus. *Neuroreport* 13, 2017–2021.

Wright, J.J., Liley, D.T.J. (1996) Dynamics of the brain at global and microscopic scales: neural networks and the EEG. *Behav. Brain Sci.* 19, 285–320.

Wu, J.-Y., Cohen, L.B., Falk, C.X. (1994) Neuronal activity during different behaviors in aplysia: a distributed organization? *Science* 263, 820–823.

Yamanishi, J., Kawato, M., Suzuki, R. (1980) Two coupled oscillators as a model for the coordinated finger tapping by both hands. *Biol. Cybern.* 37, 219–225.

Yang, Y., Raine, A. (2009) Prefrontal structural and functional brain imaging findings in antisocial, violent, and psychopathic individuals: a meta-analysis. *Psychiatry Res. Neuroimaging* 174, 81–88.

Yechiam, E., Kanz, J.E., Bechara, A., Stout, J.C., Busemeyer, J.R., Altmaier, E.M., Paulsen, J.S. (2008) Neurocognitive deficits related to poor decision making in people behind the bars. *Psychon. Bull. Rev.* 15(1), 44–51.

Yuste, R., MacLean, J.N., Smith, J., Lansner, A. (2005) The cortex as a central pattern generator. *Nat. Rev. Neurosci.* 6, 477–483.

Zilbovicius, M., Meresse, I., Chabane, N., Brunelle, F., Samson, Y., Boddaert, N. (2006) Autism, the superior temporal sulcus and social perception. *Trends Neurosci.* 29, 359–366.

Zillmer, R., Brunel, N., Hansel, D. (2009) Very long transients, irregular firing, and chaotic dynamics in networks of randomly connected inhibitory integrate-and-fire neurons. *Phys. Rev. E* 79, 031909.

Zorzano, M.P., Vazquez, L. (2003) Emergence of synchronous oscillations in neural networks excited by noise. *Physica D* 179, 105–114.

Index